Reflexive Structures

Luis E. Sanchis

Reflexive Structures

An Introduction to Computability Theory

Springer-Verlag
New York Berlin Heidelberg
London Paris Tokyo

Luis E. Sanchis
School of Computer and Information Science
Syracuse University
Syracuse, New York 13210, USA

Library of Congress Cataloging-in-Publication Data
Sanchis, Luis E.
Reflexive structures.
Bibliography: p.
1. Computable functions. 2. Recursive functions.
I. Title.
QA9.59.S26 1988 511.3 88-4614

Softcover reprint of the hardcover 1st edition 1988

Typeset by Asco Trade Typesetting Ltd., Hong Kong.

9 8 7 6 5 4 3 2 1

ISBN 978-1-4612-8386-7 ISBN 978-1-4612-3878-2 (eBook)
DOI 10.1007/978-1-4612-3878-2

To Freide

I don't know if you've noticed it, but it's rummy how nothing in this world ever seems to be absolutely perfect.

P. G. Wodehouse
The Inimitable Jeeves

Preface

Computability theory deals ostensibly with a collection of objects, the recursive functions, which satisfy a number of structural properties. Elementary books in the field are concerned primarily with the identification of such structures and with applications. In this book three kinds of properties are considered, which are universally thought of as crucial in the theory. They are closure properties (particularly under recursion and minimalization), reflexivity properties (usually in the form of universal or indexing functions), and enumeration properties (particularly in the form of the normal form theorem). The purpose is to develop these structures in a reasonably general setting, but avoiding the introduction of abstract axiomatic domains. Natural numbers and functions on natural numbers are dealt with exclusively. While the approach is certainly related to general recursion theory, and many ideas are derived from there, the presentation of a concrete theory is conceptually organized around Church's thesis.

There is a natural hierarchy among these structures, which plays a role in the presentation. The first three chapters involve only closure properties and consequences. The results obtained in this way are very general and apply to many classes of functions, in some cases to the class of all functions or the class of all total functions. At some stage the notion of minimal closure becomes crucial, and proper computability theory appears. In fact, Church's thesis can be formulated purely in terms of minimal closure under adequate operations. The normal form property, or enumeration theorem, can be introduced at this stage, but it is not of much use without reflexivity properties.

Reflexivity is defined and studied in Chapter 4, and it is the central piece of this work. It is formalized via interpreters, which appears to be an elegant and economical technique for introducing universal functions and indexing. As usual, the recursion theorem in different forms provides the crucial tool to deal with these structures. Traditional applications to classification of sets and predicates, in particular the theory of productive and creative sets, are developed in some detail. A general selector theorem is proved, which is later applied in Chapter 5.

The last chapter is devoted to stronger forms of enumeration, namely, hyperenumeration (universal function quantification) and hyperhyperenumeration (existential-universal function quantification). These operations are nonfinitary and require special techniques that involve rudimentary knowledge about ordinals. The traditional approach introduced by Kleene, via ordinal notations, is followed. Results by Spector and others are included.

Considerable attention is given to inductive definitions and the correlated proofs by induction. In Chapter 3 finitary induction is formalized and in Chapter 5 nonfinitary induction is covered. Inductive proofs over nonfinitary inductions are a delicate matter, particularly the proper handling of inductive assumptions. This process is formalized so that it becomes almost mechanical.

Notation and terminology have been a constant problem in the execution of this work. Changes and innovations have been kept to a minimum—a statement that is sure to be received with skepticism by some readers. Because the material is being dealt with in a very general setting, it was necessary to introduce some notation that departs from the standard in the literature. At any rate, universal and practical notations are not being rejected, even when new notations are introduced. Whenever this book is used as a text, it is expected that the instructor will provide the necessary information regarding the standard notation in the literature. Furthermore, historical and bibliographical references are given in the notes, which should help the reader to place the notation presented here in the right perspective.

The choice of reflexive structures to support the presentation of computability theory is well known in the professional literature, but it has not been used before in an elementary text. It is hoped that instructors, students, and persons interested in computability will find this approach rewarding, particularly in terms of the amount of new material included in this introductory, and relatively short, volume.

A work of this type depends on the contributions of many persons, and some of them are mentioned in the references at the end of the volume. I would like to mention here the true pioneers: K. Gödel, S. C. Kleene, E. L. Post, and A. Turing. Among modern researchers I have a particular debt to J. Fenstad, H. Friedman, P. G. Hinman, Y. N. Moschovakis, and L. P. Sasso.

I am very grateful to the School of Computer and Information Science at Syracuse University for generous support given in different forms, particularly for the sabbatical leave granted in 1984. In the preparation of the final version of the manuscript, I received valuable help from my student Dilia Rodriguez.

This work was partially supported by CASE, The New York Center for Advanced Technology in Computer Applications and Software Engineering at Syracuse University.

Syracuse, New York Luis E. Sanchis

Contents

CHAPTER 1

Functions and Predicates

This chapter introduces the fundamental objects to be discussed in this work: numerical functions and numerical predicates. Initially, we are concerned with different procedures that can be used to specify functions and predicates and the manner in which such procedures can be combined. The main tool in our discussion is the notion of minimal closure, which characterizes the class of all functions and predicates that can be generated by using a given set of specification rules.

§1. Definitions

As usual we take the notion of *set* as an undefined primitive, but no attempt is made to reduce functions to sets. This type of reduction is fashionable in modern mathematics, but we think it is an essentially artificial approach that distorts the proper description of the mathematical activity. Furthermore, in the context of computability theory where functions are determined by algorithms, it appears more natural to take functions as a relatively independent primitive.

Let A and B be two sets or collections of objects. A *single-valued applicative rule* from A to B is a rule that can be applied to elements of A to obtain a unique value that is an element of B, with the understanding that in some cases the rule may fail to produce such a value. This means that the application of the rule to $x \in A$ may result in two different situations: in one case the rule fails to produce a value, and we say that the rule is *undefined* for x; in the other case the rule produces a unique value, which is an element of B, and we say the rule is *defined* for x. The rule as such is any valid mathematical procedure, and eventually an algorithm, but at this stage we do not impose any type of restriction. Still, the concept of applicative rule involves some subtleties that must be accounted for properly. The rule must be single-valued in the sense that a unique value is produced whenever defined, and this imposes a restriction on the procedures that can safely be used. Usually these are operational procedures involving the execution of some operations, for example, substitution. In some cases the single-valuedness property is not obvious and some argument is required.

EXAMPLE 1.1.1. Let $A = B =$ the set of natural numbers, and consider the rule which when applied to $n \in A$ produces the value $n - 1$. Clearly, this is a single-value applicative rule, but note that it is undefined when applied to $0 \in A$.

EXAMPLE 1.1.2. We consider again $A = B =$ the natural numbers, and let p be some fixed prime number. We associate with $n \in A$ a number $m \in B$ such that p^m divides n but p^{m+1} does not divide n. The single-valuedness property follows by an easy mathematical argument. The rule is undefined for $n = 0$.

Assume a given single-valued applicative rule from A to B. If we abstract this rule, retaining only the sets A and B, the elements of A for which the rule is defined, and the values associated with every element for which it is defined, we obtain a *mapping from A to B*. We call this process *extensional abstraction*. It is essentially a process in which the details of the applicative rule are disregarded but the final action of the rule is preserved. It is also a process of identification, which applies to different applicative rules from a set A to a set B. More precisely, we identify all applicative rules that are defined for the same elements of A and produce the same values whenever defined. The object obtained by this identification is a mapping from A to B. Note that the sets A and B are part of the applicative rule and are preserved in the process of extensional abstraction.

This process of extensional abstraction is extremely useful in all branches of mathematics. The point is that an applicative rule creates an object, a mapping, which can be created by many other applicative rules. Such a mapping has many properties that are independent of any particular rule and it is convenient and economical to identify such properties. For example, the properties of being 1-1 or onto are independent and there is no reason to formulate them in terms of applicative rules.

On the other hand, extensional abstraction does not imply the complete elimination of applicative rules. This is particularly true in computability theory, where functions are classified precisely in terms of computational procedures.

If h is a mapping from A to B, then h can be applied to any element $x \in A$, and the result of this application is denoted by hx. If h is undefined for x, then the expression hx is undefined, and we write $hx\uparrow$. If h is defined for x and produces the value v, we say that hx is defined with value v and write $hx \simeq v$.

The expression $hx\downarrow$ means that h is defined for x; that is, there is a v such that $hx \simeq v$.

It is convenient to generalize this notation to arbitrary expressions that can be defined or undefined, depending on the values of some parameters. If U and V are such expressions we put:

(i) $U\downarrow$ means that U is defined.
(ii) $U\uparrow$ means that U is undefined.
(iii) $U \simeq V$ means that either both U and V are undefined, or both are defined with the same value.
(iv) $U \not\simeq V$ means that either both U and V are undefined, or both are defined with different value.

We shall use the standard equality $=$ with the usual meaning, but only in some restricted contexts. More precisely, we write $U = V$ only if both U and V are totally defined expressions, that is, defined for all possible values of the parameters. If either U or V is possibly undefined, then $U = V$ is not a legitimate expression. The reason for this restriction will be explained later in connection with the rule of substitution in predicates.

As explained above a mapping from a set A to a set B must be introduced via some applicative rule. This process we call a *specification* for the function. In theory specifications can be given in any manner that is mathematically satisfactory. In practice we shall rely on specifications involving very definite procedures, which are described later in this chapter. We anticipate a general pattern of such procedures as follows. If U is an expression that depends on some variable x ranging over a set A, and is undefined or defined with values in a set B depending on the values of x, we introduce a mapping h from A to B by the specification

$$hx \simeq U,$$

which means that hx is defined if and only if U is defined, and whenever defined the value of hx is the value of U. Such a specification we call an *explicit specification.*

If A and B are sets, there is a unique mapping from A to B which is undefined for every $x \in A$. Such a mapping can be specified by an empty rule, which assigns no value to the elements of A, or by a rule that imposes a condition that is impossible to satisfy. This mapping is *totally undefined*, and it is denoted by UD_A, since the reference to the set B is clearly unnecessary. It follows that $\mathsf{UD}_A x\uparrow$ for all $x \in A$.

In general, mappings from a set A to a set B are partially defined, which means that they are defined for some elements of A and undefined for others. The mapping UD_A represents an extreme case of this situation. The other extreme case is represented by mappings that are defined for all elements of A. We say that such mappings are *total.*

If we allow the set A to be empty, there is exactly one mapping from A to any set B, which is the totally undefined mapping UD_A. In this case UD_A is also a total mapping.

The general theory of mappings has been formalized to a great extent, usually in the context of set theory, but in most applications only total mappings are considered. The extension to nontotal mappings offers no particular problems, but the reader is advised to be alert to the possibility of a mapping not being defined for some argument. The notation introduced above is adequate to convey this information. For example, we say that a mapping h from A to B is *1-1* if whenever $hx \simeq v$ and $hx' \simeq v$ then $x = x'$. It is implicit in this notation that x, x' are elements of A, and v is an element of B. Similarly, we say that h is *onto* if for every $v \in B$ there is $x \in A$ such that $hx \simeq v$.

A particular type of mapping will play an important role in our discussion. We denote by $\mathsf{BOOL} = \{\mathsf{true}, \mathsf{false}\}$ the set that contains the two boolean values $\mathsf{true} = \mathsf{T}$ and $\mathsf{false} = \mathsf{F}$. A total mapping from A to BOOL is called a *predicate* on A. Note the condition that a predicate is a total mapping, which must be taken into account when the specification for a predicate is written. To enforce this condition we shall impose restrictions on this process, for example, requiring that only total mappings be used, even if they are not predicates.

Equality between boolean values will be written in the form $x \equiv y$, which we consider a version of the standard equality $x = y$, but restricted to boolean values. Hence $x \equiv y$ is legitimate when x and y are boolean values and it is true when x and y are the same value; otherwise it is false. In general, if U and V are boolean expressions we assume they are total, that is, defined for all values of the variables, and write $U \equiv V$ to indicate that they take the same values.

The fact that a boolean expression may take (and actually takes one of) two values can be used to avoid the functional notation altogether. For example, instead of writing $U \equiv \mathsf{T}$ we shall write simply U as an assertion, and instead of writing $U \equiv \mathsf{F}$ we may write simply not U as an assertion. We take advantage of these alternatives but in many cases we return to the functional notation that occasionally provides a sharper description for some arguments.

There are many useful operations that can be defined on the set BOOL, usually called *boolean operations*. For future reference we identify three of them.

Negation. This is a unary operation denoted in the form $\sim x$, where x is a boolean value. We define $\sim x \equiv \mathsf{T}$ if and only if $x \equiv \mathsf{F}$.

Disjunction. This is a binary operation denoted in the form $x \vee y$, where x and y are boolean values. We put $x \vee y \equiv \mathsf{T}$ if and only if $x \equiv \mathsf{T}$ or $y \equiv \mathsf{T}$.

Conjunction. This is a binary operation denoted in the form $x \wedge y$, where x and y are boolean values. We put $x \wedge y \equiv \mathsf{T}$ if and only if $x \equiv \mathsf{T}$ and $y \equiv \mathsf{T}$.

Note that in these definitions we take advantage of the fact that a boolean variable may take only two values. From this it follows that in order to determine the value it is sufficient to give a necessary and sufficient condition for one of the values, say the value T, and whenever the condition fails the variable takes the other value.

EXAMPLE 1.1.3. We take A = the set of all natural numbers and introduce a predicate P on A such that $Px \equiv \mathrm{T}$ if and only if $x = 0$. Next we can use P to specify a new predicate P' such that $P'x \equiv {\sim}Px$ (which is an example of explicit specification). Note that $(Px \vee P'x) \equiv \mathrm{T}$ holds for all x, but we express this simply in the standard form: $Px \vee P'x$ holds for all x.

We are discussing mappings at a very general level, and this makes it difficult to introduce nontrivial constructions. Still there are some operations that can be introduced here and that have important applications. We consider first the operation of *composition* that applies to two mappings, say f from A to B and g from B to C. The composition of f and g, in this order, is denoted in the form $g \circ f = h$, where h is a mapping from A to C given by the following explicit specification:

$$hx \simeq gfx.$$

The expression on the right is assumed to be undefined if fx is undefined or if $fx \simeq v$ and gv is undefined.

Another important operation is the *union* of the mappings f and g, both from A to B, which is denoted in the form $f \cup g = h$, where h is a mapping from A to B given by the specification

$$hx \simeq v \text{ if and only if } fx \simeq v \text{ or } gx \simeq v.$$

Note that this specification is not single-valued when f and g are arbitrary mappings, for it is possible that $fx \simeq v$ and $gx \simeq v'$, where $v \neq v'$. To make sure the operation is well defined we must impose some restrictions. We say that the mappings f and g, both from A to B, are *consistent* if whenever $fx \simeq v$ and $gx \simeq v'$ then $v = v'$. It follows that the union operation is well defined when applied to consistent mappings.

Closely related to the union operation is the relation of extension between mappings. If f and g are mappings from A to B, we say that g is an *extension* of f if whenever $fx \simeq v$ then $gx \simeq v$. Clearly, this relation is reflexive, transitive, and antisymmetric. Note that if g is an extension of f then f and g are consistent. Furthermore, $f \cup g = g$ if and only if g is an extension of f.

This section has been intended as an informal discussion of several basic concepts, namely, applicative rules, mappings, and specification rules. A more formal treatment of specification rules is given in the next section in connection with numerical functions and predicates, but several general principles can be discussed here.

In dealing with predicates the most important feature is totality. For example, an explicit specification of a predicate P on a set A takes the form

$$Px \equiv U,$$

where U is a boolean expression (i.e., takes values T or F) and depends on a variable x that takes values in A. It is essential here that the expression U be defined for all possible values of the variable x.

The use of the symbol $\equiv$ (essentially equality between boolean values) can be extended. We have already noted that a boolean expression U can be considered as an assertion, and to assert U is equivalent to asserting $U \equiv \mathsf{T}$. Now assume U and V are two boolean expressions. The assertion $U \equiv V$ is equivalent to the two conditions $U \equiv \mathsf{T}$ and $V \equiv \mathsf{T}$, or $U \equiv \mathsf{F}$ and $V \equiv \mathsf{F}$, and these conditions can be written informally as follows:

U is equivalent to V

U if and only if V.

In the specification of mappings the main consideration is to ensure single-valuedness. Explicit specifications are very important in this connection. In some extreme cases we have to give specifications where single-valuedness becomes problematic. For example, it is possible to specify a mapping h from A to B in the form

$$hx \simeq v \equiv \cdots x \cdots v \cdots,$$

where the expression on the right is a relation between x and v, and the connective $\equiv$ can be read "if and only if." This specification is legitimate only if the given relation is single-valued, and this usually requires some argument or proof.

Let h be a mapping from A to B and f a mapping from B to A. We say that f is a *left inverse* of h if whenever $hx \simeq y$ then $fy \simeq x$. We say that f is a *right inverse* of h if f is total and whenever $fy \simeq x$ then $hx \simeq y$. We say that f is an inverse of h if f is both a left inverse and a right inverse of h.

These concepts are classical in the theory of mappings. They are closely related to the concepts of 1-1 and onto. These relations are developed in the exercises.

Exercises

1.1.1. Which are the mappings from A to B when B is the empty set?

1.1.2. Let f and h be two mappings from A to B. Use the definition of extensional abstraction to determine the conditions that are necessary and sufficient for $f = h$, that is, for f and h to be identical mappings.

1.1.3. Let f and h be mappings from A to B. Assume f is total and h is an extension of f. Prove that $f = h$.

1.1.4. Let h be a mapping from A to B where B is nonempty. Prove that the following conditions are equivalent:
(a) h is a total mapping.
(b) The only extension of h is h itself.

1.1.5. Give an example that shows that the equivalence in Exercise 1.1.4 does not hold when B is the empty set.

1.1.6. Let A and B be sets, and assume $\mathscr{C}$ is a class of mappings from A to B such that any two mappings in $\mathscr{C}$ are consistent. Prove that there is a unique mapping h_0 from A to B which satisfies the following conditions: (a) h_0 is an extension of every mapping in $\mathscr{C}$. (b) If h is a mapping from A to B which is an extension of every mapping in $\mathscr{C}$, then h is an extension of h_0.

1.1.7. Let h be a mapping from A to B. Assume f is a left inverse of h and f' is a right inverse of h. Prove that $f = f'$.

1.1.8. Let h be a mapping from A to B. Prove that the following conditions are equivalent:
(a) h is 1-1.
(b) h has a left inverse.

1.1.9. Let h be a mapping from A to B. Prove that if h has a right inverse then h is onto.

1.1.10. Discuss the conditions under which the converse of Exercise 1.1.9 holds; that is, if h is a mapping which is onto then h has a right inverse.

1.1.11. Let h be a mapping from A to B. Prove that the following conditions are equivalent:
(a) h is onto and 1-1.
(b) h is onto and has a left inverse.
(c) h has a right inverse and a left inverse.
(d) h has an inverse.

Notes

Material in this section belongs to what is now called discrete mathematics. We have made a mild attempt to avoid a complete reduction of mappings to sets, which is the fashionable approach. On the other hand, we mention in the text the possibility of some mappings being determined purely by single-valued relations.

We emphasize from the beginning that mappings are partial operations, possibly undefined for some arguments. This is a crucial aspect of some definitions, which have been introduced to force the reader to be alert to this phenomenon. The definitions of onto and 1-1 must be considered in this context, and the exercises present some classical results in this more general setting. At the same time we have tried to identify the role of the axiom of choice in the proofs of some of these results (see Exercise 1.1.10).

§2. Numerical Functions

We denote by $\mathbb{N} = \{0, 1, 2, \ldots\}$ the set that contains all the *natural numbers*, which we call simply *numbers*. These are the only numbers to be considered in this work. To denote elements of $\mathbb{N}$ we use the following *numerical variables*: $i, j, k, m, n, u, v, w, x, y$, and z, with subscripts or superscripts if necessary.

A finite sequence of elements of $\mathbb{N}$ of length k, $k \geq 0$, is called a *k-tuple*. The sequence may contain repetitions so the same number may appear in different places. For example, $(2, 3, 2)$ is a 3-tuple, or a triple, $(3, 2, 1)$ is also a 3-tuple, $(2, 2)$ is a 2-tuple, or pair, (1) is a 1-tuple. There is a unique 0-tuple or empty tuple, which is denoted with the symbol $(\ )$.

We need variables to denote arbitrary k-tuples. In principle we can use the obvious notation with numerical variables. For example, $(x_1, \ldots, x_k)$ denotes an arbitrary k-tuple, or (x_1, x_2, x_3) denotes an arbitrary triple, and so on. This notation is used whenever it is necessary to make explicit the elements of a k-tuple. In many cases this is not necessary and we can use a more compact notation. Letters of the form $\mathbf{x}, \mathbf{y}, \mathbf{z}, \ldots$ are considered abbreviations of lists of variables, more precisely of distinct variables, including as a possibility an empty list. In most cases the letter $\mathbf{x}$ represents the list $x_1, \ldots, x_k$ for some $k \geq 0$, and $\mathbf{y}$ represents $y_1, \ldots, y_m$ for some $m \geq 0$, and similarly for $\mathbf{z}$, or any other letter constructed in the same form. But other relations are admissible. For example, if x, u, w are given variables, we may say that $\mathbf{x}$ contains x, u, and w, which means that $\mathbf{x}$ is a list of the form z, y, w, u, x, or any other combination in which the variables x, y, w appear in some places. Note that the length of the list is not referred to in the notation $\mathbf{x}$. Such information must be provided by the context in which the notation is used. In general, if $\mathbf{x}$ is a list of variables which expands, say, to $x_1, \ldots, x_k$, then $(\mathbf{x})$ is equivalent to $(x_1, \ldots, x_k)$, which can be used to denote an arbitrary k-tuple.

If $k \geq 0$, then $\mathbb{N}^k$ denotes the set of all k-tuples. For example, $\mathbb{N}^0 = \{(\ )\}$, which is a set that contains exactly one element, the empty tuple $(\ )$, and $\mathbb{N}^1 = \{(0), (1), (2), \ldots\}$, which we often identify with the set $\mathbb{N}$. The set $\mathbb{N}^2$ can be generated in several ways; for example, $\mathbb{N}^2 = \{(0,0), (1,0), (0,1), (2,0), (1,1), (0,2), \ldots\}$.

Elements of a set $\mathbb{N}^k$ are denoted using the notation introduced above. For example, $(\mathbf{x}) \in \mathbb{N}^k$ means that $(\mathbf{x})$ is a k-tuple that depends on the values of the variables occurring in the list $\mathbf{x}$.

A mapping from $\mathbb{N}^k$ to $\mathbb{N}$ is said to be a *k-ary numerical function*, or simply a *k-ary function*. A predicate on $\mathbb{N}^k$ is said to be a *k-ary numerical predicate*, or simply a *k-ary predicate*. If h is a k-ary function and $(\mathbf{x}) \in \mathbb{N}^k$, then $h(\mathbf{x})$ is the application of h to $(\mathbf{x})$, which can be defined or undefined, as explained in the preceding section. If P is a k-ary predicate and $(\mathbf{x}) \in \mathbb{N}^k$, then $P(\mathbf{x})$ is the application of P to $(\mathbf{x})$, and this expression is always defined; that is, $P(\mathbf{x}) \equiv \mathsf{T}$ or $P(\mathbf{x}) \equiv \mathsf{F}$.

EXAMPLE 1.2.1. Assume P is a 3-ary predicate, and U_1, U_2, and U_3 are total numerical expressions involving the variables x_1 and x_2; that is, they are defined for any values of x_1 and x_2. Using explicit specification we introduce a new binary predicate P' such that

$$P'(x_1, x_2) \equiv P(U_1, U_2, U_3).$$

Note that the condition that U_1, U_2, and U_3 are total is essential if we want P' to be a total mapping. We could overcome this difficulty by setting $P'(x_1, x_2) \equiv \mathsf{F}$ whenever the expression on the right is undefined, but this is not a good solution (consider the situation that arises when the predicate P' is negated) and we prefer to enforce the restriction that substitution in predicates is permissible only with total expressions.

EXAMPLE 1.2.2. Consider the binary predicate $\mathsf{P}_=$ given by the explicit specification

$$\mathsf{P}_=(x_1, x_2) \equiv x_1 = x_2.$$

If numerical expressions U_1 and U_2 are substituted, we get the equivalent expressions

$$\mathsf{P}_=(U_1, U_2) \equiv U_1 = U_2,$$

and of course we assume U_1 and U_2 are total. This means that the connective $=$ can be used only between total expressions. If U_1 and U_2 are nontotal, we can use the connectives $\simeq$ or $\not\simeq$ to obtain

$$U_1 \simeq U_2$$

$$U_1 \not\simeq U_2.$$

Note that these are total boolean expressions; in fact both are true when U_1 and U_2 are both undefined. This machinery is sufficient to deal with nontotal expressions.

We know that for every $k \geq 0$ there is a k-ary totally undefined function, which in this context we denote by UD_k. It follows that $\mathsf{UD}_k(\mathbf{x})\uparrow$ for all $(\mathbf{x}) \in \mathbb{N}^k$. Occasionally we write UD for UD_1.

In the preceding section we described an ordering relation between mappings, which has interesting applications in terms of numerical functions. If both f and g are k-ary functions, and g is an extension of f, we write $f \subseteq_k g$. The class of all k-ary functions is a partial order under this relation, with a bottom element UD_k, but there is no top element, so this structure is not a complete lattice.

The extension relation applied to predicates is trivial, for these are total mappings. Here we consider a different ordering, which also has interesting applications. If P and Q are both k-ary predicates and $P(\mathbf{x})$ implies $Q(\mathbf{x})$ for every $(\mathbf{x}) \in \mathbb{N}^k$ (i.e., whenever $P(\mathbf{x}) \equiv \mathsf{T}$ then $Q(\mathbf{x}) \equiv \mathsf{T}$), we say that P is *included* in Q, and write $P \subseteq_k Q$. This is also a partial order with a bottom element, the predicate $\mathsf{F}_k(\mathbf{x}) \equiv \mathsf{F}$ for all $(\mathbf{x}) \in \mathbb{N}^k$, and a top element, the predicate $\mathsf{T}_k(\mathbf{x}) \equiv \mathsf{T}$ for all $(\mathbf{x}) \in \mathbb{N}^k$. This structure is in fact a complete lattice.

The predicate ordering can be described in a different but equivalent way. If P is a k-ary predicate we can identify P with a set, actually with a subset of $\mathbb{N}^k$, by the relation

$$(\mathbf{x}) \in P \equiv P(\mathbf{x}).$$

With this understanding the relation $P \subseteq_k Q$ is equivalent to the usual inclusion relation between sets. Now we have $\mathsf{F}_k = \varnothing$, and $\mathsf{T}_k = \mathbb{N}^k$.

In the case of a unary predicate we carry the notation a bit further, because here we can identify $\mathbb{N}^1$ with $\mathbb{N}$; hence a unary predicate becomes a subset of $\mathbb{N}$. This has a notational consequence, for if A is a unary predicate, and we consider A as a subset of $\mathbb{N}^1 = \mathbb{N}$, then we do not write $(x) \in A$, where $x \in \mathbb{N}$, but rather $x \in A$. In fact unary predicates will be treated systematically as subsets of $\mathbb{N}$.

Now assume P is a k-ary predicate, $k \geq 0$. We associate with the predicate P two k-ary functions. The first is a total function denoted χ_P such that

$$\chi_P(\mathbf{x}) = 0 \quad \text{if } P(\mathbf{x})$$

$$\chi_P(\mathbf{x}) = 1 \quad \text{otherwise.}$$

The function χ_P is called the *characteristic function* of P. The second function is in most cases nontotal and it is denoted by ψ_P. The specification of ψ_P is as follows:

$$\psi_P(\mathbf{x}) \simeq v \equiv P(\mathbf{x}) \wedge v = 0.$$

The relation on the right is clearly single valued so the function ψ_P is well defined. Clearly, we have $\psi_P(\mathbf{x}) \simeq 0$ in case $P(\mathbf{x})$, and $\psi_P(\mathbf{x})\uparrow$ otherwise. The function ψ_P is called the *partial characteristic function* of P.

Now assume h is a k-ary function, $k \geq 0$. With the function h we associate three numerical predicates: a k-ary predicate D_h, a unary predicate R_h, and a $(k+1)$-ary predicate G_h. The specifications of these predicates are as follows:

$$\mathsf{D}_h(\mathbf{x}) \equiv h(\mathbf{x})\downarrow$$

$$y \in \mathsf{R}_h \equiv \text{there is } (\mathbf{x}) \in \mathbb{N}^k \text{ such that } h(\mathbf{x}) \simeq y$$

$$\mathsf{G}_h(\mathbf{x}, y) \equiv h(\mathbf{x}) \simeq y.$$

The predicate D_h is called the *domain* of h. The predicate R_h is called the *range* of h. The predicate G_h is the *graph predicate* of h. If the set D_h is finite we say that h is *finite*.

Most of the time we shall be concerned with classes of numerical functions and classes of numerical predicates. A class $\mathscr{C}$ of functions may contain functions of different arities, may be finite or infinite, and eventually may be empty. If a function h is an element of the class $\mathscr{C}$, we say that h is $\mathscr{C}$-computable. This notation is intended to suggest that different forms of computability can be described via classes of functions. The idea is interesting only when the class $\mathscr{C}$ satisfies special properties to be discussed later.

Let $\mathscr{C}$ be a class of functions, and P a k-ary predicate. We say that P is *$\mathscr{C}$-decidable* if the function χ_P is $\mathscr{C}$-computable. If the function ψ_P is $\mathscr{C}$-computable, we say that P is *partially $\mathscr{C}$-decidable.*

A class $\mathscr{P}$ of predicates may contain predicates of different arities. For example, if $\mathscr{C}$ is a class of functions, then $\mathscr{C}_{\mathsf{d}}$ is the class of all predicates that are $\mathscr{C}$-decidable, and $\mathscr{C}_{\mathsf{pd}}$ is the class of all predicates that are partially $\mathscr{C}$-decidable.

EXAMPLE 1.2.3. Let TO = the class of all total functions. It follows that only total functions are TO-computable. For example, the functions UD_k, $k \geq 0$, are not TO-computable. The class TO_{d} contains all predicates, but the class of $\mathsf{TO}_{\mathsf{pd}}$ contains only those predicates of the form T_k, $k \geq 0$, that is, predicates such that $\mathsf{T}_k(\mathbf{x}) \equiv \mathsf{T}$ for all $(\mathbf{x}) \in \mathbb{N}^k$ (or equivalently $\mathsf{T}_k = \mathbb{N}^k$).

If $\mathscr{P}$ is a class of predicates, then $\mathscr{P}_{\mathsf{g}}$ denotes the class of all functions h such that G_h is in the class $\mathscr{P}$. We shall find later that this is a very interesting construction when the class $\mathscr{P}$ satisfies adequate closure properties.

The subscript notation is used very often in this work and later will be extended. It is very useful because it has a number of convenient features. For example, it is monotonic and whenever $\mathscr{C}$ and $\mathscr{C}'$ are classes of functions such that $\mathscr{C} \subseteq \mathscr{C}'$ then $\mathscr{C}_{\mathsf{d}} \subseteq \mathscr{C}'_{\mathsf{d}}$ and $\mathscr{C}_{\mathsf{pd}} \subseteq \mathscr{C}'_{\mathsf{pd}}$. It can be concatenated in the obvious way. For example, we can write $\mathscr{C}_{\mathsf{dg}}$, which denotes the class of all functions h such that G_h is $\mathscr{C}$-decidable.

To complete this section we introduce a number of functions that we call *initial functions*. This does not refer to any particular property, but rather to the fact that we identify these functions in advance. Later we shall find that some are actually initial in some constructions. The only purpose here is to identify a number of functions that appear frequently in this work.

All the functions described here are total, and the specifications are given using standard equality. Some of them are well known in general mathematics and the specification reduces to the introduction of some notation. Others are not widely known, in which case a complete specification is given.

The functions are not independent, in the sense that some of them can be specified from others by simple substitution. All of them are introduced with prefix notation, although in practice some will be used with infix notation.

(a) The identity functions. For every $k \geq 1$, and i such that $1 \leq i \leq k$, the k-ary function I_i^k such that $\mathsf{I}_i^k(\mathbf{x}) = x_i$.
(b) The constant functions. For every $k \geq 0$ and $m \geq 0$, the k-ary function C_m^k such that $\mathsf{C}_m^k(\mathbf{x}) = m$.

(c) The successor function. This is a unary function denoted by the symbol σ. The specification is $\sigma(x) = x + 1$.

(d) The conditional function. This is a 3-ary function denoted by the symbol cd. The specification is as follows:

$$\begin{aligned} \mathsf{cd}(x, y, z) &= x \quad \text{if } z = 0 \\ &= y \quad \text{if } z \neq 0 \end{aligned}$$

(e) The sign function. This is a unary function denoted by the symbol sg. The specification is $\mathsf{sg}(x) = 0$ if $x = 0$, $\mathsf{sg}(x) = 1$ if $x \neq 0$.

(f) The cosign function. This is a unary function denoted by the symbol csg. The specification is $\mathsf{csg}(x) = 0$ if $x \neq 0$, $\mathsf{csg}(x) = 1$ if $x = 0$.

(g) The addition function. This is the usual arithmetical binary addition, which we denote by the prefix symbol add, although in applications the standard infix notation will be preferred. We assume this function is known. The specification is simply $\mathsf{add}(x, y) = x + y$.

(h) The product function. This is the usual arithmetical binary function multiplication, which we denote by the prefix symbol pr, although in applications the standard infix notation will be preferred. The specification is $\mathsf{pr}(x, y) = x \times y$.

(i) The exponentiation function. This is the usual arithmetical binary power operation, which we denote by the prefix symbol exp, although in some cases the standard notation is used. The specification is $\mathsf{exp}(x, y) = x^y$ (note that $\mathsf{exp}(0, 0) = 1$).

(j) The substraction function. This is an extension of the usual (partially defined) arithmetical substraction, which is denoted by the symbol sb. We put $\mathsf{sb}(x, y) = x - y =$ the standard substraction of y from x if $x \geq y$, and $\mathsf{sb}(x, y) = 0$ if $x < y$. On occasions we use the notation $\mathsf{sb}(x, y) = x \dot{-} y$.

(k) The predecessor function. This is the unary function denoted by the symbol pd such that $\mathsf{pd}(x) = \mathsf{sb}(x, 1)$.

(l) The equality function. This is a binary function denoted by the symbol ε. The specification is $\varepsilon(x, y) = 0$ if $x = y$, $\varepsilon(x, y) = 1$ if $x \neq y$. Note that ε is the characteristic function of the predicate $\mathsf{P}_=$ introduced in Example 1.2.2.

We have already mentioned that the initial functions are not independent. We note the following relations:

$$\begin{aligned} \mathsf{C}^k_{m+1}(\mathbf{x}) &= \sigma(\mathsf{C}^k_m(\mathbf{x})) \\ \mathsf{sg}(x) &= \mathsf{cd}(0, 1, x) \\ \mathsf{csg}(x) &= \mathsf{cd}(1, 0, x) \\ \mathsf{csg}(x) &= \mathsf{sb}(1, x) \\ \mathsf{sg}(x) &= \mathsf{csg}(\mathsf{csg}(x)) \\ \mathsf{cd}(x, y, z) &= (x \times \mathsf{csg}(z)) + (y \times \mathsf{sg}(z)) \\ \varepsilon(x, y) &= \mathsf{sg}(\mathsf{sb}(x, y) + \mathsf{sb}(y, x)). \end{aligned}$$

A subtler relation between functions and predicates is given by the so-called selector property. If P is a $(k + 1)$-ary predicate, and f is a k-ary function, we say that f is a *selector function* for P if the following two conditions are satisfied:

(i) If there is y such that $P(\mathbf{x}, y)$ then $f(\mathbf{x})\downarrow$.
(ii) If $f(\mathbf{x}) \simeq y$ then $P(\mathbf{x}, y)$.

EXAMPLE 1.2.4. Consider $\mathsf{P}_=$, the binary predicate of Example 1.2.2. It follows that the identify function I_1^1 is the only selector function for the predicate $\mathsf{P}_=$.

EXAMPLE 1.2.5. Consider the binary predicate P such that

$$P(x, y) \equiv x \neq y.$$

There are infinitely many selector functions for the predicate P. In particular, the function σ is one of the selector functions.

For every $(k + 1)$-ary predicate there is at least one selector function, and in some cases only one. If $\mathscr{C}$ is a class of functions and $\mathscr{P}$ is a class of predicates, we say that $\mathscr{C}$ has the *selector property* relative to $\mathscr{P}$ if every $(k + 1)$-ary predicate in $\mathscr{P}$ has at least one $\mathscr{C}$-computable selector function.

Although selector functions are not uniquely determined, it is convenient to use a standard notation, which, given a predicate, denotes an arbitrary selector function. If P is a $(k + 1)$-ary predicate we put

$$\sigma y P(\mathbf{x}, y),$$

to denote an indeterminate selector function for P.

EXAMPLE 1.2.6. If f is a k-ary function, then G_f is a $(k + 1)$-ary predicate. It follows that f is the only selector function for the predicate G_f.

If $\mathscr{C}$ is a class of functions which has the selector property relative to $\mathscr{C}_{\mathsf{d}}$, we say that $\mathscr{C}$ has the d-*selector property*. In a similar way we define the pd-*selector property*.

The consistency relation on mappings considered in the preceding section applies to numerical functions. If f and g are k-ary functions, then f is *consistent* with g if whenever $f(\mathbf{x}) \simeq v$ and $g(\mathbf{x}) \simeq v'$ then $v = v'$. In this case the union $f \cup g$ is well defined. Note that $f \subseteq_k f \cup g, g \subseteq_k f \cup g$. Furthermore, if $f \subseteq_k h$ and $g \subseteq_k h$, then $f \cup g \subseteq_k h$.

If $\mathscr{C}$ is a class of k-ary functions, we say that $\mathscr{C}$ is *consistent* if any two functions in $\mathscr{C}$ are consistent. If $\mathscr{C} = \varnothing$, then $\mathscr{C}$ is certainly consistent. If $\mathscr{C}$ contains only total k-ary functions, then $\mathscr{C}$ is consistent if and only if $\mathscr{C}$ has at most one element.

If $\mathscr{C}$ is a consistent class of k-ary functions, we define $\mathsf{Sup}\,\mathscr{C} = h$, where h is the k-ary function such that $h(\mathbf{x}) \simeq v$ if and only if there is a $\mathscr{C}$-computable function h' such that $h'(\mathbf{x}) \simeq v$. Note that $h' \subseteq_k \mathsf{Sup}\,\mathscr{C}$ whenever h' is $\mathscr{C}$-computable. Furthermore, if $h' \subseteq_k h''$ whenever h' is $\mathscr{C}$-computable, then $\mathsf{Sup}\,\mathscr{C} \subseteq_k h''$.

If $\mathscr{C}$ is the class of all functions in the enumeration $h_0, h_1, \ldots, h_n, \ldots$ and all these functions are consistent k-ary functions, we write

$$\mathsf{Sup}\,\mathscr{C} = h_0 \cup h_1 \cup \cdots \cup h_n \cup \cdots.$$

There is a dual operation that we mention for completeness. If $\mathscr{C}$ is a nonempty class of k-ary functions, we can introduce a k-ary function h such that $h(\mathbf{x}) \simeq v$ if and only if $h'(\mathbf{x}) \simeq v$ whenever h' is a $\mathscr{C}$-computable function. We write $\mathsf{Inf}\,\mathscr{C} = h$. Note that $\mathsf{Inf}\,\mathscr{C} \subseteq_k h'$ if h' is $\mathscr{C}$-computable. Furthermore, if $h'' \subseteq_k h'$ whenever h' is $\mathscr{C}$-computable then $h'' \subseteq_k \mathsf{Inf}\,\mathscr{C}$.

Exercises

1.2.1. Describe the 0-ary functions and the 0-ary predicates.

1.2.2. Let $\mathscr{C}$ be the class that contains only the constant functions. Describe the elements of the classes $\mathscr{C}_{\mathsf{d}}$, $\mathscr{C}_{\mathsf{pd}}$, $\mathscr{C}_{\mathsf{dg}}$, and $\mathscr{C}_{\mathsf{pdg}}$.

1.2.3. Describe the elements of the classes TO_{d}, $\mathsf{TO}_{\mathsf{pd}}$, $\mathsf{TO}_{\mathsf{dg}}$, and $\mathsf{TO}_{\mathsf{pdg}}$.

1.2.4. Let $\mathscr{C}$ be the class that contains all the constant functions. Prove that $\mathscr{C}$ has the selector property relative to $\mathscr{C}_{\mathsf{pd}}$ but does not have the selector property relative to $\mathscr{C}_{\mathsf{d}}$.

1.2.5. Prove that if $\mathscr{C}$ is a class of functions which has the selector property relative to a class of predicates $\mathscr{P}$, then $\mathscr{P}_{\mathsf{g}} \subseteq \mathscr{C}$.

1.2.6. Let A be a numerical set, that is, a unary predicate. Describe the possible selector functions for A.

1.2.7. Let $\mathscr{P}$ be the class that contains all predicates F_{k+1}, $k \geq 0$. Prove that there is a class $\mathscr{C}$ which has the selector property relative to $\mathscr{P}$, and whenever $\mathscr{C}'$ is another class which has the selector property relative to $\mathscr{P}$ then $\mathscr{C} \subseteq \mathscr{C}'$.

1.2.8. Let h be a k-ary function. Prove that there is a class $\mathscr{C}$ that contains only finite k-ary functions, $\mathscr{C}$ is consistent, and $\mathsf{Sup}\,\mathscr{C} = h$.

1.2.9. Let $\mathscr{C}$ be a class of k-ary functions. A function h is an upper bound of $\mathscr{C}$ if $h' \subseteq_k h$ whenever h' is $\mathscr{C}$-computable. Similarly, we say that h is a lower bound of $\mathscr{C}$ if $h \subseteq_k h'$ whenever h' is $\mathscr{C}$-computable. Let $\mathscr{C}_1$ be the class of all upper bounds of $\mathscr{C}$, and $\mathscr{C}_2$ the class of all lower bounds of $\mathscr{C}$. Prove the following:
(a) $\mathscr{C}$ is consistent if and only if $\mathscr{C}_1$ is nonempty. If $\mathscr{C}$ is consistent the $\mathsf{Sup}\,\mathscr{C} = \mathsf{Inf}\,\mathscr{C}_1$.
(b) $\mathscr{C}$ is nonempty if and only if $\mathscr{C}_2$ is consistent. If $\mathscr{C}$ is nonempty then $\mathsf{Inf}\,\mathscr{C} = \mathsf{Sup}\,\mathscr{C}_2$.

Notes

The fact that computability theory deals with numerical functions does not mean that it is a part of number theory in the standard mathematical sense. The set $\mathbb{N}$ enters in computability theory as a kind of universal domain, among other possible domains. The universality refers to the very strong encoding properties of $\mathbb{N}$. In fact, any other reasonable computational domain (which can be considered as built by concatenation from a given alphabet) can be encoded in $\mathbb{N}$. This requires only the possibility of encoding sequences of numbers, and many techniques are known to be satisfactory for this purpose. The one we use in this work is explained in Section 6 of this chapter.

The distinction between numerical functions and numerical predicates is to some extent arbitrary. A predicate is officially a total function, but by using the partial characteristic function it becomes a nontotal operation. On the other hand, any function that whenever defined takes the value 0 can be considered as a total function where undefined is now represented by the value 1. The two versions are completely equivalent in general mathematics but may fail to be so from a computational point of view.

§3. Finitary Rules

In this section we discuss a number of specification rules involving functions and predicates. Typically, a specification rule assumes some given functions and predicates and describes a new function or predicate. All the rules presented in this section share a common property: they are finitary in a sense we proceed to explain.

Suppose a rule specifies a function h from given functions $f_1, f_2, \ldots$ and predicates $P_1, P_2, \ldots$. The number of given functions and predicates is always finite, but it may change from one application of the rule to another. Assume that $h(\mathbf{x})$ is defined for some input $(\mathbf{x})$. The value of $h(\mathbf{x})$ in general will depend on several values of the given functions and predicates. The rules are *finitary* in the sense that the number of values required from the given functions and predicates in order to evaluate $h(\mathbf{x})$ is always finite.

For example, assume we have a specification in which the unary function h is described from a given function f as follows:

$$h(x) \simeq f(x) + f(x + 1).$$

This means that whenever $f(x)$ or $f(x + 1)$ is undefined then $h(x)$ is undefined. If $h(x)$ is defined, then we need two values of f in order to evaluate $h(x)$, namely, $f(x)$ and $f(x + 1)$. So the rule is finitary.

Let us now consider an example in which the specification is not finitary. Suppose h is a unary function that is specified from a function f as follows: $h(x) \simeq 0$ if for every $y \geq x$, $f(y) \simeq 0$; $h(x) \simeq 1$ if for some $y \geq x$, $f(y) \not\simeq 0$. Otherwise the value of $h(x)$ is undefined. Clearly, in order to determine the value of $h(x)$ we may need an infinite number of values of f, although in some cases a finite number may be sufficient. So this rule is not finitary.

Although the preceding explanation is not intended as a formal definition, it should be sufficient for the reader to check that all rules described below are in fact finitary. A formal version of this property, namely, continuity, applies to functional transformations, and will be discussed in the next chapter.

We proceed now to describe the finitary specification rules. In most cases they involve nontotal functions, so the specification will be given using the symbol $\simeq$. We assume in all rules that $k \geq 0$.

Partial Substitution. Let f be a given $(k + 1)$-ary function, and g a given k-ary function. We specify a k-ary function h as follows:

$$h(\mathbf{x}) \simeq f(\mathbf{x}, g(\mathbf{x})).$$

We stipulate that when $g(\mathbf{x})$ is undefined then $h(\mathbf{x})$ is undefined. If $g(\mathbf{x}) \simeq y$ but $f(\mathbf{x}, y)$ is undefined, then $h(\mathbf{x})$ is also undefined. In all the other cases $h(\mathbf{x})$ is defined and the value is given by the right side expression.

Adjoining of Variables. Let f be a given k-ary function. We specify two new $(k + 1)$-ary functions h_1 and h_2 such that

$$h_1(\mathbf{x}, y) \simeq f(\mathbf{x})$$

$$h_2(y, \mathbf{x}) \simeq f(\mathbf{x}).$$

Full Substitution. Let f be a m-ary function, $m \geq 0$, and $g_1, \ldots, g_m$ be m k-ary functions. We specify a new k-ary function h such that

$$h(\mathbf{x}) \simeq f(g_1(\mathbf{x}), \ldots, g_m(\mathbf{x})).$$

The conditions for $h(\mathbf{x})$ to be defined are similar to those in partial substitution. All expressions $g_1(\mathbf{x}), \ldots, g_m(\mathbf{x})$ must be defined, and if $g_i(\mathbf{x}) \simeq y_i$, $i = 1, \ldots, m$, then $f(y_1, \ldots, y_m)$ must be defined.

Predicate Substitution. Let Q be a given m-ary predicate, $m \geq 0$, and let $g_1, \ldots, g_m$ be k-ary total functions. We specify a new k-ary predicate P as follows:

$$P(\mathbf{x}) \equiv Q(g_1(\mathbf{x}), \ldots, g_m(\mathbf{x})).$$

Since all functions are total the right side expression is self-explanatory.

Boolean Operations. Let Q_1 and Q_2 be given k-ary predicates. We specify three new predicates as follows:

$$P_1(\mathbf{x}) \equiv \sim Q_1(\mathbf{x})$$

$$P_2(\mathbf{x}) \equiv Q_1(\mathbf{x}) \vee Q_2(\mathbf{x})$$

$$P_3(\mathbf{x}) \equiv Q_1(\mathbf{x}) \wedge Q_2(\mathbf{x}).$$

The meaning of these boolean operations was explained in Section 1. We shall use the following notation: $P_1 = \bar{Q}_1$, $P_2 = Q_1 \vee Q_2$, and $P_3 = Q_1 \wedge Q_2$. If Q_1 and Q_2 are unary predicates, we may interpret them as numerical sets, and then P_1, P_2, and P_3 are also numerical sets. Under this interpretation we change the notation slightly and write $P_2 = Q_1 \cup Q_2$ and $P_3 = Q_1 \cap Q_2$. These are of course the usual set operations of union and intersection. For the complement of a set we use the same notation as the negation of the corresponding predicate.

Bounded Sum. Let f be a given $(k+1)$-ary function. We specify a new $(k+1)$-ary function h as follows:

$$h(\mathbf{x}, z) \simeq \sum_{y<z} f(\mathbf{x}, y).$$

This specification is understood as follows: If $z = 0$ then $h(\mathbf{x}, z) \simeq 0$. If $z > 0$, and $f(\mathbf{x}, i) \simeq y_i$ for $i = 0, 1, \ldots, z-1$, then $h(\mathbf{x}, z) \simeq y_0 + \cdots + y_{z-1}$. If for some $i < z$, $f(\mathbf{x}, i)\uparrow$ then $h(\mathbf{x}, z)\uparrow$.

Bounded Product. Let f be a given $(k+1)$-ary function. We specify a $(k+1)$-ary function h as follows:

$$h(\mathbf{x}, z) \simeq \prod_{y<z} f(\mathbf{x}, y).$$

This specification is understood essentially as in bounded sum, with the following changes: if $z = 0$ we say $h(\mathbf{x}, z) \simeq 1$; if $z > 0$ we replace addition by multiplication.

Existential Bounded Quantification. Let Q be a given $(k+1)$-ary predicate. We specify a new $(k+1)$-ary predicate P as follows:

$$P(\mathbf{x}, z) \equiv \exists y < zQ(\mathbf{x}, y)$$
$$\equiv \text{there is some } y < z \text{ such that } Q(\mathbf{x}, y) \equiv \mathsf{T}.$$

It follows from this specification that $P(\mathbf{x}, 0) \equiv \mathsf{F}$.

Universal Bounded Quantification. Let Q be a given $(k+1)$-ary predicate. We specify a new $(k+1)$-ary predicate P as follows:

$$P(\mathbf{x}, z) \equiv \forall y < zQ(\mathbf{x}, y)$$
$$\equiv \text{for all } y < z,\ Q(\mathbf{x}, y) \equiv \mathsf{T}.$$

It follows from this specification that $P(\mathbf{x}, 0) \equiv \mathsf{T}$.

Bounded Minimalization. Let Q be a given $(k+1)$-ary predicate. We specify a new $(k+1)$-ary function h as follows:

$$h(\hat{\mathbf{x}}, z) \simeq \text{the least } y < z \text{ such that } Q(\mathbf{x}, y) \equiv \mathsf{T}$$
$$\simeq z \text{ if there is no such } y.$$

Note that h is always a total function. The notation for this specification will be as follows:

$$h(\mathbf{x}, z) = \mu y < zQ(\mathbf{x}, y).$$

Unbounded Minimalization with Predicates. Let Q be a given $(k+1)$-ary predicate. We specify a new k-ary function h as follows:

$$h(\mathbf{x}) \simeq \text{the least } y \text{ such that } Q(\mathbf{x}, y) \equiv \mathsf{T}.$$

This must be understood in the sense that $h(\hat{\mathbf{x}})$ is undefined in case there is no

y such that $Q(\mathbf{x}, y) \equiv \mathsf{T}$. Note that although the predicate Q is by definition a total function, the application of this operation may result in a nontotal function. The notation for this type of specification is as follows:

$$h(\mathbf{x}) \simeq \mu y Q(\mathbf{x}, y).$$

Unbounded Minimalization with Functions. Let f be a given $(k + 1)$-ary function. We specify a new k-ary function h as follows:

$$h(\mathbf{x}) \simeq \text{the least } y \text{ such that } f(\mathbf{x}, y) \simeq 0, \text{ and}$$
$$\text{for all } y' < y,\ f(\mathbf{x}, y') \not\simeq 0.$$

Again if the least y with the property described on the right side of the specification does not exist then $h(\mathbf{x})$ is undefined. The notation for this specification is as follows:

$$h(\mathbf{x}) \simeq \mu y(f(\mathbf{x}, y) \simeq 0).$$

Total Definition by Cases. Let $Q_1, \ldots, Q_m$ be m k-ary predicates, $m \geq 1$, and $f_1, \ldots, f_m, f_{m+1}$ be $m + 1$ total k-ary functions. We specify a k-ary function h as follows:

$$\begin{aligned} h(\mathbf{x}) &\simeq f_1(\mathbf{x}) && \text{if } Q_1(\mathbf{x}) \\ &\simeq f_2(\mathbf{x}) && \text{if } Q_2(\mathbf{x}) \\ &\ \ \vdots && \ \ \vdots \quad \vdots \\ &\simeq f_m(\mathbf{x}) && \text{if } Q_m(\mathbf{x}) \\ &\simeq f_{m+1}(\mathbf{x}) && \text{otherwise.} \end{aligned}$$

Note that we do not impose any restriction on the given predicates. In order to evaluate $h(\mathbf{x})$ we look for the first i such that $Q_i(\mathbf{x}) \equiv \mathsf{T}$ and put $h(\mathbf{x}) \simeq f_i(\mathbf{x})$. If there is no such i, we put $h(\mathbf{x}) \simeq f_{m+1}(\mathbf{x})$. Since all given functions are total a value for $h(\mathbf{x})$ is always obtained, so h is a total function.

Partial Definition by Cases. Let $f_1, \ldots, f_m$ and $g_1, \ldots, g_m$ be given k-ary functions. We specify a new k-ary function h as follows:

$$\begin{aligned} h(\mathbf{x}) &\simeq f_1(\mathbf{x}) && \text{if } g_1(\mathbf{x}) \simeq 0 \\ &\simeq f_2(\mathbf{x}) && \text{if } g_2(\mathbf{x}) \simeq 0 \\ &\ \ \vdots && \\ &\simeq f_m(\mathbf{x}) && \text{if } g_m(\mathbf{x}) \simeq 0. \end{aligned}$$

To evaluate $h(\mathbf{x})$ we look for the first i such that $g_i(\mathbf{x}) \simeq 0$, and whenever $j < i$ then $g_j(\mathbf{x}) \not\simeq 0$, and put $h(\mathbf{x}) \simeq f_i(\mathbf{x})$. If there is no such i then $h(\mathbf{x})$ is undefined. Hence $h(\mathbf{x})$ may be undefined, even if all given functions are total.

Besides being finitary there is another related property that these rules share. They are all *monotonic*, relative to the ordering $\subseteq_k$ between functions.

For example, assume a k-ary function h is specified by one of the finitary rules from given functions $f_1, f_2, \ldots$ and some predicates, where f_1 is k_1-ary, f_2 is k_2-ary, and so on. If we replace $f_1, f_2, \ldots$ by $f_1', f_2', \ldots$, of the same arity, and apply the rule with these new functions and the same predicates, we obtain a new k-ary function h'. The rule is monotonic in the sense that whenever $f_1 \subseteq_{k_1} f_1'$, $f_2 \subseteq_{k_2} f_2'$, ... then $h \subseteq_k h'$. The reader is urged to check that this condition is satisfied by all finitary rules because it has important applications in this work.

Some of the finitary rules have the property that whenever in a particular application all the given functions are total, then the specified function is also total. Such rules are said to be *elementary*. There are three nonelementary rules, namely, unbounded minimalization with predicates, unbounded minimalization with functions, and partial definition by cases. The transition from elementary to nonelementary rules is a characteristic aspect of computability theory. This point will be elaborated later, but note that given initial functions of the kind introduced in the preceding section, nonelementary rules are required to generate nontotal functions.

EXAMPLE 1.3.1. Assume f is a total binary function such that $f(x, y) = x + y + 1$. Using unbounded minimalization with functions we introduce a unary function h such that

$$h(x) \simeq \mu y(f(x, y) \simeq 0).$$

Clearly h is the unary undefined function UD.

EXAMPLE 1.3.2. Using the initial binary function ε we use partial definition by cases to introduce a binary function ε' such that

$$\varepsilon'(x, y) \simeq \varepsilon(x, y) \quad \text{if } \varepsilon(x, y) \simeq 0.$$

In this application of the rule there is only one case, and although the given function ε is total, the new function ε' is nontotal. In fact, $\varepsilon'(x, y)$ is defined if and only if $x = y$. It is clear that while ε is the characteristic function of the equality predicate $\mathsf{P}_=$ (see Section 2), ε' is the partial characteristic function of the same predicate.

EXERCISES

1.3.1. Let h be a unary function that is specified from a binary predicate P, using unbounded minimalization in the form

$$h(x) \simeq \mu y P(x, y).$$

Assume $h(x)$ is defined for some x. How many values of the predicate P are required to evaluate $h(x)$?

1.3.2. Let f be a 0-ary function. Which functions can be specified from f using full substitution?

1.3.3. Show that any application of partial substitution or adjoining of variables can be replaced by full substitution with identity functions.

1.3.4. Give an example in which a specification by full substitution with nontotal functions produces a total function.

1.3.5. Consider a binary function h that is specified from a binary predicate P, using bounded minimalization in the form

$$h(x, z) = \mu y < zP(x, y).$$

Give a simple rule to evaluate $h(x, z + 1)$ using the value $h(x, z)$.

1.3.6. Give a specification for the function pr using bounded sum and an identity function.

1.3.7. Give a specification for the function exp using bounded product and an identity function.

1.3.8. Consider a set A (i.e., a unary predicate) and the two unary functions h_1 and h_2 such that

$$h_1(y) \simeq \sum_{v<y} \chi_A(v)$$

$$h_2(y) \simeq \sum_{v<y} \psi_A(v)$$

Which is the information about the set A given by the functions h_1 and h_2?

1.3.9. Consider a $(k + 1)$-ary predicate Q. Using bounded sum and bounded product we introduce two $(k + 1)$-ary total functions h_1 and h_2 such that

$$h_1(\mathbf{x}, z) = \prod_{y<z} \chi_Q(\mathbf{x}, y)$$

$$h_2(\mathbf{x}, z) = \mathsf{sg}\left(\sum_{y<z} \chi_Q(\mathbf{x}, y)\right).$$

(a) Show that there are $(k + 1)$-ary predicates P_1 and P_2 such that h_1 is the characteristic function of P_1 and h_2 is the characteristic function of P_2.
(b) Which is the relation between P_1 and Q? Which is the relation between P_2 and Q?

1.3.10. Prove that the following rules are finitary and monotonic: partial substitution, unbounded minimalization with functions, and partial definition by cases.

Notes

The finitary rules described in this section are an integral part of computability theory. Most of them were identified in the classical paper by Gödel [7], the first attempt at a mathematical theory of computation. Some of these rules are derived from logic and general mathematics, since they do not involve the structure of the set $\mathbb{N}$. Others, like minimalization, are very much dependent on such structure.

Recursive rules are not considered in this section, although we shall show later that minimalization is a form of recursion. Recursion, which is discussed in Chapter 2, is also finitary, monotonic, and continuous. Nonfinitary rules are introduced in Chapter 5.

§4. Closure Properties

Each of the rules described in the preceding section induces an associated notion of closure, which applies to classes of functions, or classes of predicates, depending on the particular rule. For example, we say that a class $\mathscr{C}$ of functions is *closed under full substitution*, if whenever a function h is specified by full substitution from functions $f, g_1, \ldots, g_m$, which are $\mathscr{C}$-computable, then h is also $\mathscr{C}$-computable. Similar definitions can be formulated for partial substitution, adjoining of variables, bounded sum, bounded product, unbounded minimalization with functions, and partial definition by cases, for in each of these rules only functions are involved.

Others rules apply to predicates, and produce predicates, in which case the closure properties apply to classes of predicates. For example, we say that a class $\mathscr{P}$ of predicates is *closed under disjunction*, if whenever two k-ary predicates P_1 and P_2 are in $\mathscr{P}$, then $P_1 \vee P_2$ is also in $\mathscr{P}$. Similar definitions can be formulated for negation, conjunction, and bounded quantification.

There are other rules that apply to functions and predicates, and the definition must take into account this situation. For example, we say that a class $\mathscr{C}$ of functions is *closed under bounded minimalization* (*unbounded minimalization with predicates*), if whenever a function h is specified by bounded minimalization (unbounded minimalization) from a predicate Q that is $\mathscr{C}$-decidable, then h is $\mathscr{C}$-computable.

Finally, we consider the two rules that apply simultaneously to functions and predicates. We say that a class $\mathscr{C}$ of functions is *closed* under *total definition by cases*, if whenever a function h is specified by total definition by cases from functions that are $\mathscr{C}$-computable, and predicates that are $\mathscr{C}$-decidable, then h is $\mathscr{C}$-computable. We say that a class $\mathscr{P}$ of predicates is *closed under substitution with $\mathscr{C}$-computable* functions, if whenever a predicate P is obtained by predicate substitution from a predicate that is in $\mathscr{P}$, and functions that are $\mathscr{C}$-computable, then P is also in $\mathscr{P}$.

In applications, we usually consider some subset of the finitary rules and also require some initial functions. A class $\mathscr{C}$ of functions is *closed under basic operations* if the identity functions and the constant functions are $\mathscr{C}$-computable, and $\mathscr{C}$ is closed under full substitution.

Theorem 1.4.1. *Let $\mathscr{C}$ be closed under basic operations. Then $\mathscr{C}$ is closed under partial substitution and adjoining of variables; $\mathscr{C}_{\mathrm{d}}$ and $\mathscr{C}_{\mathrm{pd}}$ are closed under substitution with total $\mathscr{C}$-computable functions.*

PROOF. To prove closure under partial substitution consider a specification of the form

$$h(\mathbf{x}) \simeq f(\mathbf{x}, g(\mathbf{x})),$$

where h is a k-ary function. If $k = 0$, this is full substitution. Otherwise we write another specification using full substitution in the form

$$h(\mathbf{x}) \simeq f(\mathsf{I}_1^k(\mathbf{x}), \ldots, \mathsf{I}_k^k(\mathbf{x}), g(\mathbf{x})).$$

Closure under adjoining of variables is handled in a similar way. Finally, to prove $\mathscr{C}_\mathsf{d}$ is closed under substitution with $\mathscr{C}$-computable functions, consider a specification of the form

$$P(\mathbf{x}) \equiv Q(g_1(\mathbf{x}), \ldots, g_m(\mathbf{x})),$$

where Q is $\mathscr{C}$-decidable and $g_1, \ldots, g_m$ are total $\mathscr{C}$-computable functions. It follows that

$$\chi_P(\mathbf{x}) = \chi_Q(g_1(\mathbf{x}), \ldots, g_m(\mathbf{x})),$$

so P is $\mathscr{C}$-decidable. The closure of $\mathscr{C}_\mathsf{pd}$ is proved in a similar way. □

There is a good motivation to introduce basic operations, namely, to make sure closure applies under general substitution. This is a fairly obvious notion that can easily be explained with examples, but we prefer to give a formal treatment, which is applied later in Chapter 2.

We define a class of symbolic constructions, or terms, in which the initial symbols are numerical variables, numerals, and function symbols, each of a fixed arity. These are assumed to be uninterpreted symbols. We define *basic terms* by the following rules:

BT 1. A variable or a numeral is a basic term.
BT 2. If f is a function symbol of arity $n \geq 1$, and $U_1, \ldots, U_n$ are basic terms, then $f(U_1, \ldots, U_n)$ is a basic term.

Examples of basic terms are

x	BT 1
y	BT 1
10	BT 1
$f(x, y, 10)$	BT 2
$g(f(x, y, 10), y)$	BT 2,

where f is assumed to be a function symbol of arity 3, g a function symbol of arity 2, x and y are numerical variables, and 10 is a numeral.

Now let U be a basic term. An *interpretation* for U is obtained by associating with every function symbol in U a function of the same arity. Once an interpretation for U is available, and the variables in U have been assigned a numerical value, we can attempt the evaluation of U. Such an evaluation may fail to produce a value, since it is possible that some of the functions are nontotal. If such is the case we say that U is undefined (for that interpretation and assignment). Otherwise the evaluation produces a unique value.

Let U be a basic term with a fixed interpretation for the function symbols, and let $\mathbf{x}$ be a list of variables of length $k \geq 0$, containing all variables in U, and perhaps variables that are not in U. We specify a k-ary function h in the form

$$h(\mathbf{x}) \simeq U,$$

which is an explicit specification of h from the term U (with the given interpretation).

We say that a class $\mathscr{C}$ of functions is *closed under basic explicit specifications*, if whenever a function h is given an explicit specification from a term U, and the interpretation of each function symbol in U is a $\mathscr{C}$-computable function, then h is $\mathscr{C}$-computable.

Theorem 1.4.2. *A class $\mathscr{C}$ of functions is closed under basic operations if and only if it is closed under basic explicit specifications.*

PROOF. If $\mathscr{C}$ is closed under basic operations, and $h(\mathbf{x}) \simeq U$, we prove that h is $\mathscr{C}$-computable by induction in the construction of U. Rule BT 1 gives an identity function or a constant function, and rule BT 2 reduces, via the induction hypothesis, to full substitution. In the other direction, it is clear that if $\mathscr{C}$ is closed under basic explicit specifications, then $\mathscr{C}$ contains the identity functions, the constant functions, and it is closed under full substitution, which is just a special case of basic explicit specification. □

Basic terms are used throughout this book, and usually the interpretation is given explicitly by using symbols that denote functions previously introduced.

We say that a class $\mathscr{C}$ of functions is *closed under elementary operations*, if the functions C_0^0, σ, cd, and ε, are $\mathscr{C}$-computable, and $\mathscr{C}$ is closed under partial substitution, adjoining of variables, bounded sum, and bounded product.

Theorem 1.4.3. *Let $\mathscr{C}$ be a class closed under elementary operations. Then*

(i) *$\mathscr{C}$ is closed under basic operations.*
(ii) *$\mathscr{C}_\mathrm{d}$ is closed under boolean operations and bounded quantification.*
(iii) *$\mathscr{C}$ is closed under total definition by cases.*
(iv) *$\mathscr{C}$ is closed under bounded minimalization.*
(v) *All initial functions are $\mathscr{C}$-computable.*

PROOF. To prove (i) we need first to show that the identity functions and the constant functions are $\mathscr{C}$-computable. From the function C_0^0, using partial substitution with σ, we get all functions C_m^0, $m \geq 0$. For example, $\mathsf{C}_1^0(\ ) = \sigma(\mathsf{C}_0^0(\ ))$, and so on. And from C_m^0 we get all functions C_m^k, $k \geq 0$, using adjoining of variables. Concerning the identity functions we use partial substitution to introduce

$$\mathsf{I}_1^2(x, y) = \mathsf{cd}(x, y, \mathsf{C}_0^2(x, y))$$
$$\mathsf{I}_1^1(x) = \mathsf{I}_1^2(x, \mathsf{C}_0^1(x)).$$

From I_1^1 we get all identify functions using adjoining of variables.

We need also to prove closure under full substitution. Without loss of generality we consider a typical example of full substitution:

$$h(x_1, x_2) \simeq f(g_1(x_1, x_2), g_2(x_1, x_2)).$$

Using adjoining of variables we introduce functions f' and g_2' such that

$$f'(x_1, x_2, y_1, y_2) \simeq f(y_1, y_2)$$
$$g_2'(x_1, x_2, y_1) \simeq g_2(x_1, x_2).$$

Finally, we apply partial substitution to get

$$h_1(x_1, x_2, y_1) \simeq f'(x_1, x_2, y_1, g_2'(x_1, x_2, y_1))$$
$$\simeq f(y_1, g_2(x_1, x_2))$$
$$h(x_1, x_2) \simeq h_1(x_1, x_2, g_1(x_1, x_2))$$
$$\simeq f(g_1(x_1, x_2), g_2(x_1, x_2)).$$

This shows that h can be generated by a sequence of applications of adjoining of variables and partial substitution, starting with the $\mathscr{C}$-computable functions f, g_1, and g_2; hence h is also $\mathscr{C}$-computable. This completes the proof of part (i).

To prove that $\mathscr{C}_\mathrm{d}$ is closed under boolean operations assume that Q_1 and Q_2 are $\mathscr{C}$-decidable predicates, $P_1 = \bar{Q}_1$, $P_2 = Q_1 \wedge Q_2$, and $P_3 = Q_1 \vee Q_2$. It follows that

$$\chi_{P_1}(\mathbf{x}) = \mathsf{cd}(1, 0, \chi_{Q_1}(\mathbf{x}))$$
$$\chi_{P_2}(\mathbf{x}) = \mathsf{cd}(\chi_{Q_2}(\mathbf{x}), 1, \chi_{Q_1}(\mathbf{x}))$$
$$\chi_{P_3}(\mathbf{x}) = \mathsf{cd}(0, \chi_{Q_2}(\mathbf{x}), \chi_{Q_1}(\mathbf{x})),$$

which shows that P_1, P_2, and P_3 are $\mathscr{C}$-decidable. To prove closure under existential bounded quantification note that if

$$P(\mathbf{x}, z) \equiv \exists y < z Q(\mathbf{x}, y),$$

where Q is $\mathscr{C}$-decidable, then

$$\chi_P(\mathbf{x}, z) = \prod_{y<z} \chi_Q(\mathbf{x}, y),$$

so P is also $\mathscr{C}$-decidable. Universal bounded quantification can be obtained using negation and existential bounded quantification.

Closure under total definition by cases is easily derivable by induction on the number of cases. If there is only one case, the specification takes the form

$$\begin{aligned} h(\mathbf{x}) &= f_1(\mathbf{x}) \quad \text{if } Q(\mathbf{x}) \\ &= f_2(\mathbf{x}) \quad \text{otherwise,} \end{aligned}$$

where f_1, f_2 are total $\mathscr{C}$-computable functions, and Q is $\mathscr{C}$-decidable predicate. It follows that

$$h(\mathbf{x}) = \mathsf{cd}(f_1(\mathbf{x}), f_2(\mathbf{x}), \chi_Q(\mathbf{x}));$$

hence h is $\mathscr{C}$-computable. If the specification involves $m + 1$ cases $Q_1, \ldots, Q_m$, Q_{m+1}, it can be reduced to a specification with only one case in the form

$$\begin{aligned} h(\mathbf{x}) &= f_1(\mathbf{x}) \quad \text{if } Q_1(\mathbf{x}) \\ &= h'(\mathbf{x}) \quad \text{otherwise,} \end{aligned}$$

where h' can be specified using only m cases. By the induction hypothesis it follows that h' is $\mathscr{C}$-computable; hence h is also $\mathscr{C}$-computable.

The proof of closure under bounded minimalization is more involved. Assume that a $(k + 1)$-ary function h is specified in the form

$$h(\mathbf{x}, z) = \mu y < z Q(\mathbf{x}, y),$$

where Q is $\mathscr{C}$-decidable. Using total definition by cases we introduce a $(k + 1)$-ary function f such that

$$\begin{aligned} f(\mathbf{x}, y) &= \mathsf{C}_0^{k+1}(\mathbf{x}, y) \quad \text{if } \exists v < y Q(\mathbf{x}, v) \\ &= \chi_Q(\mathbf{x}, y) \qquad \text{otherwise.} \end{aligned}$$

We claim that $h(\mathbf{x}, z) = \sum_{y<z} f(\mathbf{x}, y)$. If $h(\mathbf{x}, z) = z$ then there is no $y < z$ such that $Q(\mathbf{x}, y)$ holds; hence $f(\mathbf{x}, y) = 1$ for all $y < z$, and the equation holds. If $h(\mathbf{x}, z) = v < z$, then for $y < v$ we have $f(\mathbf{x}, y) = 1$, and for $y \geq v$ we have $f(\mathbf{x}, y) = 0$; hence $\sum_{y<z} f(\mathbf{x}, y) = v$. From (iii) above it follows that f is $\mathscr{C}$-computable, then h is also $\mathscr{C}$-computable.

To complete the proof we must show that all initial functions are $\mathscr{C}$-computable. This follows from the following specifications:

$$\begin{aligned} \mathsf{sg}(x) &= \mathsf{cd}(0, 1, x) \\ \mathsf{csg}(x) &= \mathsf{cd}(1, 0, x) \\ \mathsf{add}'(x, y, z) &= \sum_{v<z} \mathsf{cd}(x, y, v) \\ \mathsf{add}(x, y) &= \mathsf{add}'(x, y, 2) \\ \mathsf{pr}(x, y) &= \sum_{v<y} \mathsf{I}_1^2(x, v) \\ \mathsf{exp}(x, y) &= \prod_{v<y} \mathsf{I}_1^2(x, y) \\ \mathsf{sb}(x, y) &= \mu v < x(\exists w < y(y + v = x + w)) \\ \mathsf{pd}(x) &= \mathsf{sb}(x, 1) \end{aligned}$$

□

EXAMPLE 1.4.1. Let $\mathscr{C} = \{\sigma\}$. Then $\mathscr{C}$ is not closed under full substitution, because the function $h(x) = \sigma(\sigma(x)) = x + 2$ is not $\mathscr{C}$-computable. But $\mathscr{C}$ is closed under partial substitution, because no application of this rule is possible using only the function σ.

EXAMPLE 1.4.2. Let $\mathscr{C}$ = the class of all initial functions. Then $\mathscr{C}$ is not closed under partial substitution, because the function $h(x) = \mathsf{add}(x, \sigma(x)) = 2 \times x + 1$ is not $\mathscr{C}$-computable.

EXAMPLE 1.4.3. Let $\mathscr{C}$ be the class of all constant and identity functions. Then $\mathscr{C}$ is closed under basic operations.

EXERCISES

1.4.1. Show that TO (= the class of all total functions) is not closed under unbounded minimalization with predicates.

1.4.2. A total k-ary function h is *linearly bounded* if there is a number m such that $h(\mathbf{x}) \leq \max\{\mathbf{x}\} + m$ holds for all $(\mathbf{x}) \in \mathbb{N}^k$. We denote by LB the class of all linearly bounded functions. Prove that LB is closed under basic operations but is not closed under elementary operations.

1.4.3. Let $\mathscr{C}_1$ and $\mathscr{C}_2$ be two classes of functions closed under elementary operations. Prove that $\mathscr{C}_1 \cap \mathscr{C}_2$ is also closed under elementary operations. Is the same property true for $\mathscr{C}_1 \cup \mathscr{C}_2$?

1.4.4. Let P be a unary predicate (i.e., a set) and let $\mathscr{C}$ be a class closed under elementary operations. The unary numerical function c_P is such that

$$\mathsf{c}_P(y) = \text{the number of } x\text{'s such that } x \in P \text{ and } x < y.$$

For example, if P = the set of even numbers, then $\mathsf{c}_P(2) = 1$, $\mathsf{c}_P(5) = 3$. Prove P is $\mathscr{C}$-decidable if and only if c_P is $\mathscr{C}$-computable.

1.4.5. Let P be an infinite $\mathscr{C}$-decidable set, where $\mathscr{C}$ is closed under elementary operations. The unary function en_P enumerates the elements of P in increasing order (so $\mathsf{en}_P(0)$ is the smallest element of P, $\mathsf{en}_P(1)$ the next, etc.). Assume there is a total $\mathscr{C}$-computable unary function g such that $\mathsf{en}_P(x) < g(x)$ for all x. Prove that en_P is $\mathscr{C}$-computable.

1.4.6. Let Q be a $\mathscr{C}$-decidable $(k + 1)$-ary predicate, where $\mathscr{C}$ is closed under elementary operations. Let h be the $(k + 1)$-ary function such that

$$h(\mathbf{x}, z) \simeq \text{the greatest } y < z \text{ such that } Q(\mathbf{x}, y)$$
$$\simeq z \text{ if there is no such } y.$$

Prove that h is $\mathscr{C}$-computable.

1.4.7. Let f be a total $\mathscr{C}$-computable $(k + 1)$-ary function, where $\mathscr{C}$ is closed under elementary operations. Let h_1 and h_2 be $(k + 1)$-ary functions such that

$$h_1(\mathbf{x}, z) = \max\{f(\mathbf{x}, 0), \ldots, f(\mathbf{x}, z)\}$$
$$h_2(\mathbf{x}, z) = \min\{f(\mathbf{x}, 0), \ldots, f(\mathbf{x}, z)\}.$$

Prove that h_1 and h_2 are $\mathscr{C}$-computable.

1.4.8. Let $\mathscr{C}$ be closed under elementary operations, and let P be some k-ary predicate. Prove the following conditions are equivalent:
(a) P is $\mathscr{C}$-decidable.
(b) There is a $\mathscr{C}$-decidable $(k + 1)$-ary predicate Q such that

$$\psi_P(\mathbf{x}) \simeq \mu y Q(\mathbf{x}, y).$$

(c) The graph predicate G_{ψ_P} is $\mathscr{C}$-decidable.

1.4.9. Let $\mathscr{C}$ be a class closed under elementary operations, and assume that $\mathscr{C}_{\mathrm{pd}} \nsubseteq \mathscr{C}_{\mathrm{d}}$. Prove that there is a k-ary $\mathscr{C}$-computable function h such that for no $(k + 1)$-ary $\mathscr{C}$-decidable predicate Q the relation

$$h(\mathbf{x}) \simeq \mu y Q(\mathbf{x}, y),$$

holds for all $(\mathbf{x}) \in \mathbb{N}^k$.

1.4.10. Let $\mathscr{C}$ be a class closed under elementary operations. Prove that if h is a total $\mathscr{C}$-computable function, then G_h is $\mathscr{C}$-decidable.

1.4.11. Give an example of a class $\mathscr{C}$ closed under elementary operations, such that $\mathscr{C}_{\mathrm{d}} \nsubseteq \mathscr{C}_{\mathrm{pd}}$, and $\mathscr{C}_{\mathrm{pd}}$ is not closed under negation.

1.4.12. Let $\mathscr{C}$ be closed under elementary operations. Prove the following:
(a) $\mathscr{C}_{\mathrm{pd}}$ is closed under conjunction and universal bounded quantification.
(b) If P_1 is a $\mathscr{C}$-decidable k-ary predicate, and P_2 is a partially $\mathscr{C}$-decidable k-ary predicate, then $P_1 \vee P_2$ is partially $\mathscr{C}$-decidable.

1.4.13. Let $\mathscr{C}$ be the class of all identity and constant functions. Prove that $\mathscr{C}$ is closed under basic operations and bounded minimalization.

1.4.14. Let $\mathscr{C}$ be a class closed under basic operations and bounded minimalization. Assume the functions add and sb are $\mathscr{C}$-computable. Prove that the functions σ, pd, sg, csg, ε, and cd are $\mathscr{C}$-computable, the predicates $\leq$ and $<$ are $\mathscr{C}$-decidable, and $\mathscr{C}_{\mathrm{d}}$ is closed under boolean operations and bounded quantification.

Notes

Closure under given operations is a traditional technique in mathematics. In this work we pay considerable attention to closure properties, which we try to identify in advance in their proper generality. The idea, of course, is to have such properties available for application whenever possible.

By means of basic operations we formalize a very general notion of substitution, which is a very useful but rather weak construction. Basic terms play an important role in recursive specifications.

Elementary operations provide a substantial amount of computational power. In particular, they are sufficient to generate the functions required by the encoding of Section 6. These operations were identified by Kalmar and have recently been incorporated in several presentations of computability theory. For example, see Monk [14] and Tourlakis [29]. More references and information are given in Rose [19].

§5. Minimal Closure

If $\mathscr{C}$ is an arbitrary class of functions, we can always find an extension of $\mathscr{C}$ which is closed under elementary operations. The intersection of all classes that are extensions of $\mathscr{C}$ and are closed under elementary operations is also an extension of $\mathscr{C}$, and it is easy to show that it is closed under elementary operations. Such a minimal extension of $\mathscr{C}$ we call the *elementary closure* of $\mathscr{C}$. The above definition is certainly simple, but it is unsatisfactory on several grounds. For example, it does not make explicit how the elements of the closure of $\mathscr{C}$ are generated from the elements of $\mathscr{C}$. For this reason we give a more constructive definition, involving restricted set-theoretical methods.

Let $\mathscr{C}$ be a class of functions. An *elementary $\mathscr{C}$-derivation* is a sequence of functions $h_1, \ldots, h_m$, $m \geq 1$, where for each i, $1 \leq i \leq m$, at least one of the following conditions is satisfied:

(a) h_i is either $\mathscr{C}$-computable, or one of the functions C_0^0, σ, cd, or ε.
(b) There is j such that $1 \leq j < i$ and h_i can be obtained from h_j by either adjoining of variables, bounded sum, or bounded product.
(c) There are j and j' such that $1 \leq j, j' < i$ and h_i is obtained from h_j and $h_{j'}$ by partial substitution.

Note that if $h_1, \ldots, h_m$ and $h'_1, \ldots, h'_{m'}$ are two elementary $\mathscr{C}$-derivations, then the concatenation $h_1, \ldots, h_m, h'_1, \ldots, h'_{m'}$ is also an elementary $\mathscr{C}$-derivation. Furthermore, a nonempty initial segment of an elementary $\mathscr{C}$-derivation is again an elementary $\mathscr{C}$-derivation.

A function h is *elementary $\mathscr{C}$-derivable* if there is an elementary $\mathscr{C}$-derivation $h_1, \ldots, h_m$, where $h = h_m$. This is equivalent to requiring that h be an element (not necessarily the last) in some elementary $\mathscr{C}$-derivation. The class of all functions that are elementarily $\mathscr{C}$-derivable is denoted by $\mathsf{EL}(\mathscr{C})$. We shall show that $\mathsf{EL}(\mathscr{C})$ is the elementary closure of $\mathscr{C}$ referred to above.

Theorem 1.5.1. *Let $\mathscr{C}$ be a class of functions. Then*

(i) $\mathscr{C} \subseteq \mathsf{EL}(\mathscr{C})$.
(ii) $\mathsf{EL}(\mathscr{C})$ *is closed under elementary operations.*
(iii) *If $\mathscr{C}'$ is an extension of $\mathscr{C}$ closed under elementary operations, then* $\mathsf{EL}(\mathscr{C}) \subseteq \mathscr{C}'$.

Proof. If h is $\mathscr{C}$-computable, then h is, by itself, an elementary $\mathscr{C}$-derivation; hence h is in $\mathsf{EL}(\mathscr{C})$. A similar argument shows that the function C_0^0, σ, cd, and ε are in $\mathsf{EL}(\mathscr{C})$. To complete the proof of (ii) we need only to show that $\mathsf{EL}(\mathscr{C})$ is closed under adjoining of variables, partial substitution, bounded sum, and bounded product. For example, assume that a function h is obtained by partial substitution, from functions f and g that are in $\mathsf{EL}(\mathscr{C})$. This means there are elementary $\mathscr{C}$-derivations $h_1, \ldots, h_m = f$ and $h'_1, \ldots, h'_{m'} = g$. Then $h_1, \ldots, h_m$, $h'_1, \ldots, h'_{m'}, h$ is also an elementary $\mathscr{C}$-derivation; hence h is in $\mathsf{EL}(\mathscr{C})$. A similar argument is used for the other rules.

To prove (iii) assume that $\mathscr{C} \subseteq \mathscr{C}'$ and that $\mathscr{C}'$ is closed under elementary operations. We prove by induction on m that whenever $h_1, \ldots, h_m$ is an elementary $\mathscr{C}$-derivation then h_i is $\mathscr{C}'$-computable, for $i = 1, \ldots, m$. This is clear when $m = 1$. If the property is true for m, and $h_1, \ldots, h_m, h_{m+1}$ is an elementary $\mathscr{C}$-derivation, then by the induction hypothesis we have that h_i is $\mathscr{C}'$-computable, for $i = 1, \ldots, m$. Furthermore, h_{m+1} is either $\mathscr{C}$-computable, or it is obtained by one of the defining rules from functions that are $\mathscr{C}'$-computable. In either case h_{m+1} is $\mathscr{C}'$-computable. □

Let $\mathscr{C}$ be a class of functions. If h is a function in $\mathsf{EL}(\mathscr{C})$, we say that *h is elementary* in $\mathscr{C}$. If c is a function, we put $\mathsf{EL}(c) = \mathsf{EL}(\{c\})$, and the functions in $\mathsf{EL}(c)$ are said to be *elementary in c*. Finally, we put $\mathsf{EL} = \mathsf{EL}(\varnothing)$, and the functions in EL are said to be *elementary*.

This notation extends to predicates via the characteristic function. A predicate P is *elementary* (in c) (in $\mathscr{C}$) if χ_P is elementary (in c) (in $\mathscr{C}$).

Note that Theorem 1.5.1 applies to $\mathsf{EL} = \mathsf{EL}(\varnothing)$. Since every class is an extension of $\varnothing$, we say only that EL is closed under elementary operations, and whenever $\mathscr{C}$ is a class closed under elementary operations, then $\mathsf{EL} \subseteq \mathscr{C}$. These two conditions characterize uniquely the class EL.

Theorem 1.5.2. *Let $\mathscr{C}$ and $\mathscr{C}'$ be classes of functions. Then*

(i) $\mathsf{EL}(\mathscr{C}) \subseteq \mathsf{EL}(\mathscr{C}')$ *if and only if* $\mathscr{C} \subseteq \mathsf{EL}(\mathscr{C}')$.
(ii) *If* $\mathscr{C} \subseteq \mathscr{C}'$ *then* $\mathsf{EL}(\mathscr{C}) \subseteq \mathsf{EL}(\mathscr{C}')$.
(iii) *If* $\mathscr{C} \subseteq \mathsf{EL}(\mathscr{C}')$ *and* $\mathscr{C}$ *is finite, there is a finite class* $\mathscr{C}'' \subseteq \mathscr{C}'$, *such that* $\mathscr{C} \subseteq \mathsf{EL}(\mathscr{C}'')$.

PROOF. To prove (i) note that if $\mathsf{EL}(\mathscr{C}) \subseteq \mathsf{EL}(\mathscr{C}')$, we have also $\mathscr{C} \subseteq \mathsf{EL}(\mathscr{C})$; hence $\mathscr{C} \subseteq \mathsf{EL}(\mathscr{C}')$. Conversely, if $\mathscr{C} \subseteq \mathsf{EL}(\mathscr{C}')$, since $\mathsf{EL}(\mathscr{C}')$ is closed under elementary operations, it follows from Theorem 1.5.1 (iii) that $\mathsf{EL}(\mathscr{C}) \subseteq \mathsf{EL}(\mathscr{C}')$. Now (ii) follows from (i) because if $\mathscr{C} \subseteq \mathscr{C}'$, then $\mathscr{C} \subseteq \mathsf{EL}(\mathscr{C}')$.

To prove (iii) assume $\mathscr{C}$ is finite and $\mathscr{C} \subseteq \mathsf{EL}(\mathscr{C}')$. For each function in $\mathscr{C}$ there is an elementary $\mathscr{C}'$-derivation, involving a finite number of functions in $\mathscr{C}'$. Since $\mathscr{C}$ is finite the class $\mathscr{C}''$ of all functions involved in all derivations is finite, and clearly $\mathscr{C} \subseteq \mathsf{EL}(\mathscr{C}'')$. □

Corollary 1.5.2.1. *Let c and c' be arbitrary functions. Then* $\mathsf{EL}(c) \subseteq \mathsf{EL}(c')$ *if and only if c is elementary in c'.*

PROOF. Immediate from Theorem 1.5.2 (i), taking $\mathscr{C} = \{c\}$ and $\mathscr{C}' = \{c'\}$. □

Theorem 1.5.3. *Let $\mathscr{C}$ be a class of functions. The following conditions are equivalent*:

(i) $\mathscr{C}$ *is closed under elementary operations.*
(ii) $\mathsf{EL}(\mathscr{C}) \subseteq \mathscr{C}$.
(iii) $\mathsf{EL}(\mathscr{C}) = \mathscr{C}$.

PROOF. To prove the implication from (i) to (ii) assume that $\mathscr{C}$ is closed under elementary operations. Since $\mathscr{C} \subseteq \mathscr{C}$ it follows from Theorem 1.5.1 (iii) that $\mathsf{EL}(\mathscr{C}) \subseteq \mathscr{C}$. The implication from (ii) to (iii) follows from Theorem 1.5.1 (i), and the implication from (iii) to (i) from Theorem 1.5.1. (ii). □

Let h be a k-ary function and $\mathscr{C}$ a class of functions. We say that h is *bounded* by $\mathscr{C}$ if there is a $\mathscr{C}$-computable function h' such that $h \subseteq_k h'$. The class of all functions that are bounded by $\mathscr{C}$ is denoted by $\mathsf{BD}(\mathscr{C})$. This means that whenever $\mathscr{C}'$ is a class of functions such that $\mathscr{C}' \subseteq \mathsf{BD}(\mathscr{C})$, then every $\mathscr{C}'$-computable function is bounded by $\mathscr{C}$.

The construction $\mathsf{BD}(\mathscr{C})$ is useful to obtain classes that are closed under given operations and satisfy special properties.

Theorem 1.5.4. *Let $\mathscr{C}$ be a class of functions. Then*

(i) $\mathscr{C} \subseteq \mathsf{BD}(\mathscr{C})$.
(ii) *If $\mathscr{C}$ is closed under one of the finitary rules, then $\mathsf{BD}(\mathscr{C})$ is also closed under the same rule.*
(iii) *If $\mathscr{C}$ is closed under elementary operations then $\mathsf{BD}(\mathscr{C})$ is also closed under elementary operations.*

PROOF. Part (i) is immediate from the definition. The proof of (ii) depends on the fact that the finitary rules are monotonic. For example, let h be a k-ary function that is specified by partial substitution from functions f and g. This means that

$$h(\mathbf{x}) \simeq f(\mathbf{x}, g(\mathbf{x})),$$

where f and g are functions in $\mathsf{BD}(\mathscr{C})$; hence there are $\mathscr{C}$-computable functions f' and g', such that $f \subseteq_{k+1} f'$, $g \subseteq_k g'$, so by using partial substitution we get a $\mathscr{C}$-computable function h' such that

$$h'(\mathbf{x}) \simeq f'(\mathbf{x}, g'(\mathbf{x})).$$

Since partial substitution is monotonic it follows that $h \subseteq_k h'$, so h is in $\mathsf{BD}(\mathscr{C})$ and $\mathsf{BD}(\mathscr{C})$ is closed under partial substitution. The same argument can be used for any of the finitary rules.

The proof of (iii) is immediate from (ii), for in this case $\mathsf{BD}(\mathscr{C})$ is closed under all elementary rules, and since $\mathscr{C} \subseteq \mathsf{BD}(\mathscr{C})$ it follows that $\mathsf{BD}(\mathscr{C})$ contains all the initial functions. □

Corollary 1.5.4.1. *Let $\mathscr{C}$ and $\mathscr{C}'$ be classes of functions such that $\mathscr{C} \subseteq \mathsf{BD}(\mathscr{C}')$. Then*

(i) *If $\mathscr{C}$ contains only total functions, then $\mathscr{C} \subseteq \mathscr{C}'$.*
(ii) *If $\mathscr{C}'$ is closed under elementary operations then $\mathsf{EL}(\mathscr{C}) \subseteq \mathsf{BD}(\mathscr{C}')$.*
(iii) *If $\mathscr{C}'$ is closed under elementary operations, and h is a total function elementary in $\mathscr{C}$, then h is $\mathscr{C}'$-computable.*

PROOF. Part (i) is clear, for in the case $h \subseteq_k h'$, and h is total, then $h = h'$. Part (ii) follows from Theorem 1.5.1 (iii), for $\mathsf{BD}(\mathscr{C}')$ is closed under elementary operations. To prove (iii) note that if h is a total function elementary in $\mathscr{C}$, then by (ii) h is in $\mathsf{BD}(\mathscr{C}')$; hence by (i) h is in $\mathscr{C}'$. $\square$

EXAMPLE 1.5.1. Let $\mathscr{C} = \mathsf{BD}(\mathsf{EL})$. Since EL is closed under elementary operations, it follows that $\mathscr{C}$ is also closed under elementary operations. If h is a total $\mathscr{C}$-computable function, that is, elementary in $\mathscr{C}$, it follows from Corollary 1.5.4.1 (iii) that h is in EL, that is, h is elementary. Hence $\mathscr{C}_{\mathsf{d}} = \mathsf{EL}_{\mathsf{d}}$.

EXAMPLE 1.5.2. Let $\mathscr{C} = \mathsf{EL}(\varepsilon')$ (see Example 1.3.2). Since $\mathscr{C}$ is closed under elementary operations, we have $\mathsf{EL} \subseteq \mathscr{C}$. On the other hand, ε' is bounded by EL (for $\varepsilon' \subseteq_2 \mathsf{C}_0^2$); hence by Corollary 1.5.4.1 (ii) it follows that $\mathscr{C} = \mathsf{EL}(\varepsilon') \subseteq \mathsf{BD}(\mathsf{EL})$, and from the same Corollary, part (iii), it follows that the total functions in $\mathscr{C}$ are exactly the elementary functions; hence $\mathscr{C}_{\mathsf{d}} = \mathsf{EL}_{\mathsf{d}}$. Furthermore, if P is a $\mathscr{C}$-decidable predicate, that is, χ_P is $\mathscr{C}$-computable, then $\psi_P(\mathbf{x}) \simeq \varepsilon'(\chi_P(\mathbf{x}), 0)$; hence P is also partially $\mathscr{C}$-decidable, and $\mathscr{C}_{\mathsf{d}} \subseteq \mathscr{C}_{\mathsf{pd}}$.

EXERCISES

1.5.1. Prove that TO ($=$ the class of all total functions) is closed under elementary operations, and $\mathsf{EL} \subseteq \mathsf{TO}$.

1.5.2. Let $\mathscr{C}$ be a class of functions such that whenever $\mathscr{C}' \subseteq \mathscr{C}$, and $\mathscr{C}'$ is finite, then $\mathsf{EL}(\mathscr{C}') \subseteq \mathscr{C}$. Prove that $\mathscr{C}$ is closed under elementary operations.

1.5.3. Let $\mathscr{C}$ be a class of functions, such that whenever $\mathscr{C}' \subseteq \mathscr{C}$, and $\mathscr{C}'$ is finite, then $\mathsf{BD}(\mathscr{C}') \subseteq \mathscr{C}$. Prove that $\mathsf{BD}(\mathscr{C}) \subseteq \mathscr{C}$.

1.5.4. Let $\mathscr{C}$ and $\mathscr{C}'$ be classes of functions such that $\mathscr{C} \subseteq \mathscr{C}'$. Prove the following:
(a) $\mathsf{BD}(\mathscr{C}) \subseteq \mathsf{BD}(\mathscr{C}')$.
(b) $\mathsf{EL}(\mathsf{BD}(\mathscr{C})) \subseteq \mathsf{BD}(\mathsf{EL}(\mathscr{C}'))$.

1.5.5. Prove that the class $\mathsf{BD}(\mathsf{EL})$ contains all the finite functions.

1.5.6. Let P be a predicate and h a total function elementary in ψ_P. Prove that h is elementary.

1.5.7. Let $\mathscr{C}$ be a denumerable class of functions (i.e., the elements of $\mathscr{C}$ can be arranged in an infinite list h_0, h_1, ..., possible with repetitions). Prove that $\mathsf{EL}(\mathscr{C})$ is denumerable.

1.5.8. Prove that if h is a unary total function then $\mathsf{BD}(h)$ is not denumerable.

Notes

Minimal closure is the logical extension of closure properties and pervades all branches of mathematics. We do not introduce explicitly minimal closure for basic operations, since it is clearly determined by the identity and constant functions. Minimal closure for the elementary operations determines the elementary functions, which include the initial functions and also the encoding functions of Section 6.

There are two equivalent ways of defining minimal closure. One is from above, which is strongly set theoretical and very economical, and another is from below, which is more constructive and also more involved. We use the latter, which is given in some detail.

Although we do not insist on the point, it should be clear that the definition of minimal closure is essentially an inductive definition of the type introduced in Chapter 3, and the minimal closure itself if the minimal fixed point of the induction (cf. Theorem 1.5.1).

§6. More Elementary Functions and Predicates

We know that all the initial functions are elementary, and we know also that the equality predicate is elementary, because the function ε is elementary. We want to expand this list by introducing new elementary functions and predicates. For this purpose we can use any of the elementary rules, with functions and predicates already known to be elementary.

Usually the specification of a particular function or predicate involves more than one rule, with substitution in most cases being one of them. Rather than writing the sequence of partial specifications, one for each rule being applied, we describe the total specification by a self-explanatory expression, from which the actual sequence of partial specifications can be extracted. This technique was used in the proof of Theorem 1.4.3, in the specification of the function sb, where the total definition by cases, bounded minimalization, and substitution are combined in one step. Anyway, the reader is urged to carry out the actual decoding in one or two of the applications given below.

We use two kinds of expressions: numerical expressions that take a numerical value (when defined) and boolean expressions that take a boolean value (and are always defined). Numerical expressions can be substituted in boolean expressions, provided they contain only total functions, and the result is another boolean expression. On the other hand, boolean expressions can be used to obtain numerical expressions, via bounded minimalization.

Substitution in upper bounds requires some care, whenever nontotal functions are involved (which is not the case in the examples below). For example, we can write a specification in the form

$$h(\mathbf{x}) \simeq \sum_{y<V} f(\mathbf{x}, y),$$

where V is a numerical expression. This is understood as the following sequence of specifications:

$$f'(\mathbf{x}, z) \simeq \sum_{y<z} f(\mathbf{x}, y)$$

$$g(\mathbf{x}) \simeq V$$

$$h(\mathbf{x}) \simeq f'(\mathbf{x}, g(\mathbf{x})).$$

Note that whenever V is undefined then $h(\mathbf{x})$ is undefined. On the other hand, if $V \simeq 0$, then $h(\mathbf{x}) \simeq 0$, even if f is the totally undefined function.

Each of the functions and predicates below is introduced with a prefix notation, the arity being given by the number of arguments. In some cases an alternate notation is introduced (in parentheses to the right), which is sometimes infixed, and it is used in most applications. Some functions and predicates are self-explanatory and require no elaboration. Others are outside the standard mathematical practice and are discussed below.

(A) $\mathsf{P}_<(x, y) \equiv \exists v < y(x + v + 1 = y) \qquad (x < y)$

(B) $\mathsf{P}_\leq(x, y) \equiv x < y \vee x = y \qquad (x \leq y)$

(C) $\mathsf{Dv}(x, y) \equiv \exists v \leq x(v \times y = x)$

(D) $\mathsf{Pr}(x) \equiv 1 < x \wedge \forall z < x(z = 1 \vee {\sim}\mathsf{Dv}(x, z))$

(E) $\mathsf{p}(y) = \mu v \leq \mathsf{exp}(2, \mathsf{exp}(2, y))\left(\mathsf{Pr}(v) \wedge y = \sum_{z<v} \chi_{\bar{\mathsf{Pr}}}(z)\right) \qquad (\mathsf{p}_y)$

(F) $\mathsf{sp}(x, y) = \mu v \leq x {\sim} \mathsf{Dv}(x + 1, \mathsf{exp}(\mathsf{p}(y), v + 1)) \qquad (\ (x)_y)$

(G) $\mathsf{lh}(x) = \mu v \leq x\left(x + 1 = \prod_{y<v} \mathsf{exp}(\mathsf{p}(y), \mathsf{sp}(x, y))\right) \qquad (\ell x).$

The meaning of the predicates $\mathsf{P}_\leq$ and $\mathsf{P}_<$ is clear. The predicate $\mathsf{Dv}(x, y)$ holds exactly when the number y divides the number x (but note that $\mathsf{Dv}(0, 0) \equiv \mathsf{T}$). The predicate Pr defines the property of being a prime number in the usual sense.

The function p enumerates the prime numbers in increasing order. To show that this is the case, assume the primes in increasing order are given by the sequence $p_0, p_1, \ldots, p_y, \ldots$. This means that $p_0 = 2, p_1 = 3, p_2 = 5$, and so on. We want to show that $\mathsf{p}(y) = p_y$ holds for all y. It is clear that if $p_y = v$, then v is a prime and the number of primes that precede v is y. Hence, if P is the unary predicate such that $P(z) \equiv {\sim}\mathsf{Pr}(z)$, then v is the unique number such that

$$\mathsf{Pr}(v) \wedge y = \sum_{z<v} \chi_P(z).$$

It follows that in order to show that $\mathsf{p}(y) = p_y$ we need only to prove that $p_y \leq \mathsf{exp}(2, \mathsf{exp}(2, y))$ holds for all y. The proof of this relation is by induction on y.

The case $y = 0$ is trivial, since $p_0 = 2$ and $\mathsf{exp}(2, \mathsf{exp}(2, 0)) = 2$. Assume the relation holds for all $y' \leq y$, to prove it for $y + 1$. From the Euclid theorem we know that

$$p_{y+1} \leq p_0 \times p_1 \times \cdots \times p_y + 1.$$

Using the induction hypothesis we have

$$\begin{aligned} p_{y+1} &\leq \mathsf{exp}(2, \mathsf{exp}(2,0)) \times \cdots \times \mathsf{exp}(2, \mathsf{exp}(2,y)) + 1 \\ &= \mathsf{exp}\left(2, \sum_{z \leq y} \mathsf{exp}(2,z)\right) + 1 \\ &= \mathsf{exp}(2, \mathsf{exp}(2, y+1) - 1) + 1 \\ &\leq \mathsf{exp}(2, \mathsf{exp}(2, y+1) - 1) + \mathsf{exp}(2, \mathsf{exp}(2, y+1) - 1) \\ &= \mathsf{exp}(2, \mathsf{exp}(2, y+1)). \end{aligned}$$

The meaning of the function $\mathsf{sp}(x, y) = (x)_y$ depends on the fact that for every number x, the number $x + 1$ has a unique factorization, and $(x)_y$ is the exponent of the prime $\mathsf{p}(y) = \mathsf{p}_y$ in such factorization. The unique factorization of $x + 1$ contains only a finite number of primes with nonzero exponent. The function $\mathsf{lh}(x) = \ell(x)$ denotes the least y such that for $y' \geq y$, $(x)_{y'} = 0$. In particular, $\ell(0) = 0$.

We have now a basic stock of elementary functions, and a few more are introduced in the exercises. They are used in several ways, but mainly as encoding structures. For example, we encode a k-tuple of numbers $(x_1, \ldots, x_k)$, $k > 0$, in the number

$$\langle x_1, \ldots, x_k \rangle = (\mathsf{exp}(2, x_1 + 1) \times \cdots \times \mathsf{exp}(\mathsf{p}_{k-1}, x_k + 1)) \dot{-} 1.$$

This is a proper encoding because if $x = \langle x_1, \ldots, x_k \rangle$ we can obtain from x the elements of the tuple, using elementary functions. For example, $x_1 = \mathsf{pd}((x)_0)$, $x_2 = \mathsf{pd}((x)_1)$, and so on. The general decoding function is given by $[x]_y = \mathsf{pd}((x)_{\mathsf{pd}(y)})$.

The previous notation for writing k-tuples in the form $(\mathbf{x})$ can be extended to the encoding. Hence $\langle \mathbf{x} \rangle$ is the encoding of the tuple $(\mathbf{x})$. In particular we put $\langle\ \rangle = 0$ as the encoding of $(\ )$.

Not only the elements of a tuple can be obtained from the encoding. We can derive the length of the tuple using the function ℓ. In fact, if $x = \langle x_1, \ldots, x_k \rangle$, then $\ell(x) = k$. In particular, $\ell(\langle\ \rangle) = \ell(0) = 0$.

Note that while every tuple is encoded by a unique number, it is not the case that every number is the encoding of some tuple. We shall see that this does not create any particular problem. Still, this is not the only way in which numbers are used to encode structures, and other procedures are described in the literature.

EXERCISES

1.6.1. Prove that the following functions are elementary:
   (a) $\mathsf{q}(x, y)$ = quotient when x is divided by y.
   (b) $\mathsf{rm}(x, y)$ = remainder when x is divided by y.
   (c) $\mathsf{gcd}(x, y)$ = the greatest common divisor of x and y.
   (d) $\mathsf{lcm}(x, y)$ = the least common multiple of x and y.
   (e) $x!$ = the factorial of x.

1.6.2. Specify an elementary binary function f such that whenever $x = \langle x_1, \dots, x_k \rangle$, $k \geq 0$, and $y = \langle y_1, \dots, y_m \rangle$, $m \geq 0$, then $f(x, y) = \langle x_1, \dots, x_k, y_1, \dots, y_m \rangle$.

1.6.3. Specify an elementary unary function g such that whenever $x = \langle x_1, \dots, x_k, x_{k+1} \rangle$, $k \geq 0$, then $g(x) = \langle x_1, \dots, x_k \rangle$.

1.6.4. Let $\mathscr{C}$ be a class of functions closed under basic operations and bounded minimalization. Assume that $\mathscr{C}$ contains the functions add and sb and, for every $m \geq 0$, contains the binary function g_m such that

$$g_m(x, y) = x \times \mathsf{exp}(\mathsf{p}_m, y).$$

Prove that for every $k \geq 0$, the k-ary function $f_k(\mathbf{x}) = \langle \mathbf{x} \rangle$, is $\mathscr{C}$-computable, and for every $m \geq 0$, the unary function $d_m(x) = (x)_m$ is also $\mathscr{C}$-computable.

1.6.5. Prove that the following equation holds for all x, y, and v.

$$(\mathsf{exp}(x, y) \dot{-} 1)_v = y \times (x \dot{-} 1)_v.$$

Notes

The main objective of this section is the introduction of the functions related to the enumeration of the prime numbers, prime factorization, and encoding. The real point is actually to show that these functions are elementary. The only nontrivial case is the enumeration function p, where we use an argument taken from Monk [14].

The encoding of tuples is traditional; in fact, it is derived from the encoding in Gödel [7]. Other encodings, more convenient for some applications, are discussed in the literature (see Davis [2] and Rogers [20]).

CHAPTER 2

Recursive Functions

All the specification rules in the preceding chapter are explicit, in the sense that the function (or predicate) being specified is determined entirely by a procedure involving only the given functions and predicates. Now we consider recursive specifications, in which a function h is specified by a rule involving the function h itself. Such a procedure is essentially ambiguous, because in general there are many functions satisfying the specification. This ambiguity can be disposed of only by elimination of the recursion, and this can be done in several ways. Closure under recursive specifications is used to define recursive functions. The relation with Church's thesis is discussed in some detail.

§1. Primitive Recursion

We study first a very special form of recursion, namely, *primitive recursion*, which is interpreted as describing an iterative evaluation. The iteration proceeds in the following form: if h is, for example, a unary function, then the specification of h determines first the value of $h(0)$, if such a value exists, and then the rule determines the value of $h(y + 1)$ for arbitrary y, assuming the value of $h(y)$ is available. To evaluate $h(y)$ for a given y, first we evaluate $h(0)$, then we evaluate $h(1)$ using $h(0)$, then we evaluate $h(2)$ using $h(1)$, and so on. This procedure is carried out as far as possible. If a value of $h(y)$ is obtained, then $h(y)$ is defined; otherwise $h(y)$ is undefined.

In the general case the function h is $(k + 1)$-ary, $k \geq 0$, and k parameters are involved. There are two given functions—a k-ary function that we call f, and a $(k + 2)$-ary function that we call g. Now the specification takes the following form:

$$h(\mathbf{x}, 0) \simeq f(\mathbf{x})$$

$$h(\mathbf{x}, y + 1) \simeq g(\mathbf{x}, y, h(\mathbf{x}, y)).$$

We say that the function h is *specified by primitive recursion from the functions f and g*. This specification is clearly recursive, because the symbol h appears in the right side of the second defining equation. But we eliminate the recursion by interpreting the two equations as describing the iterative procedure to evaluate $h(\mathbf{x}, y)$ for an arbitrary y. During the evaluation, the values of $\mathbf{x}$ remain constant but y takes values $0, 1, 2, \ldots$.

It is clear that if the given functions f and g are total, then the iteration procedure can be continued indefinitely, so that the function h is also total. In this sense primitive recursion is close to the elementary operations. If the given functions are not total then h is not necessarily total. In this case the iteration may stop after reaching some y, and from then on the function is undefined.

We say that a class $\mathscr{C}$ of functions is *closed under primitive recursion* if whenever a function h is specified by primitive recursion from $\mathscr{C}$-computable functions f and g, then h is also $\mathscr{C}$-computable. In applications the class $\mathscr{C}$ will satisfy some closure properties and this makes possible a more flexible use of primitive recursion. For instance, it is convenient to use primitive recursion with basic terms, where the function symbols are interpreted in some given class $\mathscr{C}$ of functions. In this way we may specify a $(k + 1)$-ary function h using two given basic numerical terms U and V, as follows:

$$h(\mathbf{x}, 0) \simeq U$$

$$h(\mathbf{x}, y + 1) \simeq V,$$

provided the terms U and V satisfy the following conditions:

(i) All numerical variables occurring in U are in the list $\mathbf{x}$.
(ii) All numerical variables occurring in V are in the list $\mathbf{x}, y$.
(iii) The symbol h does not occur in U and may occur in V only in the form $h(\mathbf{x}, y)$.

As before, these two equations describe an iteration procedure where the value of $h(\mathbf{x}, 0)$ is given by U, and the value of $h(\mathbf{x}, y + 1)$ is given by V, provided the value of $h(\mathbf{x}, y)$ is available. If $h(\mathbf{x}, y)$ is undefined then $h(\mathbf{x}, y + 1)$ is undefined, even if V is defined, a situation that may arise only if the symbol h does not occur in V.

EXAMPLE 2.1.1. Consider the specification of the binary function h,

$$h(x, 0) \simeq \mathsf{UD}_1(x)$$

$$h(x, y + 1) \simeq x.$$

The function h is totally undefined, because the iteration fails to obtain a value for $h(x, 0)$.

A specification of the form described above with terms U and V satisfying conditions (i), (ii), and (iii) is said to be a *basic primitive recursive specification*. The given functions are the functions in the terms U and V, that is, the interpretation of the function symbols in U and V, not including h which may appear in V. A class $\mathscr{C}$ of functions is *closed under basic primitive recursion* if whenever a function h is specified by basic primitive recursion from given $\mathscr{C}$-computable functions, h is also $\mathscr{C}$-computable.

Theorem 2.1.1. *If $\mathscr{C}$ is a class of functions closed under basic operations and primitive recursion, then $\mathscr{C}$ is closed under basic primitive recursion.*

PROOF. Assume h is specified by basic primitive recursion with terms U and V. We introduce a function f by putting $f(\mathbf{x}) \simeq U$, and a function g by putting $g(\mathbf{x}, y, z) \simeq V'$, where V' is a basic numerical term obtained from V by replacing all occurrences of $h(\mathbf{x}, y)$ in V by a new variable z. Since $\mathscr{C}$ is closed under basic operations it follows that f and g are $\mathscr{C}$-computable. Now we can specify h by primitive recursion in the form

$$h(\mathbf{x}, 0) \simeq f(\mathbf{x})$$

$$h(\mathbf{x}, y + 1) \simeq g(\mathbf{x}, y, h(\mathbf{x}, y)),$$

and since $\mathscr{C}$ is closed under primitive recursion it follows that h is $\mathscr{C}$-computable. □

We say that a class $\mathscr{C}$ of functions is *closed under primitive recursive operations* if $\mathscr{C}$ contains the identity functions, the constant functions, and the function σ, and furthermore that $\mathscr{C}$ is closed under full substitution and primitive recursion.

Theorem 2.1.2. *If $\mathscr{C}$ is closed under primitive recursive operations, then $\mathscr{C}$ is closed under basic operations, basic primitive recursion, and elementary operations.*

PROOF. Closure under basic operations follows from the definition, and closure under basic primitive recursion follows from Theorem 2.1.1. Using basic primitive recursion we obtain the following functions:

$$\mathsf{cd}(x_1, x_2, 0) \simeq x_1$$

$$\mathsf{cd}(x_1, x_2, y + 1) \simeq x_2,$$

$$\mathsf{add}(x, 0) \simeq x$$

$$\mathsf{add}(x, y + 1) \simeq \sigma(\mathsf{add}(x, y))$$

$$\mathsf{pr}(x, 0) \simeq 0$$
$$\mathsf{pr}(x, y + 1) \simeq \mathsf{add}(\mathsf{pr}(x, y), x)$$
$$\mathsf{pd}(0) \simeq 0$$
$$\mathsf{pd}(y + 1) \simeq y$$
$$\mathsf{sb}(x, 0) \simeq x$$
$$\mathsf{sb}(x, y + 1) \simeq \mathsf{pd}(\mathsf{sb}(x, y)).$$

Finally, we use substitution to give specifications of sg and ε:

$$\mathsf{sg}(x) \simeq \mathsf{cd}(0, 1, x)$$
$$\varepsilon(x, y) \simeq \mathsf{sg}(\mathsf{sb}(x, y) + \mathsf{sb}(y, x)).$$

To complete the proof we need only to consider closure under the rules. Note that partial substitution and adjoining of variables follow from Theorem 1.4.1. Now consider the following applications of bounded sum and bounded product:

$$h_1(\mathbf{x}, z) \simeq \sum_{y<z} f(\mathbf{x}, y)$$
$$h_2(\mathbf{x}, z) \simeq \prod_{y<z} f(\mathbf{x}, y),$$

where f is a given $\mathscr{C}$-computable function. The following are new specifications using basic primitive recursion:

$$h_1(\mathbf{x}, 0) \simeq 0$$
$$h_1(\mathbf{x}, y + 1) \simeq h_1(\mathbf{x}, y) + f(\mathbf{x}, y),$$
$$h_2(\mathbf{x}, 0) \simeq 1$$
$$h_2(\mathbf{x}, y + 1) \simeq h_2(\mathbf{x}, y) \times f(\mathbf{x}, y),$$

and this shows that $\mathscr{C}$ is closed under bounded sum and bounded product. This completes the proof. □

If $\mathscr{C}$ is a class of functions, then PR($\mathscr{C}$) denotes the *primitive recursive closure of* $\mathscr{C}$. A formal definition can be given as in the case of EL($\mathscr{C}$), by introducing the notion of *primitive recursive* $\mathscr{C}$*-derivation*. The details of this definition are left to the reader. As usual we can prove that PR($\mathscr{C}$) is an extension of $\mathscr{C}$ closed under primitive recursive operations. Furthermore, if $\mathscr{C}'$ is an extension of $\mathscr{C}$ closed under primitive recursive operations, then PR($\mathscr{C}$) $\subseteq \mathscr{C}'$. From Theorem 2.1.2 it follows that EL($\mathscr{C}$) $\subseteq$ PR($\mathscr{C}$).

If h is a function in the class PR($\mathscr{C}$), we say that h is *primitive recursive* in $\mathscr{C}$. If c is a function, we put PR(c) = PR($\{c\}$), and the functions in PR(c) are said to be *primitive recursive in* c. Finally, we put PR = PR($\varnothing$), and the functions in PR are said to be *primitive recursive*.

This notation extends to predicates via the characteristic function. A predicate P is *primitive recursive* (in c) (in $\mathscr{C}$), if χ_P is primitive recursive (in c) (in $\mathscr{C}$).

A number of results about the operator $\mathsf{EL}(\mathscr{C})$ were proved in Section 5 of Chapter 1, and all of them can be extended to the operator $\mathsf{PR}(\mathscr{C})$. We have already mentioned the properties of Theorem 1.5.1. In the same way, Theorem 1.5.2 and Theorem 1.5.3 are valid for primitive recursive operations.

Of particular interest is the relation between the operators $\mathsf{PR}(\mathscr{C})$ and $\mathsf{BD}(\mathscr{C})$. Again, all basic connections given in Theorem 1.5.4 and Corollary 1.5.4.1 extend to primitive recursive operations.

A number of iterations that do not appear to be primitive recursive can be reduced to primitive recursion. For example, we can specify a $(k + 1)$-ary function by an iteration in which the value $h(\mathbf{x}, 0)$ is given as a function of $\mathbf{x}$, and the value $h(\mathbf{x}, y + 1)$ is given in terms of values $h(\mathbf{x}, y')$, where $y' \leq y$. An iteration of this form is called *course-of-values iteration*. To formalize this idea, we associate with h the $(k + 1)$-ary function $\bar{h}$ such that

$$\bar{h}(\mathbf{x}, y) \simeq \left(\prod_{v \leq y} \mathsf{exp}(\mathsf{p}_v, h(\mathbf{x}, v))\right) \dot{-} 1.$$

Now let $\mathscr{C}$ be a class of functions closed under elementary operations. It is clear that if h is $\mathscr{C}$-computable, then $\bar{h}$ is also $\mathscr{C}$-computable. The converse does not follow in general, because in order that $\bar{h}(\mathbf{x}, y)$ be defined it is necessary that $h(\mathbf{x}, y')$ be defined for all $y' \leq h$. If this is the case, then $h(\mathbf{x}, y) \simeq (\bar{h}(\mathbf{x}, y))_y$.

The formalization of *course-of-values recursion* is as follows. Let f be a given k-ary function, and g a given $(k + 2)$-ary function. A new $(k + 1)$-ary function h is introduced by the equations

$$h(\mathbf{x}, 0) \simeq f(\mathbf{x})$$

$$h(\mathbf{x}, y + 1) \simeq g(\mathbf{x}, y, \bar{h}(\mathbf{x}, y)).$$

We interpret these equations as an iteration procedure to evaluate $h(\mathbf{x}, 0)$, $\bar{h}(\mathbf{x}, 0)$, $h(\mathbf{x}, 1)$, $\bar{h}(\mathbf{x}, 1), \ldots, h(\mathbf{x}, y)$, $\bar{h}(\mathbf{x}, y)$, and so on. The evaluation of $h(\mathbf{x}, y + 1)$ uses the value of $\bar{h}(\mathbf{x}, y)$, and the evaluation of $\bar{h}(\mathbf{x}, y + 1)$ uses the values of $\bar{h}(\mathbf{x}, y)$ and $h(\mathbf{x}, y + 1)$. In these conditions the relation $h(\mathbf{x}, y) \simeq (\bar{h}(\mathbf{x}, y))_y$ is valid. We say that h is *specified by course-of-values recursion* from f and g.

Theorem 2.1.3. *Let $\mathscr{C}$ be closed under primitive recursive operations. If a function h is specified by course-of-values recursion from $\mathscr{C}$-computable functions f and g, then h is also $\mathscr{C}$-computable.*

PROOF. The following is a basic primitive recursive specification of the function $\bar{h}$:

$$\bar{h}(\mathbf{x}, 0) \simeq \mathsf{exp}(2, f(\mathbf{x})) \dot{-} 1$$

$$\bar{h}(\mathbf{x}, y + 1) \simeq (\bar{h}(\mathbf{x}, y) + 1 \times \mathsf{exp}(\mathsf{p}_{y+1}, g(\mathbf{x}, y, \bar{h}(\mathbf{x}, y)))) \dot{-} 1.$$

It follows that $\bar{h}$ is $\mathscr{C}$-computable, and since $h(\mathbf{x}, y) \simeq (\bar{h}(\mathbf{x}, y))_y$ it follows that h is also $\mathscr{C}$-computable. □

In applications, course-of-values recursion is usually used rather as *basic course-of-values recursion*, that is, as an iteration of the form

$$h(\mathbf{x}, 0) \simeq U$$

$$h(\mathbf{x}, y + 1) = V,$$

where U and V are basic numerical terms, all variables of U are in the list $\mathbf{x}$, all variables of V are in the list $\mathbf{x}, y$, the symbol h does not occur in U or V, and the symbol $\bar{h}$ does not occur in U but may occur in V in the form $\bar{h}(\mathbf{x}, y)$. The given functions in the specification are the functions in U and V, excluding $\bar{h}$ of course.

Corollary 2.1.3.1. *Let $\mathscr{C}$ be closed under primitive recursive operations. If a function h is specified by basic course-of-values recursion from given functions which are all $\mathscr{C}$-computable, then h is also $\mathscr{C}$-computable.*

PROOF. Basic course-of-values recursion can be reduced to standard course-of-values recursion, exactly as in the proof of Theorem 2.1.1. □

Finally, the restriction that the function h does not appear in the term V can be relaxed in some cases. For example, an occurrence of $h(\mathbf{x}, d(y))$ is admissible if d is a total function and the condition $d(y) \leq y$ is satisfied by all y, because in this case we can replace such an occurrence by $(\bar{h}(\mathbf{x}, y))_{d(y)}$. In this way it is possible to give a course-of-values specification of h in which the symbol $\bar{h}$ does not appear.

EXAMPLE 2.1.2. An example of course-of-values recursion is the following:

$$h(0) \simeq 1$$

$$h(y + 1) \simeq \mathsf{exp}(y, \bar{h}(y)).$$

To evaluate $h(3)$ the iteration will produce values $h(0) = 1, \bar{h}(0) = 1, h(1) = 0, \bar{h}(1) = 1, h(2) = 1, \bar{h}(2) = 9, h(3) = \mathsf{exp}(2, 9)$. From Theorem 2.1.3 it follows that this function is primitive recursive.

EXAMPLE 2.1.3. An example of basic course-of-values recursion is as follows:

$$h(x, 0) \simeq f(x)$$

$$h(x, y + 1) \simeq \bar{h}(x, y) + x.$$

If for some x $f(x)$ is undefined, then $h(x, y)$ is undefined for all y.

EXAMPLE 2.1.4. In this example the explicit occurrences of $\bar{h}$ have been eliminated:

$$h(0) \simeq 0$$
$$h(y+1) \simeq \mathsf{csg}(y) + h(y) + h(\mathsf{pd}(y)).$$

This function generates the Fibonacci sequence: 0, 1, 1, 2, 3, 5, ... The occurrence of $\bar{h}(y)$ is implicit, because $\mathsf{pd}(y) \leq y$; hence $h(\mathsf{pd}(y)) = (\bar{h}(y))_{\mathsf{pd}(y)}$. From Corollary 2.1.3.1 it follows that the function h is primitive recursive.

EXAMPLE 2.1.5. A basic primitive recursive specification is said to be by *pure iteration*, if it has the form

$$h(\mathbf{x}, 0) \simeq U$$
$$h(\mathbf{x}, y+1) \simeq g(h(\mathbf{x}, y)),$$

where U is a basic term, and g is a given unary function. For example, the specification of add and sb in the proof of Theorem 2.1.2 is by pure iteration. If $m \geq 0$, consider the elementary function g_m such that

$$g_m(x, y) = x \times \mathsf{exp}(\mathsf{p}_m, y),$$

introduced in Exercise 1.6.4. A specification of g_m by pure iteration can be given as follows:

$$g_m(x, 0) \simeq x$$
$$g_m(x, y+1) \simeq g(g_m(x, y)),$$

where $g(z) = \mathsf{p}_m \times z$.

EXAMPLE 2.1.6. Some forms of simultaneous recursion can be reduced to primitive recursion. For example, consider two $(k+1)$-ary functions h_1 and h_2, which are specified from functions f_1, f_2, g_1 and g_2 as follows:

$$h_1(\mathbf{x}, 0) \simeq f_1(\mathbf{x}) \qquad h_2(\mathbf{x}, 0) \simeq f_2(\mathbf{x})$$
$$h_1(\mathbf{x}, y+1) \simeq g_1(\mathbf{x}, y, h_1(\mathbf{x}, y), h_2(\mathbf{x}, y))$$
$$h_2(\mathbf{x}, y+1) \simeq g_2(\mathbf{x}, y, h_1(\mathbf{x}, y), h_2(\mathbf{x}, y)).$$

As usual we interpret this specification as the description of an iteration such that in order to evaluate $h_1(\mathbf{x}, y)$ or $h_2(\mathbf{x}, y)$ we generate values $h_1(\mathbf{x}, 0)$, $h_2(\mathbf{x}, 0)$, $h_1(\mathbf{x}, 1)$, $h_2(\mathbf{x}, 1), \ldots, h_1(\mathbf{x}, y)$, $h_2(\mathbf{x}, y)$. Note that under this interpretation $h_1(\mathbf{x}, y)$ is not defined in case $h_2(\mathbf{x}, y)$ is not defined. The reduction to primitive recursion involves a coding function h such that

$$h(\mathbf{x}, y) \simeq \langle h_1(\mathbf{x}, y), h_2(\mathbf{x}, y)\rangle.$$

A primitive recursive specification of h is given as follows:

$$h(\mathbf{x}, 0) \simeq \langle f_1(\mathbf{x}), f_2(\mathbf{x})\rangle$$
$$h(\mathbf{x}, y+1) \simeq \langle g_1(\mathbf{x}, y, h'(\mathbf{x}, y), h''(\mathbf{x}, y)), g_2(\mathbf{x}, y, h'(\mathbf{x}, y), h''(\mathbf{x}, y))\rangle,$$

where $h'(\mathbf{x}, y) \simeq [h(\mathbf{x}, y)]_1$ and $h''(\mathbf{x}, y) \simeq [h(\mathbf{x}, y)]_2$. Finally, we get h_1 and h_2

from h in the form

$$h_1(\mathbf{x}, y) \simeq [h(\mathbf{x}, y)]_1$$
$$h_2(\mathbf{x}, y) \simeq [h(\mathbf{x}, y)]_2.$$

This construction shows that in case f_1, f_2, g_1, and g_2 are $\mathscr{C}$-computable, where $\mathscr{C}$ is closed under primitive recursive operations, then h_1 and h_2 are also $\mathscr{C}$-computable.

We have already mentioned that recursive specifications are essentially ambiguous and require some clarification. A general method, which applies to arbitrary recursive specifications, will be discussed later in this chapter. Regarding primitive recursive specifications the situation is rather simple, because we consider the equations as a procedure to evaluate $h(\mathbf{x}, 0)$, $h(\mathbf{x}, 1), \ldots$, and this iteration determines unambiguously the function h. This idea was extended to course-of-values recursion, which involves a slightly more complicated iteration. Still there are situations that look quite similar to primitive recursion but cannot be reduced to an iteration procedure. For example, consider a specification for a $(k + 1)$-ary function h, $k \geq 1$, of the form

$$h(\mathbf{x}, 0) \simeq U$$
$$h(\mathbf{x}, y + 1) \simeq V,$$

where U and V are basic numerical terms, and V is allowed to contain occurrences of the symbol h in the form $h(V_1, \ldots, V_k, y)$, where the terms $V_1, \ldots, V_k$ may contain function symbols, the variables $\mathbf{x}, y$, and also occurrences of the symbol h. In general, no iteration can be derived from such equations, and the specification must be clarified in a different way. Our approach is the following. We consider that the first equation specifies $h(\mathbf{x}, 0)$ for all $(\mathbf{x}) \in \mathbb{N}^k$, and the second equation specifies $h(\mathbf{x}, y + 1)$ also for all $(\mathbf{x}) \in \mathbb{N}^k$, assuming $h(\mathbf{x}, y)$ has been specified for all $(\mathbf{x}) \in \mathbb{N}^k$. Any specification of this kind we call *global primitive recursion*. We discuss a few examples of this general situation and show how they are reduced to primitive recursion.

EXAMPLE 2.1.7. Let d be a binary function. We specify by global primitive recursion a 3-ary function d' such that

$$d'(x_1, x_2, 0) \simeq x_2$$
$$d'(x_1, x_2, y + 1) \simeq d(x_1, d'(x_1 + 1, x_2, y)).$$

Assume d is $\mathscr{C}$-computable, where $\mathscr{C}$ is closed under primitive recursive operations. To prove that d' is also $\mathscr{C}$-computable we introduce another 3-ary function d'', using basic primitive recursion in the form

$$d''(x_1, x_2, 0) \simeq x_2$$
$$d''(x_1, x_2, y + 1) \simeq d(x_1 \dot{-} y, d''(x_1, x_2, y)).$$

It follows that d'' is $\mathscr{C}$-computable, and using induction on y we can prove that

$$d'(x_1, x_2, y) \simeq d''((x_1 + y) \dot{-} 1, x_2, y);$$

hence d' is also $\mathscr{C}$-computable.

EXAMPLE 2.1.8. In a more general example a binary function h is specified by global primitive recursion in the form

$$h(x, 0) \simeq f(x)$$
$$h(x, y + 1) \simeq g(x, y, h(d(y, x), y)),$$

where f, g, and d are given $\mathscr{C}$-computable functions, and $\mathscr{C}$ is closed under primitive recursive operations. In order to prove that h is $\mathscr{C}$-computable, we introduce a 3-ary function h', such that

$$h'(y, x, 0) \simeq f(d'(0, x, y))$$
$$h'(y, x, v + 1) \simeq g(d'(v + 1, x, y \dot{-} (v + 1)), v, h'(y, x, v)),$$

where d' is the function induced by d, proved to be $\mathscr{C}$-computable in the preceding example. Now h' is $\mathscr{C}$-computable, and using induction on y we can prove that

$$h(d'(y, x, v), y) \simeq h'(y + v, x, y)$$

holds for all v, x, y. Taking $v = 0$ we get $h(x, y) \simeq h'(y, x, y)$, so h is $\mathscr{C}$-computable.

EXERCISES

2.1.1. Let $\mathscr{C}$ be a class of functions closed under primitive recursive operations, and let h be a k-ary function that is specified using partial definition by cases in the form

$$h(\mathbf{x}) \simeq f_1(\mathbf{x}) \quad \text{if } g(\mathbf{x}) \simeq 0$$
$$\simeq f_2(\mathbf{x}) \quad \text{if } g(\mathbf{x}) \not\simeq 0,$$

where f_1, f_2, and g are $\mathscr{C}$-computable, and furthermore either f_1 or f_2 is a total function. Prove that h is $\mathscr{C}$-computable.

2.1.2. Consider a binary function h that is specified by the following iteration:

$$h(x, 0) \simeq x + 1 \qquad \text{if } x \text{ is even}$$
$$\simeq x + 2 \qquad \text{if } x \text{ is odd}$$
$$h(x, y + 1) \simeq h(x, y) + 1 \qquad \text{if } h(x, y) \text{ is even}$$
$$\simeq \mathsf{exp}(2, h(x, y)) \qquad \text{if } h(x, y) \text{ is odd.}$$

Given a primitive recursive specification for h, with given elementary functions f and g.

2.1.3. Consider a binary function h that is specified by the following iteration:

$$h(x,0) \simeq x + 1$$

$$h(x, y + 1) \simeq h(x,(y)_0) + h(x,(y)_1).$$

Given a course-of-values specification for h, with given elementary functions f and g.

2.1.4. Let $\mathscr{C}$ be a class containing the functions σ, pd, and closed under basic operations, bounded minimalization, and pure iteration. Prove that $\mathscr{C}$ is closed under primitive recursive operations.

2.1.5. Let h be a binary function given by global primitive recursion in the form

$$h(x,0) \simeq f(x)$$

$$h(x, y + 1) \simeq h(g(x), y),$$

where f and g are $\mathscr{C}$-computable functions, and $\mathscr{C}$ is closed under basic operations and pure iteration. Prove that h is $\mathscr{C}$-computable.

2.1.6. Consider the binary function h specified by global primitive recursion in the form

$$h(x,0) \simeq f(x)$$

$$h(x, y + 1) \simeq h(x + 1, y),$$

where f is $\mathscr{C}$-computable and $\mathscr{C}$ is closed under elementary operations. Prove that h is $\mathscr{C}$-computable.

2.1.7. A total $(k + 1)$-ary function h is specified by primitive recursion from total functions f and g, which are $\mathscr{C}$-computable, where $\mathscr{C}$ is closed under elementary operations. Furthermore, there is a total $\mathscr{C}$-computable $(k + 1)$-ary function h' such that $h(\mathbf{x}, y) \leq h'(\mathbf{x}, y)$ for all x, y. Prove that h is $\mathscr{C}$-computable.

2.1.8. Let $\mathscr{C}$ be a class closed under primitive recursive operations. Prove that $\mathsf{BD}(\mathscr{C})$ is also closed under primitive recursive operations.

2.1.9. Let A be a numerical set and h a total function primitive recursive in ψ_A. Prove that h is primitive recursive.

2.1.10. Let h be a binary primitive recursive function given by

$$h(x,0) \simeq x$$

$$h(x, y + 1) \simeq \mathsf{exp}(x, h(x, y)).$$

Prove that h is not elementary.

2.1.11. Consider a binary function A given by the following specification:

$$A(0, y) \simeq y + 1$$

$$A(x + 1, 0) \simeq A(x, 1)$$

$$A(x + 1, y + 1) \simeq A(x, A(x + 1, y)).$$

Determine the function A using global primitive recursion; that is, the first equation specifies the value of $A(0, y)$ for arbitrary y, the second and third equations specify the value of $A(x + 1, y)$ for arbitrary y, assuming the value of $A(x, y)$ has been specified for arbitrary y. Prove that A is not primitive recursive.

2.1.12. Consider the following specification by global primitive recursion of a binary function h, such that

$$h(x, 0) \simeq f(x)$$

$$h(x, y + 1) \simeq h(h(x, y), y),$$

where the given function f is $\mathscr{C}$-computable, and the class $\mathscr{C}$ is closed under elementary operations and pure iteration. Prove that h is $\mathscr{C}$-computable.

2.1.13. Let $\mathscr{C}$ be a class of functions closed under elementary operations and unbounded minimalization with predicates. Prove that $\mathscr{C}$ is closed under primitive recursion with given functions f and g such that both G_f and G_g are $\mathscr{C}$-decidable predicates.

2.1.14. Prove the relation $d'(x_1, x_2, y) \simeq d''((x_1 + y) \dot{-} 1, x_2, y)$ in Example 2.1.7.

2.1.15. Prove the relation $h(d'(y, x, v), y) \simeq h'(y + v, x, y)$ in Example 2.1.8.

Notes

Primitive recursion has played a central role in early presentation of the subject. For example, in Kleene [9] it is considered as a model of formal computation, which is generalized via a system of equations. More recently, the tendency has been to note that the primitive recursive functions are simply a basic class of total functions, which can be replaced by other classes, for example, the elementary functions.

While we agree in principle with this approach, we are more interested in primitive recursion than in the primitive recursive functions. The analysis of Church's thesis in Section 5 shows that primitive recursive iteration is a basic mechanism essential for the description of computations.

More information about primitive recursion can be found in Davis [2], Hinman [8], Monk [14], Rose [19], and Tourlakis [29]. Proof of Exercises 2.1.10 and 2.1.11 are given in Monk [14].

§2. Functional Transformations

Numerical functions are operations that can be applied to numerical values and produce, when defined, numerical values. Now we introduce higher-order operations, which can be applied to numerical functions, and the output is also a numerical function. They provide the right foundation for the theory of recursive specifications, to be discussed in the next section.

A k-ary *functional transformation*, $k \geq 0$, is an operation that can be applied to one k-ary numerical function and produces as output another k-ary numerical function. If F is a k-ary functional transformation, and h is a k-ary numerical function, the application of F to h is denoted by $F(h)$, and it is always defined, so there is a k-ary numerical function h' such that $F(h) = h'$. On the other hand, if $(\mathbf{x}) \in \mathbb{N}^k$, then we can write $h'(\mathbf{x})$ simply as $F(h)(\mathbf{x})$, and it is possible that this expression is undefined. A complete specification for $F(h)(\mathbf{x})$ provides a specification for $F(h)$.

For example, for any $k \geq 0$, the k-ary functional transformation ID_k is specified by writing $\mathsf{ID}_k(h)(\mathbf{x}) \simeq h(\mathbf{x})$, which means that $\mathsf{ID}_k(h) = h$ for any k-ary function h. The unary functional transformation SC is given by the specification $\mathsf{SC}(h)(x) \simeq h(x) + 1$, which means that $\mathsf{SC}(h) = h'$, where $h'(x) \simeq h(x) + 1$.

If F is a k-ary functional transformation, and h is a k-ary function such that $F(h) \subseteq_k h$, we say that h is *closed under* F. If $F(h) = h$, we say that h is a *fixed point of* F. Furthermore, if h is a fixed point of F, and whenever h' is another fixed point of F then $h \subseteq_k h'$, we say h is a *minimal fixed point* of F.

A k-ary functional transformation F is *monotonic* if whenever $h \subseteq_k h'$ then $F(h) \subseteq_k F(h')$, for arbitrary k-ary functions h and h'. This means that whenever $F(h)(\mathbf{x}) \simeq v$ for some $(\mathbf{x}) \in \mathbb{N}^k$ then $F(h')(\mathbf{x}) \simeq v$.

A k-ary functional transformation F is *continuous* if it is monotonic, and furthermore when $F(h)(\mathbf{x})\downarrow$ for some k-ary function h and $(\mathbf{x}) \in \mathbb{N}^k$, then there is finite function $h_1 \subseteq_k h$ such that $F(h_1)(\mathbf{x})\downarrow$. Since F is monotonic it follows that in case $F(h_1)(\mathbf{x}) \simeq v$ then $F(h)(\mathbf{x}) \simeq v$.

Theorem 2.2.1. *If F is a continuous k-ary functional transformation then there is a k-ary function h which is closed under F, and furthermore when h' is another k-ary function closed under F then $h \subseteq_k h'$.*

Proof. We construct a sequence $h_0, h_1, \ldots, h_n, \ldots$ of k-ary functions, where $h_0 = \mathsf{UD}_k$ and $h_{n+1} = F(h_n)$. We claim that $h_n \subseteq_k h_{n+1}$ holds for all $n \geq 0$. The case $n = 0$ is trivial. Assume that it is true for n, so $h_n \subseteq_k h_{n+1}$. Since F is monotonic it follows that $F(h_n) \subseteq_k F(h_{n+1})$; that is, $h_{n+1} \subseteq_k h_{n+2}$.

Next we put $h = h_0 \cup h_1 \cup \cdots \cup h_n \cup \cdots$, and the condition $h_n \subseteq_k h_{n+1}$ implies that h is single valued, so it is a function. From the definition it follows that $h(\mathbf{x}) \simeq v$ if and only if for some n, $h_n(\mathbf{x}) \simeq v$. The following two properties of h are crucial in the proof. First, we have $h_n \subseteq_k h$ for every $n \geq 0$. Second, if h' is a k-ary function such that $h_n \subseteq h'$ for every $n \geq 0$, then $h \subseteq_k h'$.

To prove that h is closed under F assume that $F(h)(\mathbf{x}) \simeq v$. Since F is continuous it follows that there is h_n such that $F(h_n)(\mathbf{x}) \simeq v$; hence $h_{n+1}(\mathbf{x}) \simeq v$, and this implies that $h(\mathbf{x}) \simeq v$. This proves that $F(h) \subseteq_k h$.

Now assume h' is a k-ary function closed under F. To prove $h \subseteq_k h'$ it is sufficient to prove $h_n \subseteq_k h'$ for every $n \geq 0$. We use induction on n. The case $n = 0$ is trivial. Assume that $h_n \subseteq_k h'$. It follows that $F(h_k) \subseteq_k F(h') \subseteq_k h'$; that is, $h_{n+1} \subseteq_k h'$. □

Corollary 2.2.1.1. *Let F be a continuous k-ary functional transformation. Then F has a minimal fixed point h such that whenever h' is a k-ary function closed under F then $h \subseteq_k h'$.*

PROOF. It is sufficient to prove that the function h in Theorem 2.2.1 is a fixed point. We know that $F(h) \subseteq_k h$; hence $F(F(h)) \subseteq_k F(h)$. This means that $F(h)$ is closed under F; hence $h \subseteq_k F(h)$ and $F(h) = h$. □

If $\mathscr{C}$ is a class of k-ary functions and F is a k-ary functional transformation, then $F(\mathscr{C})$ denotes the class of all functions $F(h)$ where h is $\mathscr{C}$-computable. Note that if $\mathscr{C}$ is consistent and F is monotonic, then $F(\mathscr{C})$ is also consistent.

A class $\mathscr{C}$ of k-ary functions is *directed* if it is nonempty and whenever h_1 and h_2 are $\mathscr{C}$-computable functions there is a $\mathscr{C}$-computable function h such that $h_1 \subseteq_k h$ and $h_2 \subseteq_k h$. It is clear that if $\mathscr{C}$ is directed then it is also consistent and $\mathsf{Sup}\,\mathscr{C}$ is defined. Furthermore, if F is a monotonic functional transformation, then $F(\mathscr{C})$ is also directed.

Theorem 2.2.2. *Let F be a monotonic k-ary functional transformation. The following conditions are equivalent:*

(i) *F is continuous.*
(ii) *If $\mathscr{C}$ is a directed class of k-ary functions then $F(\mathsf{Sup}\,\mathscr{C}) = \mathsf{Sup}\,F(\mathscr{C})$.*

PROOF. If we assume (i) it follows by monotonicity that $\mathsf{Sup}\,F(\mathscr{C}) \subseteq_k F(\mathsf{Sup}\,\mathscr{C})$ whenever $\mathscr{C}$ is directed. On the other hand, if $F(\mathsf{Sup}\,\mathscr{C})(\mathbf{x}) \simeq v$, there is a finite k-ary function h such that $h \subseteq_k \mathsf{Sup}\,\mathscr{C}$ and $F(h)(\mathbf{x}) \simeq v$. Since $\mathscr{C}$ is directed there is a $\mathscr{C}$-computable function h' such that $h \subseteq_k h'$; hence $F(h')(\mathbf{x}) \simeq v$ and $\mathsf{Sup}\,F(\mathscr{C})(\mathbf{x}) \simeq v$.

Conversely, if F satisfies (ii) and $F(h)(\mathbf{x}) \simeq v$, we take $\mathscr{C}$ = the class of all k-ary finite functions bounded by h. It follows that $\mathscr{C}$ is directed and $\mathsf{Sup}\,\mathscr{C} = h$. From (ii) it follows that there is a function h' that is $\mathscr{C}$-computable and $F(h')(\mathbf{x}) \simeq v$, so F is continuous. □

EXERCISES

2.2.1. Prove that the functional transformations ID_k and SC are continuous. Which are the minimal fixed points?

2.2.2. Consider the unary functional transformation F such that $F(h)(x) \simeq 1$ if $x = 0$, and $F(h)(x) \simeq h(x + 1)$ otherwise. Prove that F is continuous, and find all fixed points of F.

2.2.3. The composition of two k-ary functional transformations F_1 and F_2 is the k-ary functional transformation $F = F_1 \circ F_2$ such that $F(h) = F_1(F_2(h))$. Prove that if F_1 and F_2 are monotonic (continuous) then $F_1 \circ F_2$ is also monotonic (continuous).

2.2.4. Prove that the following functional transformations are continuous, and for each one find the minimal fixed point:

$$\begin{aligned} F_1(h)(x, y) &\simeq 2 \times x && \text{if } y = 0 \\ &\simeq h(2 \times x, \mathsf{pd}(y)) && \text{otherwise} \\ F_2(h)(x, y) &\simeq \mathsf{exp}(x, 2) && \text{if } y = 0 \\ &\simeq h(h(x, \mathsf{pd}(y)), \mathsf{pd}(y)) && \text{otherwise} \\ F_3(h)(x, y) &\simeq y && \text{if } x = 0 \\ &\simeq h(x, h(x, y)) && \text{otherwise} \\ F_4(h)(x, y) &\simeq y + 1 && \text{if } x = y \\ &\simeq h(x, h(\mathsf{pd}(x), y + 1)) && \text{otherwise} \\ F_5(h)(x, y) &\simeq y && \text{if } x = 0 \\ &\simeq h(\mathsf{pd}(x), y + 1) && \text{otherwise.} \end{aligned}$$

2.2.5. Let F be the unary functional transformation such that

$$\begin{aligned} F(h)(x) &\simeq 0 && \text{if there is } y > x \text{ such that } h(y)\uparrow \\ &\simeq \mathsf{UD}(x) && \text{otherwise.} \end{aligned}$$

Prove that F has no fixed point.

2.2.6. Let F be a monotonic k-ary functional transformation and $\mathscr{C}$ the class of all functions that are closed under F. Prove that $\mathscr{C}$ is nonempty and $\mathsf{Inf}\,\mathscr{C}$ is the minimal fixed point of F.

2.2.7. Let F be a continuous k-ary functional transformation in which all fixed points are total functions. Prove that F has only one fixed point.

2.2.8. Let h be the minimal fixed point of the k-ary continuous functional transformation F, and let h' be a k-ary function such that $F(h \cap h') \subseteq_k h'$. Prove that $h \subseteq_k h'$.

2.2.9. Let f and h be two k-ary functions. Prove the following:
(a) f is consistent with h if and only if there is a k-ary function h' such that $f \subseteq_k h'$ and $h \subseteq_k h'$.
(b) If f is consistent with h, $f' \subseteq_k f$, and $h' \subseteq_k h$ then f' is consistent with h'.
(c) If f is consistent with h and F is a k-ary monotonic functional transformation, then $F(f)$ is consistent with $F(h)$.
(d) Let $h = h' \cup h''$. Then f is consistent with h if and only if it is consistent with h' and h''.
(e) Assume $\mathscr{C}$ is directed and $\mathsf{Sup}\,\mathscr{C} = h$. Then f is consistent with h if and only if it is consistent with every function in $\mathscr{C}$.

2.2.10. Let F be a k-ary continuous functional transformation and f a k-ary function such that $f \subseteq_k F(f)$. Prove there is a k-ary function h such that:
(a) $f \subseteq_k h$ and $F(h) \subseteq_k h$.
(b) If $f \subseteq_k h'$, where $F(h') \subseteq_k h'$, then $h \subseteq_k h'$.

(c) $F(h) = h$.
(d) f is consistent with h.
(e) If f is consistent with a k-ary function h' such that $h' \subseteq_k F(h')$, then h is consistent with h'.

2.2.11. Let F be a continuous k-ary functional transformation, and let $\mathscr{C}_F$ be the class of k-ary functions such that $h \in \mathscr{C}_F$ if and only if h is a fixed point of F and it is consistent with every fixed point of F. Prove the following:
(a) $\mathscr{C}_F$ is directed.
(b) If $\mathsf{Sup}\,\mathscr{C}_F = f$ then $f \in \mathscr{C}_F$. If h' is a k-ary function such that $h' \subseteq_k F(h')$, then h' is consistent with f.

2.2.12. Let F_1 and F_2 be continuous k-ary functional transformations such that $F_1(h) \subseteq_k F_2(h)$ holds for all k-ary functions h. Prove that if h_1 is the minimal fixed point of F_1 and h_2 is the minimal fixed point of F_2 then $h_1 \subseteq_k h_2$.

Notes

The purpose of this section is to give a complete mathematical foundation for recursion, which is actually introduced in Section 3. It should be compared with the evaluation algorithm in Section 4, which also provides an equivalent foundation for recursion.

The proof of the minimal fixed point given in the text assumes continuity. This assumption can be avoided (see Exercise 2.2.6), but the proof requires transfinite ordinal recursion.

§3. Recursive Specifications

While the results in the preceding section provide a general foundation for recursive specifications, in applications we need concrete methods to define continuous functional transformations. We now introduce a procedure that is practical and also fairly general, at least for the applications in this book.

First we recall that in a partial definition by cases a function h is specified from m functions $f_1, \ldots, f_m$ and m functions $g_1, \ldots, g_m$ (the case functions). It is clear that instead of using functions we can use basic numerical terms, say terms $U_1, \ldots, U_m$ and $V_1, \ldots, V_m$, constructed according to rules BT 1 and BT 2 in the definition of Chapter 1. The specification will have the following form:

$$\begin{aligned} h(\mathbf{x}) &\simeq U_1 \quad \text{if } V_1 \simeq 0 \\ &\simeq U_2 \quad \text{if } V_2 \simeq 0 \\ &\;\;\vdots \qquad\quad \vdots \qquad \vdots \\ &\simeq U_m \quad \text{if } V_m \simeq 0. \end{aligned}$$

We assume here that all variables in the terms $U_1, \ldots, U_m, V_1, \ldots, V_m$ appear in the list $\mathbf{x}$, which may contain extra variables not occurring in the terms. The evaluation procedure is exactly the same described in Chapter 1. We look for the first i such that $V_i \simeq 0$, and for $i' < i$ $V_{i'} \not\simeq 0$, and put $h(\mathbf{x}) \simeq U_i$. If there is no such i, then $h(\mathbf{x})$ is undefined.

Technically, the terms $U_1, \ldots, U_m, V_1, \ldots, V_m$ contain only variables, numerals, and function symbols. In practice, we allow conditions of the form $V \simeq 0$ to be expressed using predicates, which can be eliminated via the characteristic function. For example, a condition $\mathsf{pd}(x) = y$ is equivalent to $\varepsilon(\mathsf{pd}(x), y) \simeq 0$.

We obtain a *recursive specification* if we allow the symbol h to occur inside the terms $U_1, \ldots, U_m, V_1, \ldots, V_m$. Occurrences of the symbol h must be in the form $h(W_1, \ldots, W_k)$, where $W_1, \ldots, W_k$ are basic terms, which may also contain the symbol h. We assume that all symbols are interpreted in the usual way, but h is an uninterpreted symbol. We say that h is the *main symbol* in the specification, and the other function symbols are called *auxiliary symbols*. The interpretation of the auxiliary symbols are *auxiliary functions*, or simply the given functions in the specification. The equations in the specification are called *recursive equations*, and the set of all equations, in the given order, is called a *recursive system*, or simply a *system*.

We must determine which is the function specified by a given recursive system. First we note that if the symbol h is given some interpretation inside the terms, then we have a partial definition by cases specification. The interpretation of h is a function of the required arity, not necessarily total. Let h' be the function obtained by considering the system as a partial definition by cases, so h' depends on the interpretation of the symbol h inside the terms. We say that h' is the interpretation of the symbol h outside the terms, that is, on the left side of the recursive equations. In general, we have that $h \neq h'$. If in some case we have $h = h'$, we say that h is a *solution* of the recursive system.

Since a system may have many solutions, we have not yet determined the function specified by the system. This is the ambiguity, which is an essential characteristic of recursive specifications. To eliminate the ambiguity we note that the transformation from h to h' described above is actually a functional transformation F, where $F(h) = h'$, and the solutions of the system are exactly the fixed points of F. Furthermore, the functional transformation F is continuous, because the finitary rules are monotonic and finitary, as discussed in Chapter 1. Actually, the only operations involved in a recursive specification are basic operations and partial definition by cases. Since F is continuous it has a minimal fixed point (Corollary 2.2.1.1), which is then the *minimal solution* of the system. Such a minimal solution is, by definition, the function specified by the recursion system.

The choice of the minimal solution appears to be natural, for such a solution can be computed from the system, in a manner we shall describe later. Still, a system in general has many solutions, and others may be of interest in some situations.

Example 2.3.1. Consider the following system in which there is only one equation with two auxiliary functions, ε and σ:

$$h(x) \simeq h(x) + 1 \quad \text{if } \varepsilon(x, x) \simeq 0.$$

As a transformation this system describes the functional transformation SC discussed in the preceding section. This system has only one solution, namely, the function UD. Note that the condition $\varepsilon(x, x) \simeq 0$ means that the equation holds for every x. Such universal conditions can be omitted.

Example 2.3.2. This example has only one equation, and the auxiliary functions are ε and σ. We omit the universal condition.

$$h(x) \simeq h(x + 1).$$

The function UD is a solution of this system, so it is the minimal solution. All the constant unary functions are also solutions.

Example 2.3.3. The next example has two equations, and the main symbol is unary.

$$\begin{aligned} h(x) &\simeq x \dot{-} 10 && \text{if } x > 100 \\ &\simeq h(h(x + 11)) && \text{otherwise.} \end{aligned}$$

This is a typical case in which the conditions are expressed using predicates. The condition $x > 100$ is of course equivalent to $\chi_{>}(x, 100) \simeq 0$, and "otherwise" is equivalent to $\mathsf{csg}(\chi_{>}(x, 100)) \simeq 0$, where $\chi_{>}$ is the characteristic function of the predicate $>$. This system has only one solution, which is a total function, and can be specified using total definition by cases in the form

$$\begin{aligned} h(x) &\simeq 91 && \text{if } x \leq 100 \\ &\simeq x \dot{-} 10 && \text{otherwise.} \end{aligned}$$

We say that a class $\mathscr{C}$ of functions is *closed under recursive specifications*, if whenever a function h is given by a recursive specification in which all auxiliary functions are $\mathscr{C}$-computable, then h is also $\mathscr{C}$-computable. This is a fairly strong closure property, with many consequences. For example, a partial definition by cases is actually a (trivial) recursive specification in which the main symbol does not occur in the defining terms, so closure under recursive specifications implies closure under partial definition by cases. Furthermore, if U is a basic numerical term, and h is a function that is explicitly given by $h(\mathbf{x}) \simeq U$, then we can write a partial definition by cases for h in the form

$$h(\mathbf{x}) \simeq U \quad \text{if } 0 \simeq 0;$$

hence closure under recursive specifications implies closure under basic operations.

Theorem 2.3.1. *Let $\mathscr{C}$ be a class of functions closed under recursive specifications, and assume $\mathscr{C}$ contains the functions σ,* pd, *and* csg. *Then $\mathscr{C}$ is closed* *under primitive recursive operations, elementary operations, and unbounded minimalization with functions.*

PROOF. Consider a primitive recursion specification of the form

$$h(\mathbf{x}, 0) \simeq f(\mathbf{x})$$
$$h(\mathbf{x}, y + 1) \simeq g(\mathbf{x}, y, h(\mathbf{x}, y)).$$

We introduce a recursive specification with the same symbols, but also including pd and csg, as follows:

$$\begin{aligned} h(\mathbf{x}, y) &\simeq f(\mathbf{x}) & &\text{if } y \simeq 0 \\ &\simeq g(\mathbf{x}, \mathsf{pd}(y), h(\mathbf{x}, \mathsf{pd}(y))) & &\text{if } \mathsf{csg}(y) \simeq 0. \end{aligned}$$

The recursive specification has only one solution, since if h' and h'' are two solutions, it is easy to prove by induction on y that $h'(\mathbf{x}, y) \simeq h''(\mathbf{x}, y)$; hence $h' = h''$. On the other hand, it is clear that the function h given by the primitive recursion is a solution of the recursive specification; hence it is the minimal solution. This proves that $\mathscr{C}$ is closed under primitive recursion and hence under primitive recursive operations.

From the closure under primitive recursive operations it follows that $\mathscr{C}$ is closed under elementary operations, by Theorem 2.1.2.

Now consider a specification by unbounded minimalization with functions of the form

$$h(\mathbf{x}) \simeq \mu y(f(\mathbf{x}, y) \simeq 0).$$

Let us recall that in order that $h(\mathbf{x}) \simeq y$ we require that $f(\mathbf{x}, y) \simeq 0$, and furthermore that $f(\mathbf{x}, y') \not\simeq 0$ for all $y' < y$. We introduce a $(k + 1)$-ary function g given by the following recursive specification:

$$\begin{aligned} g(\mathbf{x}, v) &\simeq 0 & &\text{if } f(\mathbf{x}, v) \simeq 0 \\ &\simeq g(\mathbf{x}, v + 1) + 1 & &\text{if } \mathsf{csg}(f(\mathbf{x}, v)) \simeq 0. \end{aligned}$$

To prove that this system has only one solution, we assume g' and g'' are two solutions, and show that $g' \subseteq_k g''$. For this purpose we assume $g'(\mathbf{x}, v) \simeq y$ and prove $g''(\mathbf{x}, v) \simeq y$, by induction on y. The case $y = 0$ follows from the first equation, and the case $y + 1$ follows easily using the induction hypothesis and the second equation.

Having proved that the recursive specification has only one solution, we identify such a solution as the function g such that

$$g(\mathbf{x}, v) \simeq y \equiv f(\mathbf{x}, v + y) \simeq 0 \wedge \forall y' < y f(\mathbf{x}, v + y') \not\simeq 0.$$

It is clear that $g(\mathbf{x}, v)$ is undefined when $f(\mathbf{x}, v)$ is undefined, and whenever $f(\mathbf{x}, v)$ is defined the value of $g(\mathbf{x}, v)$ is determined according to the recursive equations.

If f is $\mathscr{C}$-computable, then g is also $\mathscr{C}$-computable. It follows that h is $\mathscr{C}$-computable, because $h(\mathbf{x}) \simeq g(\mathbf{x}, 0)$. □

If $\mathscr{C}$ is a class of functions that contains the functions σ, pd, and csg, and furthermore if $\mathscr{C}$ is closed under recursive specifications, we say that $\mathscr{C}$ is *closed under recursive operations*. If $\mathscr{C}$ is an arbitrary class, then $\mathsf{RC}(\mathscr{C})$ is the minimal closure of $\mathscr{C}$ under recursive operations; that is $\mathsf{RC}(\mathscr{C})$ is closed under recursive operations, and whenever $\mathscr{C} \subseteq \mathscr{C}'$ and $\mathscr{C}'$ is closed under recursive operations then $\mathsf{RC}(\mathscr{C}) \subseteq \mathscr{C}'$. The formal definition of $\mathsf{RC}(\mathscr{C})$ can be given using the techniques we have discussed before in connection with elementary operations.

The functions in $\mathsf{RC}(\mathscr{C})$ are said to be *recursive* in $\mathscr{C}$. If $\mathscr{C} = \{c\}$ we write $\mathsf{RC}(c) = \mathsf{RC}(\{c\})$, where c is a given function. We say that the functions in $\mathsf{RC}(c)$ are *recursive* in c. Finally, if $\mathscr{C} = \varnothing$ we write $\mathsf{RC} = \mathsf{RC}(\varnothing)$. The functions in RC are said to be *recursive*.

This notation extends in the usual way to predicates. A predicate P is *recursive in* $\mathscr{C}$ if the function χ_P is recursive in $\mathscr{C}$, and *partially recursive in* $\mathscr{C}$ if the function ψ_P is recursive in $\mathscr{C}$. The predicates in RC_{d} are *recursive predicates*, and the predicates in $\mathsf{RC}_{\mathsf{pd}}$ are *partially recursive* predicates.

Corollary 2.3.1.1. *If $\mathscr{C}$ is closed under recursive operations, then $\mathscr{C}$ is closed under elementary operations, primitive recursive operations, unbounded minimalization, and partial definition by cases.*

PROOF. Immediate from Theorem 2.3.1. □

A converse of this result, where primitive recursive operations, unbounded minimalization, and partial definition by cases imply closure under recursive operations, is proved in the next section.

Computability theory is primarily concerned with the classes RC and $\mathsf{RC}(c)$. Still there are many results that can be proved using closure under recursive operations and do not require minimal closure. A classical example is given in the next theorem. If h is a total unary function that is 1-1 and onto we say that h is a *permutation*. Furthermore, if h is $\mathscr{C}$-computable we say that h is a *$\mathscr{C}$-permutation*. In particular, an RC-permutation is called a *recursive permutation*.

Theorem 2.3.2. *Let $\mathscr{C}$ be a class closed under recursive operations. Let A_1 and A_2 be sets, and let f_1 and f_2 be total* 1-1 *unary $\mathscr{C}$-computable functions such that*

$$x \in A_1 \equiv f_1(x) \in A_2$$

$$x \in A_2 \equiv f_2(x) \in A_1.$$

Then there is a $\mathscr{C}$-permutation h such that $h(A_1) = A_2$.

PROOF. We describe the construction informally and leave to the reader the formalization in terms of recursive operations. We generate an infinite sequence of pairs (y_0, z_0), (y_1, z_1), $\ldots$, (y_n, z_n), $\ldots$ that satisfies the following conditions:

(a) For every $x \in \mathbb{N}$ there is exactly one $n \geq 0$ such that $x = y_n$.
(b) For every $x \in \mathbb{N}$ there is exactly one $n \geq 0$ such that $x = z_n$.
(c) $y_n \in A_1 \equiv z_n \in A_2$, for all $n \geq 0$.

We put $y_0 = 0$ and $z_0 = f_1(0)$. Assume now that the sequence has been generated up to a pair (y_n, z_n). To generate (y_{n+1}, z_{n+1}) we consider two cases.

Case 1. n is even. We take y_{n+1} = the least x such that $x \neq y_i$ for all $i = 0, \ldots, n$. We want to find z_{n+1} such that $z_{n+1} \neq z_i$ for all $i = 0, \ldots, n$, and furthermore $y_{n+1} \in A_1 \equiv z_{n+1} \in A_2$. This value is obtained by iterating the function f_1 as follows. Let $z'_1 = f_1(y_{n+1})$. If $z'_1 \neq z_i$ for $i = 0, \ldots, n$, we put $z_{n+1} = z'_1$. Otherwise there is a unique i such that $z'_1 = z_i$, and we put $z'_2 = f_1(y_i)$, and check for z'_2 in the same form as we checked for z'_1. If $z'_2 = z_i$ for some $i = 0, \ldots, n$, we take $z'_3 = f_1(y_i)$. This iteration generates values z'_1, $z'_2, \ldots$ that are all different, so eventually a value z'_m is obtained such that $z'_m \neq z_i$ for all $i = 0, \ldots, n$. We put $z_{n+1} = z'_m$.

Case 2. n is odd. We take z_{n+1} = the least x such that $x \neq z_i$ for all $i = 0, \ldots, n$. To obtain y_{n+1} we iterate f_2 starting with $f_2(z_{n+1})$ as we did in case 1 with the function f_1, so eventually a value y'_m is obtained such that $y_i \neq y'_m$ for all $i = 0, \ldots, n$, and we put $y_{n+1} = y'_m$.

Note that the fact that both functions f_1 and f_2 are 1-1 ensures that new values are generated in both iterations. Furthermore, the manner in which f_1 and f_2 relate the sets A_1 and A_2 ensures that the condition $y_{n+1} \in A_1 \equiv z_{n+1} \in A_2$ is satisfied.

The permutation h is obtained by putting $h(x) = z_n$, where n is the unique index such that $x = y_n$. From conditions (a), (b), and (c) above it follows that h is 1-1 and $f_1(A_1) = A_2$.

In order to show that h is $\mathscr{C}$-computable, we must generate h from f_1 and f_2 by recursive operations. In this application primitive recursive operations and unbounded minimalization are sufficient. For example, assume that $y = \langle y_0, \ldots, y_n \rangle$. We introduce a binary elementary function d such that

$$d(y, x) = (\mu v < \ell y [y]_{v+1} = x) + 1.$$

Note that if $d(y, x) = n + 2 = \ell y + 1$, it follows that $x \neq y_i$ for all $i = 0, \ldots, n$. On the other hand, if $d(y, x) = v + 1 \leq n + 1$, then $x = y_v$. If $z = \langle z_0, \ldots, z_n \rangle$, then $d(z, x)$ has similar properties.

In order to describe the iteration in case 1 we introduce a 3-ary $\mathscr{C}$-computable function g_1 such that

$$g_1(y, z, 0) \simeq \mu x [y]_{d(y,x)} = \ell y + 1$$

$$g_1(y, z, v + 1) \simeq [y]_{d(z, f_1(g_1(y,z,v)))}.$$

The iteration halts when $d(z, f_1(g_1(y, z, v))) = \ell z + 1$.

The iteration in case 2 is described by a 3-ary computable function g_2 that is similar to g_1 but uses f_2 rather than f_1. Finally, the main process that generates the pairs (y_0, z_0), (y_1, z_1), $\ldots$, (y_n, z_n), $\ldots$ is described by a unary $\mathscr{C}$-computable function f such that

$$f(n) \simeq \langle\langle y_0, \ldots, y_n\rangle, \langle z_0, \ldots, z_n\rangle\rangle.$$

The specification of f from g_1 and g_2 and the specification of h from f are left to the reader. □

Exercises

2.3.1. Prove that all solutions in the recursive system of Example 2.3.3 are total functions.

2.3.2. Consider a system with only one equation, and assume the system is not trivial, in the sense that the main symbol occurs in at least one of the terms of the system. Prove that UD is the minimal solution.

2.3.3. Prove that the function A in Exercise 2.1.11 is recursive.

2.3.4. Let $\mathscr{C}$ be a class of functions closed under recursive operations. Prove that $\mathsf{BD}(\mathscr{C})$ is also closed under recursive operations.

2.3.5. Let $\mathscr{C}$ be a class of functions such that $\mathscr{C} \subseteq \mathsf{BD}(\mathsf{RC})$. Prove that if h is a total function recursive in $\mathscr{C}$ then h is recursive.

2.3.6. Prove that $\mathsf{EL}(\varepsilon') \subset \mathsf{RC}$ (see Example 1.3.2 and Exercise 2.1.10).

2.3.7. Complete the proof of Theorem 2.3.2 with the formal specification of the functions g_2, f, and h.

2.3.8. Prove that if a function h is recursive in a class $\mathscr{C}$ of functions then there is a finite class $\mathscr{C}' \subseteq \mathscr{C}$ such that h is recursive in $\mathscr{C}'$.

2.3.9. Let $\mathscr{C}$ be a class of functions closed under recursive operations, and let f be a k-ary $\mathscr{C}$-computable function. Prove that G_f is a partially $\mathscr{C}$-decidable predicate.

Notes

Recursion usually appears in computability theory in the form of the first recursion theorem (see Kleene [9]) but it is not as widely used as the second recursion theorem (the recursion theorem in Chapter 4), which is more flexible and can be applied in situations well beyond ordinary recursion. In this work we use recursion in the construction of the universal interpreter in Chapter 4.

The form of recursion we introduce in this chapter is fairly restricted. For example, it would be possible to allow terms that are more general than basic terms, including bounded sum and bounded product. We exclude this extension because it requires a more complicated evaluation algorithm. Another possibility is to consider systems of recursive equations which introduce several functions at the same time. While the evaluation algorithm is not affected by this extension, the mathematical foundations must be arranged to take care of more complicated functional transformations.

Although the definition of recursive functions via recursive specifications is not standard in the literature, it appears to be at least a reasonable use of the term "recursive."

The proof of Theorem 2.3.2 is taken from Myhill [16].

§4. Recursive Evaluation

In this section we show that the necessary conditions in the preceding section, that is, partial definition by cases, primitive recursive operations, and unbounded minimalization, are sufficient to ensure closure under recursive specifications. For this purpose we formalize the well-known algorithm to evaluate recursive systems, using a stack to control the order of the recursive calls.

The evaluation procedure is given by a number of rules, and the detail of such rules is determined by the recursive equations of the system under consideration. This being the case, it will help the presentation to introduce first a particular example and to discuss the evaluation in this context. The generalization to arbitrary systems will then be straightforward. Our example consists of two equations with two unspecified functions f and g, and furthermore the functions σ and csg. The binary main symbol is denoted with the letter h.

$$\begin{aligned} h(x_1, x_2) &\simeq f(x_1) && \text{if } g(x_1, x_2) \simeq 0 \\ &\simeq h(h(x_1, \sigma(x_2)), x_2) && \text{if } \mathsf{csg}(g(x_1, x_2)) \simeq 0. \end{aligned}$$

It is convenient to have names for each of these terms, so we put

$$\begin{aligned} U_1 &= f(x_1) & U_1' &= g(x_1, x_2) \\ U_2 &= h(h(x_1, \sigma(x_2)), x_2) & U_2' &= \mathsf{csg}(g(x_1, x_2)). \end{aligned}$$

We say that U_1 and U_2 are *output terms* and U_1' and U_2' are *case terms*. Furthermore, it is necessary to rewrite the terms using polish notation, in which the arguments precede the operands, and no parentheses are required. We assume the reader is familiar with this type of transformation. The terms take now the following form:

$$\begin{aligned} U_1 &= x_1 f & U_1' &= x_1 x_2 g \\ U_2 &= x_1 x_2 \sigma h x_2 h & U_2' &= x_1 x_2 g\, \mathsf{csg}. \end{aligned}$$

Under this presentation the symbols in each term have a natural ordering, and, for example, we can talk about the *next symbol* after x_1 in U_1, namely, f. On the other hand, f has no next symbol in U_1, so we say that f is *final* in U_1.

We complete this preliminary organization by assigning an index to every symbol in the terms, which identifies the position of the corresponding symbol. We use numbers as indexes, the only requirement being that symbols with different positions should have different indexes. The assignment is described by rewriting the terms with the symbols replaced by the indexes.

$$U_1 = 1, 2 \qquad U_1' = 9, 10, 11$$

$$U_2 = 3, 4, 5, 6, 7, 8 \qquad U_2' = 12, 13, 14, 15.$$

From now on each of the numbers from 1 to 15 will be considered to be a *position* for the corresponding symbol. For example, 1 is a position for x_1 in U_1, 9 is a position for x_1 in U_1', 15 is position for csg in U_2', and so on.

The evaluation algorithm operates on *vectors*. There are two kinds of vectors: *complete vectors* of the form (p, v, v', w), where p is 0 or a position (i.e., a number from 1 to 15) and v, v', and w are arbitrary numbers, and *incomplete vectors* of the form $(p, v, v', *)$, where p, v, and v' are as before and $*$ is just a symbol. If V is a vector, say $V = (p, v, v', w)$ or $V = (p, v, v', *)$, we say that p is the *handle* of V, and w is the output component of V.

During the evaluation vectors are organized in a stack, which is simply a sequence of vectors of the form $(V_1, \ldots, V_m)$, $m \geq 1$. The last vector V_m is said to be at the *top* of the stack. If $m > 1$ we also say that V_{m-1}, V_m are at the top of the stack.

In order to evaluate $h(x_1, x_2)$, where x_1 and x_2 are fixed numerical values, we start with the initial stack

$$((0, x_1, x_2, *), (9, x_1, x_2, x_1)),$$

which is a stack containing two vectors, one incomplete and the other complete at the top of the stack. During the evaluation the stack changes, according to some rules that are derived from the equations and are described below. There are several possible changes. In some cases one new vector is inserted at the top of the stack, but retaining all vectors already present, so the length of the stack increases and the former top vector is now below the new top vector. In other cases one or two vectors at the top are taken off the stack and replaced by one new vector, or by two new vectors which are now at the top of the stack. In this case the length of the stack may decrease.

To describe the rules we assume that some stack has been generated, where the length is $m \geq 1$, and the top vector is V, which is always a complete vector, so we write $V = (p, v, v', w)$, the handle of V is p, and the output component of V is w. If $m > 1$ the vector below V is denoted by V', and we assume that p' is the handle of V', and w' is the output component of V' in case V' is complete.

The rules are determined by the recursive equations and depend essentially on the value p of the handle of the top vector V.

(a) If $p = 0$ evaluation halts with output value w.
(b) Assume $p > 0$, and p has a next position p'' for either x_1 or x_2 (i.e., p has one of the values 3, 6, 9, 12). In this case a new vector $V'' = (p'', v, v', w'')$

is added at the top of the stack, where $w'' = v$ if p'' is a position for x_1, and $w'' = v'$ if p'' is a position for x_2.

(c) Assume $p > 0$ and p has a next position p'' which is a position for one of the auxiliary symbols f, g, σ, or csg (i.e., p has one of the values 1, 4, 10, 13, 14). If $p = 1$ the auxiliary symbol is f, and we replace V by the vector $V'' = (2, v, v', w'')$, where $f(w) \simeq w''$. If $f(w)$ is not defined the evaluation halts without output. If $p = 10$ the auxiliary symbol is g, and it can be shown that $m > 1$, so there are at least two complete vectors at the top stack, V' and V. Both are taken off from the stack, and a new vector $V'' = (11, v, v', w'')$ is inserted at the top, where $g(w', w) \simeq w''$. If $g(w', w)$ is undefined the evaluation halts without output. A similar procedure is followed in the other cases.

(d) Assume $p > 0$ and p is a final position in one of the output terms U_1 or U_2. In this case $m > 1$, and the vector V' is incomplete. We take V off the stack and change V' to a complete vector by inserting w as an output component. Equivalently, if $V' = (p', s, s', *)$, then V' is replaced by $V'' = (p', s, s', w)$.

(e) Assume $p > 0$ and p is the final position in the case term U_1', that is, $p = 11$. If $w = 0$ we replace V by the vector $V'' = (1, v, v', v)$ (this means we start the evaluation of U_1). If $w \neq 0$ we repalce V by the vector $V'' = (12, v, v', v)$ (we start the evaluation of U_2').

(f) Assume $p > 0$ and p is the final position in the case term U_2', that is, $p = 15$. If $w = 0$ we replace V by the vector $V'' = (3, v, v', v)$. If $w \neq 0$ (this case is impossible in this example) evaluation halts without output.

(g) Assume $p > 0$ and p has a next position p'' which is a position for the main symbol h (i.e., $p = 5$ or 7). In this case $m > 1$ and there are at least two vectors at the top of the stack. Both vectors at the top are taken off the stack, and two new vectors are inserted, an incomplete vector $V'' = (p'', v, v', *)$ and a complete vector $V_1 = (9, w', w, w')$, which is now at the top of the stack. This means we start the evaluation algorithm assuming w' is the value of x_1, and w is the value of x_2. This step is called a *recursive call*.

It is clear that the action of the algorithm is controlled by the handle of the vector at the top of the stack, which is always a complete vector. The actual transformation is actually determined by the contents of the stack, and in this particular example by just the two vectors at the top of the stack. A complete formalization of the algorithm can be obtained by writing 16 rules, one for each possible value of p from 0 to 15. As an example, we write some of the rules and leave it to the reader to complete the algorithm. We continue to use the same notation, where V is the vector at the top of the stack, $V = (p, v, v', w)$, and in case there is a vector V' below V (which may be incomplete) we set p' = the handle of V' and w' = the output component of V' if V' is complete.

Rule 0: If $p = 0$ evaluation halts with output w.
Rule 4: If $p = 4$ replace V by the vector $(5, v, v', w + 1)$.
Rule 5: If $p = 5$ replace vectors V' and V by vectors $(6, v, v', *)$ and $(9, w', w, w')$.
Rule 8: If $p = 8$ then $m > 1$. Take V off the stack, and insert w as the output component of V'.
Rule 11: If $p = 11$ and $w = 0$ replace V by vector $(1, v, v', v)$. If $w \neq 0$ replace V by vector $(12, v, v', v)$.
Rule 13: If $p = 13$ then $m > 1$. Take V' and V off the stack, and insert the vector $(14, v, v', w'')$ at the top, where $g(w', w) \simeq w''$. If $g(w', w)$ is undefined evaluation halts without output.
Rule 15: If $p = 15$ and $w = 0$ replace V by the vector $(3, v, v', v)$. If $w \neq 0$ evaluation halts without output.

Note that although the rules have been derived from the recursive equations they contain no reference to them. The rules form a self-contained algorithm, which can be applied even if the equations are not available. The algorithm may halt in several ways, but in only one case an output is produced, namely, under rule 0. Other possibilities of halting are given by rules 1 and 10. Theoretically, rules 13 and 15 may result in halting, but this is actually impossible in this system.

Up to this point we have only shown how to derive an evaluation algorithm from the recursive system. We must also prove this evaluation is correct, that is, that the algorithm evaluates the right function. We call h_0 the function given by the algorithm, that is, $h_0(x_1, x_2) \simeq v$ if and only if the algorithm halts under rule 0 with output v. We must prove that h_0 is the minimal solution of the recursive system.

First let us assume that $F(h_0) = h'$ and $h'(x_1, x_2) \simeq v$. This value is determined using the equations as a partial definition by cases, where values $h_0(v_1, v_2) \simeq w$ are assumed to be given whenever they are required. Instead of assuming that such values are given we may insert the recursive evaluation itself, and the result will be a more inclusive recursive evaluation which shows that $h_0(x_1, x_2) \simeq v$. This means that $h' \subseteq_2 h_0$, that is, h_0 is closed under F.

Now let h be a function that is closed under F, that is, $F(h) \subseteq_2 h$, and assume that there is a recursive evaluation that shows that $h_0(x_1, x_2) \simeq v$. Using induction on the number of recursive calls in the evaluation, we can show that $h(x_1, x_2) \simeq v$. For example, if there is no recursive call this means $F(\mathsf{UD})(x_1, x_2) \simeq v$; hence $F(h)(x_1, x_2) \simeq v$, so $h(x_1, x_2) \simeq v$. If the evaluation includes a recursive call to determine the value of $h_0(v_1, v_2) \simeq w$, we use the induction hypothesis to assume that $h(v_1, v_2) \simeq w$; hence we have $F(h)(x_1, x_2) \simeq v$, and $h(x_1, x_2) \simeq v$. This induction shows that $h_0 \subseteq_2 h$.

The argument above shows that h_0 is the minimal solution of the recursive specification; hence the recursive algorithm is a correct evaluation of such a solution.

Our analysis has shown that the minimal solution of the recursive system is a function h_0 which can be evaluated by an algorithm involving vectors organized in a stack structure. The problem now is to find some specification of h_0 involving the finitary rules of Chapter 1. For this purpose we encode vectors and also the stack. If V is the vector (p, v, v', w), then V is encoded in the number $x = \langle p, v, v', w\rangle$. If V is the incomplete vector $(p, v, v', *)$, we encode V in the number $x' = \langle p, v, v'\rangle$. Note that $\ell x = 4$ and $\ell x' = 3$. A stack of the form $(V_1, \ldots, V_m)$ is encoded in the number $y = \langle y_1, \ldots, y_m\rangle$, where y_1 is the encoding of $V_1, \ldots, y_m$ is the encoding of V_m.

The evaluation algorithm operates on a stack, which we assume to be encoded in a number y, and produces a new stack, which we assume is encoded in a number y'. In this way the algorithm determines a numerical function tr such that $\mathsf{tr}(y) = y'$. This function can be specified by a simple translation of the rules 0 to 15, and the result is a specification by partial definition by cases in which a number of elementary functions are involved; furthermore, the functions f and g appear in the recursive specification. We do not intend to write the complete specification of tr, only some of the equations as derived from the algorithm rules given explicitly above. For this purpose it is convenient to introduce some elementary functions that are used to manipulate the encoding and decoding of vectors and stacks.

(i) $x \square y = (x + 1) \times \prod_{z<\ell y} \mathsf{exp}(\mathsf{p}_{\ell x+z}, (y)_z) \dot{-} 1$. If $x = \langle x_1, \ldots, x_n\rangle$ and $y = \langle y_1, \ldots, y_m\rangle$, then $x \square y = \langle x_1, \ldots, x_n, y_1, \ldots, y_m\rangle$.
(ii) $[x]_y = \mathsf{pd}((x)_{\mathsf{pd}(y)})$. If $x = \langle x_1, \ldots, x_n\rangle$, then $[x]_i = x_i$, $i = 1, \ldots, n$.
(iii) $\nabla x = [x]_{\ell x}$. If $x = \langle x_1, \ldots, x_n\rangle$ and $n \geq 1$, then $\nabla x = x_n$.
(iv) $x\S = \prod_{y<\mathsf{pd}(\ell x)} \mathsf{exp}(\mathsf{p}_y, (x)_y) \dot{-} 1$. $\S x = \prod_{y<\mathsf{pd}(\ell x)} \mathsf{exp}(\mathsf{p}_y, (x)_{y+1}) \dot{-} 1$. If $x = \langle x_1, \ldots, x_n, x_{n+1}\rangle$, $n \geq 0$, then $x\S = \langle x_1, \ldots, x_n\rangle$, and $\S x = \langle x_2, \ldots, x_{n+1}\rangle$.

Note that if $x = \langle x_1, \ldots, x_n\rangle$, $n \geq 0$, then $x \square \langle y\rangle = \langle x_1, \ldots, x_n, y\rangle$, and $\langle y\rangle \square x = \langle y, x_1, \ldots, x_n\rangle$.

We write below some of the equations entering the partial definition by cases of the function tr. Note that if y is the encoding of a stack in which the top vector is V, then $[\nabla y]_1$ is the handle of V, and $\nabla\nabla y$ is the output component of V.

$$
\begin{aligned}
\mathsf{tr}(y) &\simeq y\S \square \langle\langle 5, [\nabla y]_2, [\nabla y]_3, \nabla\nabla y + 1\rangle\rangle && \text{if } [\nabla y]_1 = 4\\
&\simeq y\S\S \square \langle\langle 6, [\nabla y]_2, [\nabla y]_3\rangle, \langle 9, \nabla\nabla(y\S), \nabla\nabla y, \nabla\nabla(y\S)\rangle\rangle && \text{if } [\nabla y]_1 = 5\\
&\simeq y\S\S \square \langle\langle [\nabla(y\S)]_1, [\nabla(y\S)]_2, [\nabla(y\S)]_3, \nabla\nabla y\rangle\rangle && \text{if } [\nabla y]_1 = 8\\
&\simeq y\S \square \langle\langle 1, [\nabla y]_2, [\nabla y]_3, [\nabla y]_2\rangle\rangle && \text{if } \mathsf{csg}(\nabla\nabla y) \times [\nabla y]_1 = 11\\
&\simeq y\S \square \langle\langle 12, [\nabla y]_2, [\nabla y]_3, [\nabla y]_2\rangle\rangle && \text{if } \mathsf{sg}(\nabla\nabla y) \times [\nabla y]_1 = 11\\
&\simeq y\S\S \square \langle\langle 14, [\nabla y]_2, [\nabla y]_3, g(\nabla\nabla(y\S), \nabla\nabla y)\rangle\rangle && \text{if } [\nabla y]_1 = 13\\
&\simeq y\S \square \langle\langle 3, [\nabla y]_2, [\nabla y]_3, [\nabla y]_2\rangle\rangle && \text{if } \mathsf{csg}(\nabla\nabla y) \times [\nabla y]_1 = 15.
\end{aligned}
$$

Now using primitive recursion with the function tr we define a 3-ary function ev that describes the evolution of the stack during the evaluation of $h_0(x_1, x_2)$:

$$\mathsf{ev}(x_1, x_2, 0) \simeq \langle\langle 0, x_1, x_2\rangle, \langle 9, x_1, x_2, x_1\rangle\rangle$$

$$\mathsf{ev}(x_1, x_2, y + 1) = \mathsf{tr}(\mathsf{ev}(x_1, x_2, y)).$$

The final stack can easily be detected using the elementary function $g_1(y) = [\nabla y]_1$, and the output of the evaluation can be obtained from y using the elementary function $g_2(y) = \nabla\nabla y$. It follows that the function h_0 satisfies the relation

$$h_0(x_1, x_2) \simeq g_2(\mathsf{ev}(x_1, x_2, \mu y(g_1(\mathsf{ev}(x_1, x_2, y)) \simeq 0)))$$

which involves finitary rules plus primitive recursion.

Note that in case the auxiliary functions f and g are total then the specification of tr can be given using total definition by cases, so ev is also a total function. From this it does not follow, of course, that the function h_0 is total.

Theorem 2.4.1. *Let $\mathscr{C}$ be a class of functions closed under primitive recursive operations, partial definition by cases, and unbounded minimalization with functions. Then $\mathscr{C}$ is closed under recursive operations.*

PROOF. It is sufficient to show that $\mathscr{C}$ is closed under recursive specifications. The example above can be generalized to any arbitrary recursive specification. □

EXAMPLE 2.4.1. The classes EL, PR, and RC are related by proper inclusion, that is, $\mathsf{EL} \subset \mathsf{PR} \subset \mathsf{RC}$. For example, the function UD_1 is recursive; in fact,

$$\mathsf{UD}_1(x) \simeq \mu y\colon \mathsf{F}_1(x, y),$$

but it is not primitive recursive. Note that $\mathsf{RC} = \mathsf{RC}(\mathsf{UD}_1)$. Furthermore, if c is an elementary function, then c is total and $\mathsf{RC} = \mathsf{RC}(c)$.

EXERCISES

2.4.1. Write all the rules necessary for the evaluation algorithm, relative to the example discussed in this section.

2.4.2. Write the complete partial definition by cases of the function tr, relative to the example discussed in this section.

2.4.3. Prove that during the execution of the evaluation algorithm the vector at the top of the stack is always complete.

2.4.4. Prove that if during the execution of the evaluation algorithm the handle of the top vector V is a position for a final symbol in one of the terms, then there is an incomplete vector V' immediately preceding V.

2.4.5. Write the rules for the evaluation algorithm corresponding to the recursive specification of Example 2.3.2.

2.4.6. Write the rules for the evaluation algorithm corresponding to the recursive specification of Example 2.3.3.

2.4.7. Prove that if $y = x \square \langle x' \rangle$ then $\nabla y = x'$.

Notes

The evaluation algorithm is the only one actually invoked in this work and can be considered a form of machine computation executed by a stack machine. The algorithm and the encoding are essential for the proof of the universal interpreter in Chapter 4.

Since a recursive specification is a special case of a system of equations in the sense of Kleene [9], it is possible to evaluate recursion using Kleene's rules of substitution and replacement.

§5. Church's Thesis

Up to now we have introduced a number of closure properties, and from each one a characteristic minimal closure construction has been derived. From the elementary operations we get the classes EL and EL(c), from the primitive recursive operations the classes PR and PR(c), and from recursive operations the classes RC and RC(c). We shall show now that with RC and RC(c) a maximal situation has been reached, in the sense that RC contains all computable functions, and RC(c) contains all functions that are deterministically computable relative to the function c. These assertions are known in the literature as Church's thesis.

The reader is warned that the word "computable" is used here in an entirely informal manner, and no proof of Church's thesis is actually available. Still some arguments can be given to support the claim.

When we say that a function is computable we mean, roughly, that there is a set of rules, or an algorithm, or a program that can be used to evaluate the function for a given input. While there is no formal definition of algorithm, and programs can be formalized only in the context of a given programming language, it is remarkable that the idea of algorithm has an old tradition in mathematics, where the algorithmic solution of problems has always been a matter of great importance. The idea appears to be clear enough to support some amount of mathematical elaboration. The difficulties appear when one tries to prove that a particular problem is algorithmically unsolvable. In fact, the main contribution of computability theory has been to provide a technique to deal with such situations.

Similarly, we say that a function is computable relative to a fixed function c if there is an evaluation algorithm that provides for instructions in which specific values of the function c are required. To execute such instructions it is assumed that there is an external agent, or oracle, which makes available the required information whenever it exists, that is, whenever the function c is defined. When the function c is not defined some extra assumptions have to be made. The most restricted assumption is that whenever such a situation takes place, that is, $c(y)$ is undefined for a required argument y, then the computation is suspended without output. A computation operating under such a restriction is said to be *deterministic*. In this chapter we consider only deterministic computability. A more general nondeterministic form will be discussed in the next chapter.

We first discuss Church's thesis for absolute computability. The relative deterministic case requires only minor modifications.

Church's thesis for absolute computability. A function is computable if and only if it is recursive.

In other words, this thesis means that RC is exactly the class of all computable functions. Now it is not difficult to argue that all recursive functions are computable. Such functions are generated from initial (computable) functions by recursive specifications, which can be computed by the stack algorithm described in the preceding section. Such algorithms can easily be combined to produce more general stack algorithms, so it can be claimed that every recursive function is computed by a stack algorithm in which the primitive operations are σ, pd, and csg.

The problem, of course, is to validate the second part, that is, that every computable function is recursive. This is by no means obvious or self-evident. Neither can we give a mathematical proof, because the notion of being computable is not formal. But it can be made convincing by other ways. The one we have chosen is derived from the experience obtained in the evaluation of recursive specifications. This does not mean that we assume that every computable function can be evaluated by a stack algorithm. For our purpose it is sufficient to realize that the basic technique used in dealing with recursion, that is, encoding of structures and computations, can be applied to any computable function.

Assume that an algorithm is available to evaluate a computable k-ary function h. In order to carry out the evaluation the algorithm assumes some fixed structure, which may be vectors, stacks, matrices, trees, and so on. These structures contain numerical values and symbols (like $*$ in the stack evaluation). The algorithm determines which is the initial structure, which usually depends on the input to be evaluated, and how the structure changes during the computation. The computation proceeds according to the rules of the algorithm, and may continue indefinitely, or may be suspended without output after reaching a particular situation, or may halt with output.

Our first assumption is that the computational structures determined by the algorithm can be numerically encoded, that is, represented by numbers.

We have seen this to be the case for vectors and stacks. The same technique can be applied to other structures. For example, matrices can be encoded as sequences of vectors. The encoding must be such that the decoding functions which compute the elements of the structure from the encoding are elementary.

With this understanding the rules of the algorithm induce a numerical function, since the change from structure to structure during the computation determines a corresponding change on the encodings. We call such a function the *transition function*, and it is denoted by tr. If y encodes some structure and $\mathsf{tr}(y) = y'$, then y' is the encoding of the next structure induced by the algorithm applied to y.

We shall assume that the function tr is always elementary. This is actually a bold assumption, which the reader should not take lightly. Note that it is possible to write algorithms involving nonelementary operations. We claim that whenever this is the case such operations can be eliminated via subroutines, and a final version of the algorithm can be obtained in which all transformations are elementary. This assumption is certainly validated by the daily experience of mathematicians and computer scientists. For example, the compilation of a higher-order program (say in PASCAL) can be interpreted as a translation to a program involving only elementary machine transformations.

The assumption that tr is always elementary is the crucial point in our argument. It corresponds to the intuition that while computations may be long and involved, every step is short and simple. On the other hand, we could assume that every step is primitive recursive, or even recursive, without affecting our final conclusion.

From now on our analysis proceeds exactly as in the encoding of recursive evaluations. We introduce a $(k + 1)$-ary function ev which is specified using primitive recursion, which describes the computation step by step. We set $\mathsf{ev}(\mathbf{x}, 0) =$ the encoding of the initial structure, and $\mathsf{ev}(\mathbf{x}, y + 1) = \mathsf{tr}(\mathsf{ev}(\mathbf{x}, y))$. It is reasonable to assume that a halting structure can be detected by an elementary function g_1, and there is an elementary function g_2 which computes the output from a halting structure. The final conclusion is that the function h can be expressed in the form

$$h(\mathbf{x}) \simeq g_2(\mathsf{ev}(\mathbf{x}, \mu y(g_1(\mathsf{ev}(\mathbf{x}, y)) \simeq 0))),$$

where ev is primitive recursive, and g_1, g_2 are elementary functions. It follows that h is a recursive function.

Now we turn our attention to deterministic computability relative to a function c, where we get the following result:

Church's thesis for deterministic relative computability. A function is deterministically computable relative to a function c if and only if it is recursive in c.

The validation of this thesis is entirely similar to the one for absolute computability. In one direction it is obvious that a function recursive in c is deterministically computable relative to c. In the other direction we argue again that given an algorithm to compute a function h relative to c, we can encode the computational structures and define a function tr that is determined by the algorithm. Such a function is specified using partial definition by cases, some of them involving the function c. Whenever c is invoked but undefined, the function tr is undefined, and this is consistent with the deterministic character of the computation. The function ev, which is derived from tr exactly as before, is now recursive in c. Here again we find that the halting function g_1 and the output function g_2 are elementary. Finally, we conclude that the function h can be expressed in the form

$$h(\mathbf{x}) \simeq g_2(\mathsf{ev}(\mathbf{x}, \mu y(g_1(\mathsf{ev}(\mathbf{x}, y)) \simeq 0))),$$

where ev is a function recursive in c, and g_1, g_2 are elementary functions. It follows that h is a function recursive in c.

We have already remarked that Church's thesis cannot be proved, and it is not really a mathematical theorem. The thesis provides a fundamental frame for the interpretation of computability theory, but it is never used in a formal sense. In fact, the only positive way in which it can be used is to derive that a function is recursive because it is known it is computable. The latter assumption can only mean that an algorithm is available to compute the function, in which case we can use the procedure described above to obtain a derivation by recursive operations.

The only real applications of Church's thesis are negative, to derive that a function is not computable because it is known it is not recursive. By using this technique it has been possible to prove that many mathematical problems are algorithmically unsolvable.

Exercises

2.5.1. Describe an algorithm to compute the function A in Exercise 2.1.11, and by a proper encoding prove that A is a recursive function.

2.5.2. Describe an algorithm to compute the permutation h in Theorem 2.3.2, and by a proper encoding prove that there exist a total binary function f that is primitive recursive in $\{f_1, f_2\}$ and unary elementary functions g_1 and g_2 such that

$$h(x) = g_2(f(x, \mu y(g_1(f(x, y)) \simeq 0))).$$

2.5.3. Let c be a total function. Use the analysis of Church's thesis to prove that a k-ary function h is recursive in c if and only if there is a unary function g, and a $(k+1)$-ary function f, both primitive recursive in c, such that

$$h(\mathbf{x}) \simeq g(\mu y(f(\mathbf{x}, y) \simeq 0)).$$

2.5.4. Extend the characterization in the preceding exercise to the case h is recursive in a class $\mathscr{C}$ containing only total functions.

Notes

Church's thesis is universally accepted, although some individual objections have occasionally been raised. Our presentation is a bit stronger than usual, for we claim there is a standard procedure to reduce the execution of an algorithm to a combination of primitive recursion, unbounded minimalization with functions, and partial definition by cases. To apply this reduction the algorithm must be refined to the point where all steps are elementary. While this argument is not certainly a proof, it appears to have an element of conviction to persons newly introduced to the field.

Kleene [9] gives a careful and systematic presentation of the different arguments that have been advanced to support Church's Thesis.

CHAPTER 3

Enumeration

This chapter is concerned mainly with predicates, in particular with the specification of functions via predicates. In a rather restricted form this takes place via the graph predicates of functions. In a more general setting we consider selector properties. Operations on predicates and the associated closure properties play a central role in our discussion. In particular, we consider the so-called unbounded quantification, existential and universal.

§1. Predicate Classes

We have already considered predicate classes, in particular the notation that can be used to denote the classes that are induced by functions. For example, RC_{d} denotes the class of predicates that are RC-decidable, or simply the recursive predicates, and $\mathsf{RC}_{\mathsf{pd}}$ denotes the class of all predicates that are partially RC-decidable, or simply the partially recursive predicates. On the other hand, if $\mathscr{C} = \mathsf{RC}(\mathscr{C}')$, where $\mathscr{C}'$ is a class of functions, we put $\mathsf{RC}_{\mathsf{d}}(\mathscr{C}') = \mathscr{C}_{\mathsf{d}}$ and $\mathsf{RC}_{\mathsf{pd}}(\mathscr{C}') = \mathscr{C}_{\mathsf{pd}}$. The predicates in $\mathsf{RC}_{\mathsf{d}}(\mathscr{C}')$ are said to be recursive in $\mathscr{C}'$, and the predicates in $\mathsf{RC}_{\mathsf{pd}}(\mathscr{C}')$ are said to be partially recursive in $\mathscr{C}'$. The same notation will be used with any combination of subscripts. In case $\mathscr{C} = \{c\}$, where c is a function, we identify $\mathscr{C}$ with c and write $\mathsf{RC}_{\mathsf{d}}(c)$ or $\mathsf{RC}_{\mathsf{pd}}(c)$.

We recall that subscript operations are always monotonic, a property that is used in many proofs. For example, if $\mathscr{C} \subseteq \mathscr{C}'$ then $\mathscr{C}_{\mathsf{d}} \subseteq \mathscr{C}'_{\mathsf{d}}$ and $\mathscr{C}_{\mathsf{pd}} \subseteq \mathscr{C}'_{\mathsf{pd}}$.

Theorem 3.1.1. *Let $\mathscr{C}$ be a class closed under elementary operations. Then*

(i) $\mathsf{EL}_{\mathsf{d}} \subseteq \mathscr{C}_{\mathsf{d}}$.
(ii) *$\mathscr{C}_{\mathsf{d}}$ is closed under conjunction, disjunction, negation, and bounded quantification.*

(iii) *$\mathscr{C}_{\mathsf{d}}$ is closed under substitution with total $\mathscr{C}$-computable functions.*
(iv) *If h is a total $\mathscr{C}$-computable function then G_h is $\mathscr{C}$-decidable.*
(v) *$\mathscr{C}_{\mathsf{pd}}$ is closed under conjunction and universal bounded quantification.*
(vi) *$\mathscr{C}_{\mathsf{pd}}$ is closed under substitution with total $\mathscr{C}$-computable functions.*
(vii) *If P_1 is a $\mathscr{C}$-decidable predicate, and P_2 is a partially $\mathscr{C}$-decidable predicate, then $P_1 \vee P_2$ is partially $\mathscr{C}$-decidable.*

PROOF. Part (i) is clear from the inclusion $\mathsf{EL} \subseteq \mathscr{C}$, and (ii) is a reformulation of Theorem 1.4.3 (ii). Part (iii) is obvious, since $\mathscr{C}$ is closed under basic operations. To prove (iv) assume h is a k-ary total $\mathscr{C}$-computable function. Then

$$\chi_{\mathsf{G}_h}(\mathbf{x}, v) = \varepsilon(h(\mathbf{x}), v)$$

so G_h is $\mathscr{C}$-decidable. To prove (v) assume first that $P = P_1 \wedge P_2$, where P_1 and P_2 are partially $\mathscr{C}$-decidable. It follows that

$$\psi_P(\mathbf{x}) \simeq \psi_{P_1}(\mathbf{x}) + \psi_{P_2}(\mathbf{x});$$

hence P is partially $\mathscr{C}$-decidable. On the other hand, if P is of the form

$$P(\mathbf{x}, z) \equiv \forall y < z Q(\mathbf{x}, y),$$

where Q is partially $\mathscr{C}$-decidable, then

$$\psi_P(\mathbf{x}, z) \simeq \sum_{y<z} \psi_Q(\mathbf{x}, y),$$

so P is partially $\mathscr{C}$-decidable. Part (vi) is obvious from the assumption.

To prove (vii) assume $P = P_1 \vee P_2$, where P_1 is $\mathscr{C}$-decidable, and P_2 is partially $\mathscr{C}$-decidable. It follows that

$$\psi_P(\mathbf{x}) \simeq \sum_{y<g(\mathbf{x})} f(\mathbf{x}, y),$$

where $g = \chi_{P_1}$, and $f(\mathbf{x}, y) \simeq \psi_{P_2}(\mathbf{x})$. Hence P is partially $\mathscr{C}$-decidable. □

Now we introduce a new subscript operation that can be applied to predicate classes. If $\mathscr{P}$ is a predicate class, then $\mathscr{P}_{\mathsf{n}}$ is the class containing the negation of the predicates in $\mathscr{P}$. Note that $\mathscr{P}$ is closed under negation if and only if $\mathscr{P} = \mathscr{P}_{\mathsf{n}}$.

As usual we allow for the meaningful concatenation of subscripts. For example, $\mathscr{C}_{\mathsf{dn}}$ is the class containing all predicates that are negations of $\mathscr{C}$-decidable predicates. $\mathsf{RC}_{\mathsf{pdn}}(\mathscr{C})$ is the class of all predicates that are negations of predicates partially recursive in $\mathscr{C}$. The equation $\mathscr{P}_{\mathsf{nn}} = \mathscr{P}$ is obviously valid for any class of predicates $\mathscr{P}$. On the other hand, $\mathscr{C}_{\mathsf{n}}$ is not meaningful when $\mathscr{C}$ is a class of functions.

More operations are introduced via the well-known unbounded quantifiers, existential and universal, which are defined as follows.

Existential Quantification. If Q is a $(k + 1)$-ary predicate, then we introduce a k-ary predicate P such that

$$P(\mathbf{x}) \equiv \exists y Q(\mathbf{x}, y)$$
$$\equiv \text{there is a } y \text{ such that } Q(\mathbf{x}, y).$$

We say that P is specified from Q by *existential unbounded quantification.*

Universal Quantification. If Q is a $(k + 1)$-ary predicate, then we introduce a k-ary predicate P such that

$$P(\mathbf{x}) \equiv \forall y Q(\mathbf{x}, y)$$
$$\equiv \text{for all } y Q(\mathbf{x}, y).$$

We say that P is specified from Q by *universal unbounded quantification.*

Let $\mathscr{P}$ be a class of predicates. We say that a k-ary predicate P is $\mathscr{P}$-enumerable if there is a $(k + 1)$-ary predicate Q in $\mathscr{P}$ such that

$$P(\mathbf{x}) \equiv \exists y Q(\mathbf{x}, y).$$

The class of all predicates that are $\mathscr{P}$-enumerable is denoted by $\mathscr{P}_{\mathsf{e}}$. We can introduce new constructions by concatenation of subscripts. For example, $\mathscr{C}_{\mathsf{de}}$ is the class of all predicates that are $\mathscr{C}_{\mathsf{d}}$-enumerable, and $\mathscr{C}_{\mathsf{pde}}$ is the class of all predicates that are $\mathscr{C}_{\mathsf{pd}}$-enumerable. Some of these constructions are very important and are identified here for future reference.

If a predicate P is RC_{d}-enumerable we say that it is *recursively enumerable,* so $\mathsf{RC}_{\mathsf{de}}$ is now the class of all recursively enumerable predicates. If P is $\mathsf{RC}_{\mathsf{pd}}$-enumerable we say that it is *partially recursively enumerable.* We shall prove later that $\mathsf{RC}_{\mathsf{de}} = \mathsf{RC}_{\mathsf{pd}} = \mathsf{RC}_{\mathsf{pde}}$, so these two notions are equivalent.

Similarly, a predicate in $\mathsf{RC}_{\mathsf{de}}(\mathscr{C})$ is said to be *recursively enumerable in* $\mathscr{C}$, and a predicate in $\mathsf{RC}_{\mathsf{pde}}(\mathscr{C})$ is said to be *partially recursively enumerable* in $\mathscr{C}$. In general, $\mathsf{RC}_{\mathsf{de}}(\mathscr{C}) \subseteq \mathsf{RC}_{\mathsf{pde}}(\mathscr{C})$, with equality holding only in special cases.

We set $\mathscr{P}_{\mathsf{u}} = \mathscr{P}_{\mathsf{nen}}$, so $\mathscr{P}_{\mathsf{u}}$ contains all predicates that can be obtained by universal quantification from predicates in $\mathscr{P}$. Note the familiar relations:

$$\mathscr{P}_{\mathsf{e}} = \mathscr{P}_{\mathsf{nun}}$$
$$\mathscr{P}_{\mathsf{en}} = \mathscr{P}_{\mathsf{nu}}$$
$$\mathscr{P}_{\mathsf{un}} = \mathscr{P}_{\mathsf{ne}}.$$

The classes $\mathscr{P}$ and $\mathscr{P}_{\mathsf{n}}$ are related by well-known duality properties, which will be used in some proofs. We note that $\mathscr{P}$ is closed under conjunction (disjunction) if and only if $\mathscr{P}_{\mathsf{n}}$ is closed under disjunction (conjunction). Furthermore, $\mathscr{P}$ is closed under universal bounded quantification (existential bounded quantification) if and only if $\mathscr{P}_{n}$ is closed under existential bounded quantification (universal bounded quantification). Finally, $\mathscr{P}$ is closed under existential quantification (universal quantification) if and only if $\mathscr{P}_{\mathsf{n}}$ is closed under universal quantification (existential quantification).

EXAMPLE 3.1.1. The class EL_{d} is closed under conjunction, disjunction, negation, and bounded quantification. On the other hand, $\mathsf{EL}_{\mathsf{pd}} = \mathsf{EL}_{\mathsf{pde}} = \mathsf{EL}_{\mathsf{pdu}}$ is not closed under negation or existential bounded quantification.

EXAMPLE 3.1.2. The class TO = the class of all total functions is closed under elementary operations, and $\mathsf{TO}_{\mathsf{d}} = \mathsf{TO}_{\mathsf{de}} = \mathsf{TO}_{\mathsf{du}}$ = the class of all predicates. On the other hand, $\mathsf{TO}_{\mathsf{pd}} = \mathsf{EL}_{\mathsf{pd}} = \mathsf{EL}_{\mathsf{pdu}}$.

Theorem 3.1.2. *Let $\mathscr{P}$ be a class of predicates closed under substitution with elementary functions. Then*

(i) *$\mathscr{P}_{\mathsf{e}}$ and $\mathscr{P}_{\mathsf{u}}$ are closed under substitution with elementary functions.*
(ii) *$\mathscr{P} \subseteq \mathscr{P}_{\mathsf{e}}$ and $\mathscr{P} \subseteq \mathscr{P}_{\mathsf{u}}$.*
(iii) *$\mathscr{P}_{\mathsf{e}}$ is closed under existential quantification, and $\mathscr{P}_{\mathsf{u}}$ is closed under universal quantification.*
(iv) *If $\mathscr{P}$ is closed under conjunction, then $\mathscr{P}_{\mathsf{e}}$ and $\mathscr{P}_{\mathsf{u}}$ are closed under conjunction.*
(v) *If $\mathscr{P}$ is closed under disjunction, then $\mathscr{P}_{\mathsf{e}}$ and $\mathscr{P}_{\mathsf{u}}$ are closed under disjunction.*
(vi) *If $\mathscr{P}$ is closed under existential bounded quantification, then $\mathscr{P}_{\mathsf{e}}$ and $\mathscr{P}_{\mathsf{u}}$ are closed under existential bounded quantification.*
(vii) *If $\mathscr{P}$ is closed under universal bounded quantification, then $\mathscr{P}_{\mathsf{e}}$ and $\mathscr{P}_{\mathsf{u}}$ are closed under universal bounded quantification.*

PROOF. Part (i) is immediate from the assumption. To prove (ii) we note that using substitution with elementary functions we can adjoin a variable to any predicate, which can be eliminated using existential or universal quantification. To prove the remaining implications, we consider first $\mathscr{P}_{\mathsf{e}}$.

To prove (iii) assume P is a $(k + 1)$-ary $\mathscr{P}$-enumerable predicate; then

$$\begin{aligned} \exists v P(\mathbf{x}, v) &\equiv \exists v \exists y Q(\mathbf{x}, v, y) \\ &\equiv \exists y Q(\mathbf{x}, (y)_0, (y)_1), \end{aligned}$$

where Q is a predicate in $\mathscr{P}$, hence $\mathscr{P}_{\mathsf{e}}$ is closed under existential quantification.

To prove (iv) for $\mathscr{P}_{\mathsf{e}}$ consider predicates P_1 and P_2, which are $\mathscr{P}$-enumerable. It follows that

$$\begin{aligned} P_1(\mathbf{x}) \wedge P_2(\mathbf{x}) &\equiv \exists y Q_1(\mathbf{x}, y) \wedge \exists y Q_2(\mathbf{x}, y) \\ &\equiv \exists y (Q_1(\mathbf{x}, (y)_0) \wedge Q_2(\mathbf{x}, (y)_1)), \end{aligned}$$

where Q_1 and Q_2 are predicates in $\mathscr{P}$. Since $\mathscr{P}$ is closed under substitution with elementary functions, and also closed under conjunction, it follows that $P_1 \wedge P_2$ is $\mathscr{P}$-enumerable. The proof of disjunction in part (v) is similar.

The proof of (vi) involves only a trivial permutation of an existential quantifier with an existential bounded quantifier. Finally, we prove (vii), which requires a more involved substitution. Consider a specification by universal bounded quantification of the form

$$\begin{aligned} \forall v < z P(\mathbf{x}, v) &\equiv \forall v < z \exists y Q(\mathbf{x}, v, y) \\ &\equiv \exists y \forall v < z Q(\mathbf{x}, v, (y)_v), \end{aligned}$$

where Q is a predicate in $\mathscr{P}$. Since $\mathscr{P}$ is closed under substitution with elementary functions, and also under universal bounded quantification, it follows that $\mathscr{P}_e$ is closed under universal bounded quantification.

This completes the proof of parts (iii) to (vii) for $\mathscr{P}_e$. Since $\mathscr{P}_n$ is also closed under substitution with elementary functions the same properties hold for $\mathscr{P}_{ne}$. From this, using duality, we complete the proof for $\mathscr{P}_u$. □

EXAMPLE 3.1.3. Since $\mathsf{EL_d}$ is closed under conjunction, disjunction, and bounded quantification, it follows that the same is true for the classes $\mathsf{EL_{de}} \subseteq \mathsf{EL_{deu}} \subseteq \mathsf{EL_{deue}} \cdots$ and also for the classes $\mathsf{EL_{du}} \subseteq \mathsf{EL_{due}} \subseteq \mathsf{EL_{dueu}} \cdots$. We shall prove in Chapter 4 that all these classes are different. Note also that $\mathsf{EL_d}$ is closed under negation; hence $\mathsf{EL_{den}} = \mathsf{EL_{du}}$, $\mathsf{EL_{deun}} = \mathsf{EL_{due}}$, and so on.

EXAMPLE 3.1.4. If $\mathscr{C}$ is any class closed under elementary operations, then the inclusions $\mathscr{C}_{de} \subseteq \mathscr{C}_{deu} \subseteq \mathscr{C}_{deue}$ and $\mathscr{C}_{du} \subseteq \mathscr{C}_{due} \subseteq \mathscr{C}_{dueu} \cdots$ are valid, but these classes are not necessarily different. For example, we have $\mathsf{TO_d} = \mathsf{TO_{de}} = \mathsf{TO_{du}} = \mathsf{TO_{deu}} \cdots$.

EXAMPLE 3.1.5. If $\mathscr{C}$ is closed under elementary operations, then both $\mathscr{C}_{pde}$ and $\mathscr{C}_{pdu}$ are closed under conjunction and universal bounded quantification. Furthermore, $\mathscr{C}_{pde}$ is closed under existential quantification, and $\mathscr{C}_{pdu}$ is closed under universal quantification. We have again the inclusions $\mathscr{C}_{pde} \subseteq \mathscr{C}_{pdeu} \subseteq \mathscr{C}_{pdeue} \cdots$ and $\mathscr{C}_{pdu} \subseteq \mathscr{C}_{pdue} \subseteq \mathscr{C}_{pdueu} \cdots$, but these classes are not necessarily different. For example, $\mathsf{EL_{pd}} = \mathsf{EL_{pde}} = \mathsf{EL_{pdu}} \cdots$.

Theorem 3.1.3. *If $\mathscr{C}$ is a class of functions closed under recursive operations, then $\mathscr{C}_{pde}$ is closed under disjunction.*

PROOF. Assume P_1 and P_2 are $\mathscr{C}_{pd}$-enumerable k-ary predicates. Hence, there are $(k + 1)$-ary predicates Q_1 and Q_2, such that

$$P_1(\mathbf{x}) \equiv \exists y Q_1(\mathbf{x}, y)$$

$$P_2(\mathbf{x}) \equiv \exists y Q_2(\mathbf{x}, y),$$

and Q_1 and Q_2 are partially $\mathscr{C}$-decidable. If we introduce a $(k + 2)$-ary predicate P such that

$$P(\mathbf{x}, y, v) \equiv (Q_1(\mathbf{x}, y) \wedge v = 0) \vee (Q_2(\mathbf{x}, y) \wedge v = 1)$$

then P is also partially $\mathscr{C}$-decidable, for we can specify the function ψ_P using partial definition by cases in the form

$$\begin{aligned} \psi_P(\mathbf{x}, y, v) &\simeq \psi_{Q_1}(\mathbf{x}, y) \quad \text{if } v = 0 \\ &\simeq \psi_{Q_2}(\mathbf{x}, y) \quad \text{if } v = 1. \end{aligned}$$

Since $\mathscr{C}$ is closed under partial definition by cases, it follows that P is partially $\mathscr{C}$-decidable. Furthermore, we have

$$P_1(\mathbf{x}) \vee P_2(\mathbf{x}) \equiv \exists y \exists v P(\mathbf{x}, y, v);$$

hence $P_1 \vee P_2$ is $\mathscr{C}_{pd}$-enumerable. □

Now we introduce a new subscript operation, which in this case applies to a class of predicates and produces a class of functions. If $\mathscr{P}$ is a class of predicates, and h is a k-ary function such that the graph predicate G_h is $\mathscr{P}$-enumerable, we say that h is *$\mathscr{P}$-enumerable.* The class of all functions that are $\mathscr{P}$-enumerable is denoted by $\mathscr{P}_{\mathsf{eg}}$. We simplify this notation by putting $\mathscr{P}_{\#} = \mathscr{P}_{\mathsf{eg}}$. For example, $\mathscr{C}_{\mathsf{pd}\#}$ is the class of all k-ary functions h such that G_h is $\mathscr{C}_{\mathsf{pd}}$-enumerable; that is, there is a partially $\mathscr{C}$-decidable $(k + 2)$-ary predicate Q such that

$$\mathsf{G}_h(\mathbf{x}, v) \equiv \exists y Q(\mathbf{x}, v, y).$$

Another example is $\mathsf{RC}_{\mathsf{d}\#}(\mathscr{C})$, the class of all functions h such that G_h is recursively enumerable in $\mathscr{C}$.

If h is a function that is RC_{d}-enumerable, we say that h is *recursively enumerable.* Hence h is recursively enumerable if and only if G_h is recursively enumerable, and $\mathsf{RC}_{\mathsf{d}\#}$ denotes the class of all recursively enumerable functions. A function in $\mathsf{RC}_{\mathsf{pd}\#}$ is said to be *partially recursively enumerable.* We show later that $\mathsf{RC} = \mathsf{RC}_{\mathsf{d}\#} = \mathsf{RC}_{\mathsf{pd}\#}$, so these notions are equivalent.

A function in $\mathsf{RC}_{\mathsf{d}\#}(\mathscr{C})$ is said to be *recursively enumerable in* $\mathscr{C}$, and a function in $\mathsf{RC}_{\mathsf{pd}\#}(\mathscr{C})$ is said to be *partially recursively enumerable* in $\mathscr{C}$. In general, the following relation holds: $\mathsf{RC}_{\mathsf{d}\#}(\mathscr{C}) \subseteq \mathsf{RC}(\mathscr{C}) \subseteq \mathsf{RC}_{\mathsf{pd}\#}(\mathscr{C})$ but the inequalities $\mathsf{RC}_{\mathsf{d}\#}(\mathscr{C}) \neq \mathsf{RC}(\mathscr{C})$ and $\mathsf{RC}_{\mathsf{pd}\#}(\mathscr{C}) \neq \mathsf{RC}(\mathscr{C})$ are both possible for the same class $\mathscr{C}$.

Theorem 3.1.4. *Let $\mathscr{C}$ be a class of functions closed under elementary operations and unbounded minimalization with predicates. Then*

(i) $\mathscr{C}_{\mathsf{de}} \subseteq \mathscr{C}_{\mathsf{pd}}$.
(ii) $\mathscr{C}_{\mathsf{d}\#} \subseteq \mathscr{C} \subseteq \mathscr{C}_{\mathsf{pd}\#}$.
(iii) $\mathscr{C}_{\mathsf{d}} = \mathscr{C}_{\mathsf{de}} \cap \mathscr{C}_{\mathsf{du}}$

PROOF. To prove (i) consider a k-ary predicate P such that

$$P(\mathbf{x}) \equiv \exists y Q(\mathbf{x}, y),$$

where Q is $\mathscr{C}$-decidable. It follows that

$$\psi_P(\mathbf{x}) \simeq \mu y Q(\mathbf{x}, y) \times 0;$$

hence P is partially $\mathscr{C}$-decidable.

To prove (ii) we consider a k-ary function h that is $\mathscr{C}_{\mathsf{d}}$-enumerable; hence

$$\mathsf{G}_h(\mathbf{x}, v) \equiv \exists y Q(\mathbf{x}, v, y),$$

where Q is $\mathscr{C}$-decidable. It follows that

$$h(\mathbf{x}) \simeq g(\mu y Q(\mathbf{x}, (y)_0, (y)_1)),$$

where $g(y) = (y)_0$; hence h is $\mathscr{C}$-computable. This proves $\mathscr{C}_{\mathsf{d}\#} \subseteq \mathscr{C}$. To prove $\mathscr{C} \subseteq \mathscr{C}_{\mathsf{pd}\#}$ consider a k-ary function h that is $\mathscr{C}$-computable. Note that the binary function ε' such that

$$\varepsilon'(x, v) \simeq \mu y\colon x = v \wedge y = 0$$

is $\mathscr{C}$-computable; hence

$$\psi_{\mathsf{G}_h}(\mathbf{x}, v) \simeq \varepsilon'(h(\mathbf{x}), v),$$

which means that G_h is partially $\mathscr{C}$-decidable, and therefore G_h is $\mathscr{C}_{\mathsf{pd}}$-enumerable and h is also $\mathscr{C}_{\mathsf{pd}}$-enumerable.

To prove (iii) note that the inclusion $\mathscr{C}_{\mathsf{d}} \subseteq \mathscr{C}_{\mathsf{de}} \cap \mathscr{C}_{\mathsf{du}}$ follows from Theorem 3.1.2 (ii). To prove the converse note that $\mathscr{C}_{\mathsf{du}} = \mathscr{C}_{\mathsf{den}}$; hence if P is a predicate in $\mathscr{C}_{\mathsf{de}} \cap \mathscr{C}_{\mathsf{du}} = \mathscr{C}_{\mathsf{de}} \cap \mathscr{C}_{\mathsf{den}}$, it follows that

$$\chi_P(\mathbf{x}) = v \equiv (P(\mathbf{x}) \wedge v = 0) \vee (\bar{P}(\mathbf{x}) \wedge v = 1)$$

Since $\mathscr{C}_{\mathsf{de}}$ is closed under conjunction and disjunction it follows that χ_P is $\mathscr{C}_{\mathsf{d}}$-enumerable; hence it is $\mathscr{C}$-computable by (ii), and P is $\mathscr{C}$-decidable. This proves that $\mathscr{C}_{\mathsf{de}} \cap \mathscr{C}_{\mathsf{du}} \subseteq \mathscr{C}_{\mathsf{d}}$. □

We are interested in the relations among the classes $\mathscr{P}_{\#}$, $\mathscr{P}_{\#\mathsf{de}}$, $\mathscr{P}_{\#\mathsf{d}\#}$, $\mathscr{P}_{\#\mathsf{pd}}$, $\mathscr{P}_{\#\mathsf{pde}}$, and $\mathscr{P}_{\#\mathsf{pd}\#}$ when $\mathscr{P}$ is a given class of predicates satisfying some restrictions. We denote by $\mathscr{P}_{\mathsf{d}}$ the class of all characteristic functions of predicates in $\mathscr{P}$. As a consequence the subscript d has a double meaning, but no confusion arises because the applications are disjoint. Note that $\mathscr{P}_{\mathsf{d}}$ is a class of functions, and $\mathscr{P}_{\mathsf{dd}}$ is a class of predicates; in fact, $\mathscr{P}_{\mathsf{dd}} = \mathscr{P}$. On the other hand, if $\mathscr{C}$ is a class of functions, then $\mathscr{C}_{\mathsf{dd}} \subseteq \mathscr{C}$.

If $\mathscr{P}$ is a class of predicates, then $\mathscr{P}_{\mathsf{pd}}$ is the class of all partial characteristic functions of predicates in $\mathscr{P}$. Here again we have $\mathscr{P}_{\mathsf{pdpd}} = \mathscr{P}$ and $\mathscr{C}_{\mathsf{pdpd}} \subseteq \mathscr{C}$.

Theorem 3.1.5. *Let $\mathscr{P}$ be a class of predicates closed under substitution with elementary functions. Assume $\mathscr{P}_{\mathsf{e}}$ is closed under conjunction and $\mathsf{EL}_{\mathsf{d}} \subseteq \mathscr{P}_{\mathsf{e}}$. Then*

(i) $\mathscr{P}_{\#\mathsf{d}} \subseteq \mathscr{P}_{\#\mathsf{pd}} = \mathscr{P}_{\mathsf{e}}$.
(ii) $\mathscr{P}_{\#\mathsf{d}} \subseteq \mathscr{P}_{\#\mathsf{de}} \subseteq \mathscr{P}_{\#\mathsf{pde}} = \mathscr{P}_{\mathsf{e}}$.
(iii) $\mathscr{P}_{\#\mathsf{d}\#} \subseteq \mathscr{P}_{\#\mathsf{pd}\#} = \mathscr{P}_{\#}$.
(iv) *If $\mathscr{P}_{\mathsf{e}}$ is closed under disjunction and universal bounded quantification then $\mathscr{P}_{\#}$ is closed under recursive operations and $\mathsf{RC}_{\mathsf{pd}\#}(\mathscr{P}_{\mathsf{pd}}) = \mathscr{P}_{\#}$.*

PROOF. To prove (i) assume P is a k-ary predicate in $\mathscr{P}_{\#\mathsf{d}}$. This means that χ_P is in $\mathscr{P}_{\#}$; hence the graph predicate of χ_P is in $\mathscr{P}_{\mathsf{e}}$. It follows that we can write

$$\psi_P(\mathbf{x}) \simeq v \equiv \chi_P(\mathbf{x}) = v \wedge v = 0,$$

which means that ψ_P is in $\mathscr{P}_{\#}$; hence P is in $\mathscr{P}_{\#\mathsf{pd}}$. Now if P is in $\mathscr{P}_{\#\mathsf{pd}}$ we can write

$$P(\mathbf{x}) \equiv \psi_P(\mathbf{x}) \simeq 0,$$

which means that P is in $\mathscr{P}_{\mathsf{e}}$. Finally, if P is in $\mathscr{P}_{\mathsf{e}}$ we can write

$$\psi_P(\mathbf{x}) \simeq v \equiv P(\mathbf{x}) \wedge v = 0;$$

hence P is in $\mathscr{P}_{\#\mathsf{pd}}$.

To prove (ii) we note that $\mathscr{P}_{\#\mathsf{d}} \subseteq \mathscr{P}_{\#\mathsf{de}}$ by Theorem 3.1.2 (ii), since $\mathscr{P}_{\#\mathsf{d}}$ is closed under substitution with elementary functions. From part (i) we get

$$\mathscr{P}_{\#\mathsf{de}} \subseteq \mathscr{P}_{\#\mathsf{pde}} = \mathscr{P}_{\mathsf{ee}} = \mathscr{P}_{\mathsf{e}},$$

since by Theorem 3.1.2 (iii) $\mathscr{P}_{\mathsf{e}}$ is closed under existential quantification. Part (iii) is immediate from part (ii).

To prove part (iv) we note that from $\mathsf{EL}_{\mathsf{d}} \subseteq \mathscr{P}_{\mathsf{e}}$ we get $\mathsf{EL} \subseteq \mathscr{P}_{\#}$. The proof of closure under full substitution is straightforward. To prove closure under primitive recursion we consider the specification

$$h(\mathbf{x}, 0) \simeq f(\mathbf{x})$$
$$h(\mathbf{x}, y + 1) \simeq g(\mathbf{x}, y, h(\mathbf{x}, y)),$$

where f and g are $\mathscr{P}$-enumerable functions. It follows that

$$\mathsf{G}_h(\mathbf{x}, y, v) \equiv \exists w\colon \mathsf{G}_f(\mathbf{x}, (w)_0) \wedge \forall z < y \mathsf{G}_g(\mathbf{x}, z, (w)_z, (w)_{z+1}) \wedge (w)_y = v.$$

To prove closure under unbounded minimalization consider the specification

$$h(\mathbf{x}) \simeq \mu y(f(\mathbf{x}, y) \simeq 0),$$

where f is $\mathscr{P}$-enumerable. It follows that

$$\mathsf{G}_h(\mathbf{x}, v) \equiv \mathsf{G}_f(\mathbf{x}, v, 0) \wedge \forall y < v \exists z \mathsf{G}_g(\mathbf{x}, y, z + 1).$$

To prove closure under partial definition by cases we consider one example with three cases:

$$\begin{aligned} h(\mathbf{x}) &\simeq f_1(\mathbf{x}) \quad \text{if } g_1(\mathbf{x}) \simeq 0 \\ &\simeq f_2(\mathbf{x}) \quad \text{if } g_2(\mathbf{x}) \simeq 0 \\ &\simeq f_3(\mathbf{x}) \quad \text{if } g_3(\mathbf{x}) \simeq 0. \end{aligned}$$

It follows that

$$\begin{aligned} \mathsf{G}_h(\mathbf{x}, v) \equiv\ & \mathsf{G}_{g_1}(\mathbf{x}, 0) \wedge \mathsf{G}_{f_1}(\mathbf{x}, v) \\ & \vee \exists y\colon \mathsf{G}_{g_1}(\mathbf{x}, y + 1) \wedge \mathsf{G}_{g_2}(\mathbf{x}, 0) \wedge \mathsf{G}_{f_2}(\mathbf{x}, v) \\ & \vee \exists y_1 \exists y_2\colon \mathsf{G}_{g_1}(\mathbf{x}, y_1 + 1) \wedge \mathsf{G}_{g_2}(\mathbf{x}, y_2 + 1) \wedge \mathsf{G}_{g_3}(\mathbf{x}, 0) \wedge \mathsf{G}_{f_3}(\mathbf{x}, v). \end{aligned}$$

To complete the proof of part (iv) we compute

$$\begin{aligned} \mathscr{P}_{\mathsf{pd}} &\subseteq \mathsf{RC}(\mathscr{P}_{\mathsf{pd}}) \\ \mathscr{P} &\subseteq \mathsf{RC}_{\mathsf{pd}}(\mathscr{P}_{\mathsf{pd}}) \\ \mathscr{P}_{\#} &\subseteq \mathsf{RC}_{\mathsf{pd\#}}(\mathscr{P}_{\mathsf{pd}}). \end{aligned}$$

On the other hand, we have

$$\mathscr{P} \subseteq \mathscr{P}_{\mathsf{e}} = \mathscr{P}_{\#\mathsf{pd}}$$

$$\mathscr{P}_{\mathsf{pd}} \subseteq \mathscr{P}_{\#\mathsf{pdpd}} \subseteq \mathscr{P}_{\#}$$

$$\mathsf{RC}(\mathscr{P}_{\mathsf{pd}}) \subseteq \mathscr{P}_{\#} \quad (\mathscr{P}_{\#} \text{ is closed under recursive operations})$$

$$\mathsf{RC}_{\mathsf{pd}\#}(\mathscr{P}_{\mathsf{pd}}) \subseteq \mathscr{P}_{\#\mathsf{pd}\#} = \mathscr{P}_{\#}. \qquad \square$$

Theorem 3.1.6. *Let $\mathscr{P}$ be a class of predicates closed under substitution with elementary functions and negation. Assume $\mathscr{P}_{\mathsf{e}}$ is closed under conjunction, disjunction, and $\mathsf{EL}_{\mathsf{d}} \subseteq \mathscr{P}$. Then*

(i) $\mathscr{P} \subseteq \mathscr{P}_{\mathsf{e}} \cap \mathscr{P}_{\mathsf{u}} = \mathscr{P}_{\#\mathsf{d}}$.
(ii) $\mathscr{P}_{\#\mathsf{de}} = \mathscr{P}_{\#\mathsf{pd}} = \mathscr{P}_{\#\mathsf{pde}} = \mathscr{P}_{\mathsf{e}}$.
(iii) $\mathscr{P}_{\#\mathsf{d}\#} = \mathscr{P}_{\#\mathsf{pd}\#} = \mathscr{P}_{\#}$.
(iv) *If $\mathscr{P}_{\mathsf{e}}$ is closed under universal bounded quantification then* $\mathsf{RC}(\mathscr{P}_{\mathsf{d}}) = \mathscr{P}_{\#}$.

Proof. To prove (i) we consider a k-ary predicate P in $\mathscr{P}$, and it follows from Theorem 3.1.2 (ii) that P is in $\mathscr{P}_{\mathsf{e}} \cap \mathscr{P}_{\mathsf{u}}$. If P is a predicate in $\mathscr{P}_{\mathsf{e}} \cap \mathscr{P}_{\mathsf{u}}$ it follows that $\bar{P}$ is in $\mathscr{P}_{\mathsf{e}}$, and we can write

$$\chi_P(\mathbf{x}) = v \equiv (P(\mathbf{x}) \wedge v = 0) \vee (\bar{P}(\mathbf{x}) \wedge v = 1)$$

which means that χ_P is in $\mathscr{P}_{\#}$; hence P is in $\mathscr{P}_{\#\mathsf{d}}$. Conversely, if P is in $\mathscr{P}_{\#\mathsf{d}}$ it follows immediately that P and $\bar{P}$ are in $\mathscr{P}_{\mathsf{e}}$; hence P is in $\mathscr{P}_{\mathsf{e}} \cap \mathscr{P}_{\mathsf{u}}$.

Part (ii) follows from Theorem 3.1.5 (i) and (ii), noting that from part (i) we get $\mathscr{P}_{\mathsf{e}} \subseteq \mathscr{P}_{\#\mathsf{de}}$. Part (iii) is immediate from (ii).

To prove (iv) we note that from Theorem 3.1.5 (iv) it follows that $\mathscr{P}_{\#}$ is closed under recursive operations. Furthermore,

$$\mathscr{P}_{\mathsf{d}} \subseteq \mathsf{RC}(\mathscr{P}_{\mathsf{d}})$$

$$\mathscr{P} \subseteq \mathsf{RC}_{\mathsf{d}}(\mathscr{P}_{\mathsf{d}})$$

$$\mathscr{P}_{\#} \subseteq \mathsf{RC}_{\mathsf{d}\#}(\mathscr{P}_{\mathsf{d}}) \subseteq \mathsf{RC}(\mathscr{P}_{\mathsf{d}}) \quad \text{(Theorem 3.1.4 (ii))}.$$

In the other direction we have

$$\mathscr{P} \subseteq \mathscr{P}_{\#\mathsf{d}} \quad \text{(part (i))}$$

$$\mathscr{P}_{\mathsf{d}} \subseteq \mathscr{P}_{\#\mathsf{dd}} \subseteq \mathscr{P}_{\#}$$

$$\mathsf{RC}(\mathscr{P}_{\mathsf{d}}) \subseteq \mathscr{P}_{\#}. \qquad \square$$

Corollary 3.1.6.1. *Let $\mathscr{C}$ be a class of functions closed under elementary operations and unbounded minimalization with predicates. Then $\mathscr{C}_{\mathsf{d}\#\mathsf{d}} = \mathscr{C}_{\mathsf{d}}$.*

Proof. Immediate from Theorem 3.1.4 (iii) and Theorem 3.1.6 (i) with $\mathscr{P} = \mathscr{C}_{\mathsf{d}}$. $\square$

Example 3.1.6. Let $\mathscr{C}$ be a class of functions closed under recursive operations. From Theorem 3.1.5 we get $\mathscr{C}_{\mathsf{pd}\#\mathsf{pd}} = \mathscr{C}_{\mathsf{pd}\#\mathsf{pde}} = \mathscr{C}_{\mathsf{pde}}$ and $\mathscr{C}_{\mathsf{pd}\#\mathsf{pd}\#} =$

$\mathscr{C}_{\mathsf{pd\#}}$. Furthermore, $\mathscr{C}_{\mathsf{pd\#}}$ is closed under recursive operations and $\mathscr{C}_{\mathsf{pd\#}} = \mathsf{RC}_{\mathsf{pd\#}}(\mathscr{C}_{\mathsf{pdpd}})$.

EXAMPLE 3.1.7. The main application of Theorem 3.1.6 is with classes of the form $\mathscr{P} = \mathscr{C}_{\mathsf{d}}$, and $\mathscr{C}$ is closed under elementary operations. In this case we can also apply Theorem 3.1.5 (and also Theorem 3.1.4 and Corollary 3.1.6.1 if $\mathscr{C}$ is closed under unbounded minimalization with predicates). It follows that

$$\begin{aligned}
\mathscr{C}_{\mathsf{d\#de}} &= \mathscr{C}_{\mathsf{d\#pd}} = \mathscr{C}_{\mathsf{d\#pde}} = \mathscr{C}_{\mathsf{de}} \\
\mathscr{C}_{\mathsf{d\#d\#}} &= \mathscr{C}_{\mathsf{d\#pd\#}} = \mathscr{C}_{\mathsf{d\#}} \\
\mathscr{C}_{\mathsf{d\#}} &= \mathsf{RC}_{\mathsf{pd\#}}(\mathscr{C}_{\mathsf{dpd}}) = \mathsf{RC}(\mathscr{C}_{\mathsf{dd}}) \\
\mathscr{C}_{\mathsf{d\#d}} &= \mathscr{C}_{\mathsf{de}} \cap \mathscr{C}_{\mathsf{du}} = \mathscr{C}_{\mathsf{d}}
\end{aligned}$$

EXAMPLE 3.1.8. If we apply these results to the class RC we note that $\mathsf{RC} \subseteq \mathsf{RC}_{\mathsf{d\#}}$ because $\mathsf{RC}_{\mathsf{d\#}}$ is closed under recursive operations and $\mathsf{RC}_{\mathsf{d\#}} \subseteq \mathsf{RC}$ by Theorem 3.1.4 (ii); hence $\mathsf{RC} = \mathsf{RC}_{\mathsf{d\#}}$. It follows that

$$\begin{aligned}
\mathsf{RC}_{\mathsf{de}} &= \mathsf{RC}_{\mathsf{pd}} = \mathsf{RC}_{\mathsf{pde}} \\
\mathsf{RC} &= \mathsf{RC}_{\mathsf{pd\#}}.
\end{aligned}$$

This means that recursively enumerable, partially recursive, and partially recursively enumerable are equivalent notions when applied to predicates. Furthermore, recursive, recursively enumerable, and partially recursively enumerable are equivalent notions when applied to functions.

Some subscripts or combination of subscripts can be iterated any number of times, but in general this does not generate new classes of functions, or predicates, as the case may be. For example, $\mathscr{P}_{\mathsf{ee}} = \mathscr{P}_{\mathsf{e}}$ and $\mathscr{P}_{\mathsf{uu}} = \mathscr{P}_{\mathsf{u}}$ when $\mathscr{P}$ is closed under substitution with elementary functions. Also $\mathscr{P}_{\mathsf{\#pd\#}} = \mathscr{P}_{\mathsf{\#}}$ when $\mathscr{P}$ satisfies the conditions of Theorem 3.1.5, and $\mathscr{P}_{\mathsf{\#d\#}} = \mathscr{P}_{\mathsf{\#}}$ when $\mathscr{P}$ satisfies Theorem 3.1.6. On the other hand, $\mathscr{P}_{\mathsf{ue}}$ is known, in some cases, to generate new classes. We consider now the classes generated by iterating this procedure.

Let $\mathscr{P}$ be a class of predicates. We define classes of predicates $\mathscr{P}^k_{\mathsf{:e}}$ and $\mathscr{P}^k_{\mathsf{:u}}$, where $k \geq 0$, as follows:

$$\begin{aligned}
\mathscr{P}^0_{\mathsf{:e}} &= \mathscr{P}^0_{\mathsf{:u}} = \mathscr{P} \\
\mathscr{P}^{k+1}_{\mathsf{:e}} &= \mathscr{P}^k_{\mathsf{:ue}} \\
\mathscr{P}^{k+1}_{\mathsf{:u}} &= \mathscr{P}^k_{\mathsf{:eu}}.
\end{aligned}$$

It is clear that $\mathscr{P}^k_{\mathsf{:e}}$ describes the result of applying k alternate quantifiers to predicates in $\mathscr{P}$, the last being existential. Similarly, $\mathscr{P}^k_{\mathsf{:u}}$ describes the result of applying k alternate quantifiers to predicates in $\mathscr{P}$, the last being universal. Note that $\mathscr{P}^1_{\mathsf{:e}} = \mathscr{P}_{\mathsf{e}}$ and $\mathscr{P}^1_{\mathsf{:u}} = \mathscr{P}_{\mathsf{u}}$.

The structure $\{\mathscr{P}^k_{\mathsf{:e}}, \mathscr{P}^k_{\mathsf{:u}}\}_{k \geq 0}$ is called the *enumeration hierarchy induced* by

$\mathscr{P}$. We set $\mathscr{P}^k_{:\Delta} = \mathscr{P}^k_{:\mathrm{e}} \cap \mathscr{P}^k_{:\mathrm{u}}$ and $\mathscr{P}^{k+1}_{:\#} = \mathscr{P}^k_{:\mathrm{u}\#}$. Note that $\mathscr{P}^1_{:\Delta} = \mathscr{P}_{\mathrm{e}} \cap \mathscr{P}_{\mathrm{u}}$ and $\mathscr{P}^1_{:\#} = \mathscr{P}_{\#}$.

EXAMPLE 3.1.9. The structure $\{\mathsf{RC}^k_{\mathrm{d:e}}, \mathsf{RC}^k_{\mathrm{d:u}}\}_{k \geq 0}$ is the enumeration hierarchy induced by RC_{d}, and it is usually called the *arithmetical hierarchy*.

This hierarchy can be studied under different assumptions on the class $\mathscr{P}$. We consider only the generalization of Theorem 3.1.6.

Theorem 3.1.7. *Let $\mathscr{P}$ be a class of predicates closed under substitution with elementary functions, boolean operations, and bounded quantification. Assume* $\mathsf{EL}_{\mathrm{d}} \subseteq \mathscr{P}$ and $k \geq 0$. *Then*

(i) *$\mathscr{P}^k_{:\mathrm{e}}$ and $\mathscr{P}^k_{:\mathrm{u}}$ are closed under substitution with elementary functions, conjunction, disjunction, and bounded quantification. Furthermore,* $\mathsf{EL}_{\mathrm{d}} \subseteq \mathscr{P}^k_{:\Delta}$, $\mathscr{P}^k_{:\mathrm{e}} \cup \mathscr{P}^k_{:\mathrm{u}} \subseteq \mathscr{P}^{k+1}_{:\Delta}$, $\mathscr{P}^k_{:\mathrm{en}} = \mathscr{P}^k_{:\mathrm{u}}$, *and* $\mathscr{P}^k_{:\mathrm{un}} = \mathscr{P}^k_{:\mathrm{e}}$.

(ii) *$\mathscr{P}^{k+1}_{:\#}$ is closed under recursive operations and* $\mathscr{P}^{k+1}_{:\#} \subseteq \mathscr{P}^{k+2}_{:\#}$.

(iii) $(\mathscr{P}^k_{:\mathrm{e}} \cup \mathscr{P}^k_{:\mathrm{u}})_{\mathrm{e}} = \mathscr{P}^{k+1}_{:\mathrm{e}}$, $(\mathscr{P}^k_{:\mathrm{e}} \cup \mathscr{P}^k_{:\mathrm{u}})_{\#} = \mathscr{P}^{k+1}_{:\#}$, *and* $(\mathscr{P}^k_{:\mathrm{e}} \cup \mathscr{P}^k_{:\mathrm{u}})_{\mathrm{u}} = \mathscr{P}^{k+1}_{:\mathrm{u}}$.

(iv) $\mathscr{P}^{k+1}_{:\#\mathrm{de}} = \mathscr{P}^{k+1}_{:\#\mathrm{pd}} = \mathscr{P}^{k+1}_{:\#\mathrm{pde}} = \mathscr{P}^{k+1}_{:\mathrm{e}}$.

(v) $\mathscr{P}^{k+1}_{:\#\mathrm{d}\#} = \mathscr{P}^{k+1}_{:\#\mathrm{pd}\#} = \mathscr{P}^{k+1}_{:\#}$.

(vi) $\mathsf{RC}(\mathscr{P}^k_{:\mathrm{ed}}) = \mathscr{P}^{k+1}_{:\#}$.

(vii) $\mathsf{RC}_{\mathrm{de}}(\mathscr{P}^k_{:\mathrm{ed}}) = \mathscr{P}^{k+1}_{:\mathrm{e}}$.

(viii) $\mathscr{P}^{k+1}_{:\#\mathrm{d}} = \mathscr{P}^{k+1}_{:\Delta}$.

PROOF. Part (i) is proved by induction on k using Theorem 3.1.2. Part (ii) is straightforward using Theorem 3.1.5. To prove (iii) we note that

$$(\mathscr{P}^k_{:\mathrm{e}} \cup \mathscr{P}^k_{:\mathrm{u}})_{\mathrm{e}} = \mathscr{P}^k_{:\mathrm{ee}} \cup \mathscr{P}^k_{:\mathrm{ue}} = \mathscr{P}^{k+1}_{:\mathrm{e}}$$

since $\mathscr{P}^k_{:\mathrm{ee}} \subseteq \mathscr{P}^k_{:\mathrm{e}} \subseteq \mathscr{P}^k_{:\mathrm{ue}} = \mathscr{P}^{k+1}_{:\mathrm{e}}$ by part (i).

To prove the rest we set $\mathscr{P}^k = \mathscr{P}^k_{:\mathrm{e}} \cup \mathscr{P}^k_{:\mathrm{u}}$ and note that $\mathscr{P}^k$ satisfies all conditions of Theorem 3.1.6. Since $\mathscr{P}^k_{\mathrm{e}} = \mathscr{P}^k_{:\mathrm{e}}$ and $\mathscr{P}^k_{\#} = \mathscr{P}^{k+1}_{:\#}$ we have (iv) and (v) immediately from Theorem 3.1.6 (ii) and (iii).

We get (vi) from Theorem 3.1.6 (iv), noting that

$$\mathsf{RC}(\mathscr{P}^k_{\mathrm{d}}) = \mathsf{RC}(\mathscr{P}^k_{:\mathrm{ed}}),$$

and (vii) is immediate from (iv) and (vi). Finally, (viii) follows in a similar way from Theorem 3.1.6 (i). □

EXAMPLE 3.1.10. The preceding result applies to the arithmetical hierarchy of Example 3.1.9. In particular, from (vii) we have

$$\mathsf{RC}_{\mathrm{de}}(\mathsf{RC}^k_{\mathrm{d:ed}}) = \mathsf{RC}^{k+1}_{\mathrm{d:e}},$$

which means that the predicates obtained at some level of the arithmetical hierarchy are exactly the predicates that are recursively enumerable in predicates of the preceding level. This result is known as the Post theorem.

EXERCISES

3.1.1. Let $\mathscr{P}$ be a class of predicates closed under substitution with elementary functions and conjunction. Assume $\mathsf{EL_d} \subseteq \mathscr{P}$. Prove the following:
(a) $\mathscr{P}_{\mathsf{gd}} \subseteq \mathscr{P}_{\mathsf{gpd}} = \mathscr{P}$.
(b) $\mathscr{P}_{\mathsf{egd}} \subseteq \mathscr{P}_{\mathsf{egpd}} = \mathscr{P}_{\mathsf{e}}$.

3.1.2. Give an example of a class $\mathscr{C}$ that is closed under elementary operations, but $\mathscr{C}_{\mathsf{pd\#}}$ is not closed under elementary operations.

3.1.3. Let $\mathscr{C}$ be a class closed under elementary operations. Prove the following:
(a) $\mathscr{C}_{\mathsf{pde}}$ is closed under substitution with total $\mathscr{C}_{\mathsf{pd}}$-enumerable functions.
(b) If P_1 and P_2 are k-ary predicates such that P_1 is $\mathscr{C}_{\mathsf{d}}$-enumerable, and P_2 is $\mathscr{C}_{\mathsf{pd}}$-enumerable, then $P_1 \vee P_2$ is $\mathscr{C}_{\mathsf{pd}}$-enumerable.

3.1.4. Let $\mathscr{C} = \mathsf{EL}(\varepsilon')$ (see Example 1.3.2). Prove that $\mathscr{C}_{\mathsf{d\#}} \nsubseteq \mathscr{C}$, but $\mathscr{C} \subseteq \mathscr{C}_{\mathsf{pd\#}}$.

3.1.5. Let $\mathscr{C}$ be a class of functions closed under recursive operations. Assume P_1, P_2, Q_1, and Q_2 are k-ary predicates, such that P_1 and P_2 are partially $\mathscr{C}$-decidable, Q_1 and Q_2 are $\mathscr{C}$-decidable, and the condition $Q_1(\mathbf{x}) \equiv \mathsf{F}$ or $Q_2(\mathbf{x}) \equiv \mathsf{F}$ is satisfied for all $(\mathbf{x}) \in \mathbb{N}^k$. Prove that the k-ary predicate P, such that

$$P(\mathbf{x}) \equiv (P_1(\mathbf{x}) \wedge Q_1(\mathbf{x})) \vee (P_2(\mathbf{x}) \vee Q_2(\mathbf{x}))$$

is partially $\mathscr{C}$-decidable.

3.1.6. Let $\mathscr{C}$ be bounded by $\mathscr{C}'$, where $\mathscr{C}'$ is closed under elementary operations and unbounded minimalization with predicates. Prove that $\mathscr{C}_{\mathsf{d\#}} \subseteq \mathscr{C}'$.

3.1.7. Prove that if $\mathscr{C}$ is closed under recursive operations, then $\mathscr{C}_{\mathsf{pdue}}$ is closed under disjunction.

3.1.8. Let $\mathscr{C}$ be closed recursive operations, and assume $\mathscr{C} = \mathscr{C}_{\mathsf{d\#}}$. Prove the following:
(a) $\mathscr{C}_{\mathsf{de}} = \mathscr{C}_{\mathsf{pde}}$.
(b) $\mathscr{C}_{\mathsf{d\#}} = \mathscr{C}_{\mathsf{pd\#}}$.

3.1.9. Let c be a total function, $\mathscr{C} = \mathsf{RC}(c)$ and $\mathscr{C}' = \mathsf{RC}(\mathscr{C}_{\mathsf{pdpd}})$. Prove the following:
(a) $\mathscr{C} = \mathscr{C}_{\mathsf{d\#}} = \mathscr{C}_{\mathsf{pd\#}}$.
(b) $\mathscr{C}' \subseteq \mathscr{C} \subseteq \mathscr{C}'_{\mathsf{pd\#}}$.
(c) $\mathscr{C} = \mathscr{C}'_{\mathsf{pd\#}}$.
(d) If $\mathscr{C} = \mathscr{C}'$ then $\mathscr{C} = \mathsf{RC}$.
(e) If $\mathscr{C}' = \mathscr{C}'_{\mathsf{d\#}}$ then $\mathscr{C} = \mathsf{RC}$.

3.1.10. Let $\mathscr{C}$ be closed under recursive operations. Prove that a k-ary function h is $\mathscr{C}_{\mathsf{pd}}$-enumerable if and only if there is a $\mathscr{C}$-computable $(k + 1)$-ary function f such that

$$\mathsf{G}_h(\mathbf{x}, v) \equiv \exists y \mathsf{G}_f(\mathbf{x}, v, y).$$

3.1.11. Let h be a k-ary $\mathscr{C}_{\mathsf{d}}$-enumerable function, where $\mathscr{C}$ is closed under elementary operations. Prove that there is a total $(k + 1)$-ary $\mathscr{C}$-computable function f such that:
(a) If $f(\mathbf{x}, w) \simeq v$ and $v < w$ then $h(\mathbf{x}) \simeq v$.
(b) If $h(\mathbf{x}) \simeq v$ there is w such that whenever $w' > w$ then $f(\mathbf{x}, w') \simeq v$.

3.1.12. Prove $\mathsf{RC_d} = \mathsf{RC_{de}} \cap \mathsf{RC_{du}} = \mathsf{RC_{d\#d}}$.

3.1.13. Let $\mathscr{C}$ be a class of functions closed under recursive operations and assume A and B are sets such that $A \cap B = \varnothing$ and $A \cup B$ is $\mathscr{C}$-decidable. Prove the following:
(a) A is $\mathscr{C}$-decidable if and only if B is $\mathscr{C}$-decidable.
(b) A is $\mathscr{C}_\mathsf{d}$-enumerable if and only if $\bar{B}$ is $\mathscr{C}_\mathsf{d}$-enumerable.
(c) A is partially $\mathscr{C}$-decidable if and only if $\bar{B}$ is partially $\mathscr{C}$-decidable.

3.1.14. Let $\mathscr{C}$ be a class closed under recursive operations and h a k-ary $\mathscr{C}$-computable function. Prove the following:
(a) D_h is partially $\mathscr{C}$-decidable and R_h is $\mathscr{C}_\mathsf{pd}$-enumerable.
(b) If h is total then R_h is $\mathscr{C}_\mathsf{d}$-enumerable.

3.1.15. Let $\mathscr{C}$ be a class of functions closed under recursive operations, and assume $\mathscr{C}_\mathsf{pd}$ is closed under existential quantification. Prove that $\mathscr{C}_\mathsf{pd}$ is closed under disjunction.

Notes

Enumeration, as defined in this section, is just another name for existential quantification. The motivation here is to use enumeration to generate classes of functions via the operation $\mathscr{P}_\#$, and the main result is that such classes are closed under recursive operations whenever $\mathscr{P}$ satisfies some general closure properties. In the next section we show how truly enumeration properties are obtained via the construction $\mathscr{C}_{\mathsf{d}\#}$.

Hierarchies have been a standard part of recursive function theory, particularly the arithmetical hierarchy discussed in this section. An important extension, the hyperarithmetical hierarchy, is not studied in this work. References can be found in Davis [2] and Kleene [10].

§2. Enumeration Properties

In this section we study classes of functions that are determined by their total functions and enumeration. They satisfy important selector properties.

Theorem 3.2.1. *Let $\mathscr{C}$ be a class of functions closed under elementary operations, and assume $\mathscr{C}_{\mathsf{d}\#} \subseteq \mathscr{C}$. Then*

(i) *$\mathscr{C}$ is closed under unbounded minimalization with predicates.*
(ii) $\mathscr{C}_\mathsf{de} \subseteq \mathscr{C}_\mathsf{pd}$.
(iii) $\mathscr{C} \subseteq \mathscr{C}_{\mathsf{pd}\#}$.
(iv) *A k-ary predicate P is $\mathscr{C}$-decidable if and only if both P and $\bar{P}$ are $\mathscr{C}_\mathsf{d}$-enumerable.*
(v) *$\mathscr{C}$ has the* de*-selector property.*
(vi) *$\mathscr{C}$ is closed under primitive recursion, and course-of-values recursion, with total functions.*

(vii) *An infinite set A is $\mathscr{C}$-decidable if and only if there is a total $\mathscr{C}$-computable function f such that $A = \mathsf{R}_f$ and $f(y) < f(y + 1)$ holds for all $y \in \mathbb{N}$.*

(viii) *An infinite set A is $\mathscr{C}_\mathsf{d}$-enumerable if and only if there is a total* 1-1 *$\mathscr{C}$-computable unary function h such that $A = \mathsf{R}_h$.*

(ix) *If A is an infinite $\mathscr{C}_\mathsf{d}$-enumerable set, then A has an infinite $\mathscr{C}$-decidable subset.*

PROOF. From Theorem 3.1.5 it follows that $\mathscr{C}_{\mathsf{d}\#}$ is closed under recursive operations. To prove (i) we assume that P is a $(k + 1)$-ary $\mathscr{C}$-decidable predicate, and

$$h(\mathbf{x}) \simeq \mu y P(\mathbf{x}, y).$$

It follows that

$$\mathsf{G}_h(\mathbf{x}, v) \equiv P(\mathbf{x}, v) \wedge \forall y < v \sim P(\mathbf{x}, y);$$

hence h is $\mathscr{C}_\mathsf{d}$-enumerable, and from $\mathscr{C}_{\mathsf{d}\#} \subseteq \mathscr{C}$ it follows that h is $\mathscr{C}$-computable.

From part (i) it follows that we can apply Theorem 3.1.4; hence parts (ii), (iii), and (iv) are valid.

To prove (v) let P be a $(k + 1)$-ary $\mathscr{C}_\mathsf{d}$-enumerable predicate; hence there is a $(k + 2)$-ary $\mathscr{C}$-decidable predicate Q such that

$$P(\mathbf{x}, v) \equiv \exists y Q(\mathbf{x}, v, y).$$

A $\mathscr{C}$-computable selector function for P is given by

$$h(\mathbf{x}) \simeq g(\mu y Q(\mathbf{x}, (y)_0, (y)_1)),$$

where $g(y) = (y)_0$.

To prove (vi) note that total $\mathscr{C}$-computable functions are $\mathscr{C}_\mathsf{d}$-enumerable. Since $\mathscr{C}_{\mathsf{d}\#}$ is closed under primitive recursion and course-of-values recursion, and $\mathscr{C}_{\mathsf{d}\#} \subseteq \mathscr{C}$, it follows that $\mathscr{C}$ is also closed under the same rules, whenever they are restricted to total functions.

To prove (vii) note that if $A = \mathsf{R}_f$, where f is a total $\mathscr{C}$-computable unary function satisfying the condition $f(y) < f(y + 1)$, then

$$x \in A = \exists y \leq x f(y) = x;$$

hence A is $\mathscr{C}$-decidable. To prove the converse we assume A is $\mathscr{C}$-decidable and introduce a unary function f such that

$$f(0) \simeq \mathsf{C}_m^0(\,)$$

$$f(y + 1) \simeq g(y, f(y)),$$

where m is the least element in A, and g is the binary function such that

$$g(y, z) \simeq \mu v(v \in A \wedge z < v).$$

Clearly g is $\mathscr{C}$-computable, and since A is infinite it follows that g is total. From (vi) it follows that f is $\mathscr{C}$-computable. Clearly $A = \mathsf{R}_f$, and f satisfies the condition $f(y) < f(y + 1)$.

To prove (viii) we note that if $A = \mathsf{R}_h$, where h is a total $\mathscr{C}$-computable function, then clearly A is $\mathscr{C}_{\mathsf{d}}$-enumerable. To prove the converse assume that

$$x \in A \equiv \exists y Q(x, y),$$

where Q is $\mathscr{C}$-decidable. Let B be the $\mathscr{C}$-decidable set such that

$$x \in B \equiv Q((x)_0, (x)_1) \wedge \forall v < x((v)_0 \neq (x)_0 \vee \sim Q((v)_0, (v)_1)).$$

Since A is infinite it follows that B is also infinite, so let f be the total 1-1 $\mathscr{C}$-computable function given by (vii), so that $B = \mathsf{R}_f$. We take $h(y) = (f(y))_0$, and it follows that h is 1-1 and $A = \mathsf{R}_h$.

Finally, we prove (ix). Assume A is an infinite $\mathscr{C}_{\mathsf{d}}$-enumerable set, and introduce the function h such that

$$h(0) \simeq \mathsf{C}^0_m()$$

$$h(y + 1) \simeq \sigma v(v \in A \wedge h(y) < v),$$

using the selector function given by (v), with m some fixed element in A. If $B = \mathsf{R}_h$ it follows that B is an infinite subset of A. Furthermore, h satisfies the conditions of part (vii); hence B is $\mathscr{C}$-decidable. □

EXAMPLE 3.2.1. If we apply Theorem 3.2.1 with $\mathscr{C} = \mathsf{RC}$, we get enumeration properties for the class of recursive predicates and recursively enumerable predicates. Note the following: (a) A predicate P is recursive if and only if both P and $\bar{P}$ are recursively enumerable. (b) RC has the selector property relative to $\mathsf{RC}_{\mathsf{de}} = \mathsf{RC}_{\mathsf{pd}} = \mathsf{RC}_{\mathsf{pde}}$ (Example 3.1.8). (c) An infinite set A is recursive if and only if $A = \mathsf{R}_f$ for some total recursive function f such that $f(y) < f(y + 1)$. (d) An infinite set A is recursively enumerable if and only if $A = \mathsf{R}_h$ for some total 1-1 recursive function. (e) If A is an infinite recursively enumerable set, then A has an infinite recursive subset.

We know that whenever a class $\mathscr{C}$ is closed under recursive operations then $\mathscr{C}_{\mathsf{d}\#} \subseteq \mathscr{C} \subseteq \mathscr{C}_{\mathsf{pd}\#}$. We shall show later that it is possible to have a nontotal function c such that $\mathsf{RC}_{\mathsf{d}\#}(c) \neq \mathsf{RC}(c)$ and $\mathsf{RC}(c) \neq \mathsf{RC}_{\mathsf{pd}\#}(c)$. If a class $\mathscr{C}$ is closed under recursive operations and $\mathscr{C} = \mathscr{C}_{\mathsf{pd}\#}$, we say that $\mathscr{C}$ is closed under *enumeration operations.*

Theorem 3.2.2. *Let $\mathscr{C}$ be a class closed under recursive operations. Then*

(i) *$\mathscr{C}_{\mathsf{pd}\#}$ is closed under enumeration operations.*
(ii) *If $\mathscr{C} \subseteq \mathscr{C}'$ and $\mathscr{C}'$ is closed under enumeration operations then $\mathscr{C}_{\mathsf{pd}\#} \subseteq \mathscr{C}'$.*
(iii) *If $\mathscr{C}$ is closed under enumeration operations then $\mathscr{C}_{\mathsf{pd}} = \mathscr{C}_{\mathsf{pde}}$ and $\mathscr{C} = \mathsf{RC}_{\mathsf{pd}\#}(\mathscr{C}_{\mathsf{pdpd}})$.*

PROOF. Part (i) follows from Theorem 3.1.5 (iii) and (iv). Part (ii) follows by monotonicity and part (iii) from Theorem 3.1.5 (i) and (iv). □

If a class $\mathscr{C}$ is closed under recursive operations and $\mathscr{C} = \mathscr{C}_{d\#}$, we say that $\mathscr{C}$ is e-*total*. This notation refers to the fact that $\mathscr{C}$ is determined by its total functions and enumeration.

Theorem 3.2.3. *Let $\mathscr{C}$ be a class closed under elementary operations. Then*

(i) *$\mathscr{C}_{d\#}$ is* e-*total.*
(ii) *If $\mathscr{C}' \subseteq \mathscr{C}$ and $\mathscr{C}'$ is* e-*total then $\mathscr{C}' \subseteq \mathscr{C}_{d\#}$.*
(iii) *If $\mathscr{C}$ is* e-*total then $\mathscr{C}_{de} = \mathscr{C}_{pd} = \mathscr{C}_{pde}$, $\mathscr{C}$ is closed under enumeration operations, and $\mathscr{C} = \mathsf{RC}(\mathscr{C}_{dd})$.*

Proof. Part (i) follows from Theorem 3.1.6 (iii), and part (ii) follows by monotonicity. To prove (iii) we note that $\mathscr{C}_{pde} = \mathscr{C}_{d\#pde} = \mathscr{C}_{de}$ follows from Theorem 3.1.5; hence $\mathscr{C}_{pd\#} = \mathscr{C}_{d\#} = \mathscr{C}$. The last part follows from Theorem 3.1.6. □

Theorem 3.2.4. *Let $\mathscr{C}$ be a class containing only total functions. Then*

(i) $\mathsf{EL}_{d\#}(\mathscr{C}) = \mathsf{RC}_{d\#}(\mathscr{C}) = \mathsf{RC}(\mathscr{C}) = \mathsf{RC}_{pd\#}(\mathscr{C})$.
(ii) $\mathsf{EL}_{de}(\mathscr{C}) = \mathsf{RC}_{de}(\mathscr{C}) = \mathsf{RC}_{pd}(\mathscr{C}) = \mathsf{RC}_{pde}(\mathscr{C})$.

Proof. To prove (i) it is sufficient to show that $\mathsf{RC}_{pd\#}(\mathscr{C}) \subseteq \mathsf{EL}_{d\#}(\mathscr{C})$. Since $\mathscr{C}$ contains only total functions, it follows that $\mathscr{C} \subseteq \mathsf{EL}_{d\#}(\mathscr{C})$. Furthermore, $\mathsf{EL}_{d\#}(\mathscr{C})$ is closed under recursive operations (Theorem 3.1.5 with $\mathscr{P} = \mathsf{EL}_{d}(\mathscr{C})$); hence $\mathsf{RC}(\mathscr{C}) \subseteq \mathsf{EL}_{d\#}(\mathscr{C})$. From Theorem 3.1.5 (ii) it follows that $\mathsf{EL}_{d\#}(\mathscr{C})$ is closed under enumeration operations, and from Theorem 3.2.2 (ii) that $\mathsf{RC}_{pd\#}(\mathscr{C}) \subseteq \mathsf{EL}_{d\#}(\mathscr{C})$.

To prove (ii) we need only show that $\mathsf{RC}_{pde}(\mathscr{C}) \subseteq \mathsf{EL}_{de}(\mathscr{C})$. If P is a predicate that is partially recursively enumerable in $\mathscr{C}$, then

$$P(\mathbf{x}) \equiv \exists y Q(\mathbf{x}, y),$$

where Q is partially recursive in $\mathscr{C}$. Since $\mathsf{EL}_{d\#}(\mathscr{C}) = \mathsf{RC}(\mathscr{C})$, it follows that Q is in $\mathsf{EL}_{de}(\mathscr{C})$; hence P is also in $\mathsf{EL}_{de}(\mathscr{C})$. □

Corollary 3.2.4.1. *Let $\mathscr{C}$ be a class containing only total functions. Then*

(i) *A k-ary function h is recursive in $\mathscr{C}$ if and only if there is a $(k+1)$-ary predicate Q, elementary in $\mathscr{C}$, and a unary elementary function g such that*

$$h(\mathbf{x}) \simeq g(\mu y Q(\mathbf{x}, y)).$$

(ii) *$\mathsf{RC}(\mathscr{C})$ is the smallest class that contains $\mathscr{C}$ and it is closed under elementary operations and unbounded minimalization with predicates.*

Proof. To prove (i) note that the condition is clearly sufficient. Conversely, if h is recursive in $\mathscr{C}$, we know that $\mathsf{EL}_{d\#}(\mathscr{C}) = \mathsf{RC}(\mathscr{C})$; hence there is a $(k+2)$-ary predicate Q', elementary in $\mathscr{C}$, such that

$$\mathsf{G}_h(\mathbf{x}, v) \equiv \exists y Q'(\mathbf{x}, v, y).$$

It follows that

$$h(\mathbf{x}) \simeq g(\mu y Q'(\mathbf{x}, (y)_0, (y)_1)),$$

where $g(y) = (y)_0$. Part (ii) is immediate from (i). □

Note that the preceding results apply when $\mathscr{C} = \varnothing$, which certainly contains only total functions. In this way we get $\mathsf{EL}_{\mathsf{d}\#} = \mathsf{RC}_{\mathsf{d}\#} = \mathsf{RC} = \mathsf{RC}_{\mathsf{pd}\#}$, and also $\mathsf{EL}_{\mathsf{de}} = \mathsf{RC}_{\mathsf{de}} = \mathsf{RC}_{\mathsf{pd}} = \mathsf{RC}_{\mathsf{pde}}$. Furthermore, in Corollary 3.2.4.1 (i) the predicate Q is now elementary. Finally, RC is the smallest class closed under elementary operations and unbounded minimalization with predicates.

Theorem 3.2.5. *Let $\mathscr{C}$ be a class closed under recursive operations. The following conditions are equivalent:*

(i) *$\mathscr{C}$ is* e-*total.*
(ii) $\mathscr{C} = \mathsf{RC}(\mathscr{C}_{\mathsf{dd}})$.
(iii) *There is a class $\mathscr{C}'$ containing only total functions and* $\mathscr{C} = \mathsf{RC}(\mathscr{C}')$.
(iv) *There is a class $\mathscr{C}'$ containing only total functions and* $\mathscr{C} = \mathsf{EL}_{\mathsf{d}\#}(\mathscr{C}')$.
(v) $\mathscr{C}_{\mathsf{de}} = \mathscr{C}_{\mathsf{pd}}$
(vi) $\mathscr{C}_{\mathsf{de}} = \mathscr{C}_{\mathsf{pde}}$.
(vii) $\mathscr{C}_{\mathsf{d}\#} = \mathscr{C}_{\mathsf{pd}\#}$.

PROOF. The implication from (i) to (ii) follows from Theorem 3.1.6 with $\mathscr{P} = \mathscr{C}_{\mathsf{d}}$. The implication from (ii) to (iii) is trivial, and (iv) follows from (iii) using Theorem 3.2.4. To prove (v) from (iv) we use Theorem 3.1.6 (ii), where we set $\mathscr{P} = \mathsf{EL}_{\mathsf{d}}(\mathscr{C}')$. Finally, implications from (v) to (vi), from (vi) to (vii), and from (vii) to (i) are trivial. □

EXERCISES

3.2.1. Let $\mathscr{C}$ be closed under elementary operations. Prove that the following conditions are equivalent:
(a) $\mathscr{C}_{\mathsf{d}\#} \subseteq \mathscr{C}$.
(b) $\mathscr{C}$ has the d-selector property.
(c) If h is a function such that G_h is $\mathscr{C}$-decidable, then h is $\mathscr{C}$-computable.
(d) $\mathscr{C}$ is closed under unbounded minimalization with predicates.
(e) $\mathscr{C}$ has the de-selector property.

3.2.2. Let $\mathscr{C}$ be a class closed under recursive operations, with the pd-selector property. Prove that $\mathscr{C}$ is closed under enumeration operations.

3.2.3. Let $\mathscr{C}$ be closed under enumeration operations and P be a k-ary predicate. Prove that the following conditions are equivalent:
(a) P is $\mathscr{C}$-decidable.
(b) Both P and $\bar{P}$ are $\mathscr{C}_{\mathsf{d}}$-enumerable.
(c) Both P and $\bar{P}$ are $\mathscr{C}_{\mathsf{pd}}$-enumerable.

3.2.4. Let $\mathscr{C}$ be closed under elementary operations, and assume A and B are $\mathscr{C}_d$-enumerable sets. Prove that there are $\mathscr{C}_d$-enumerable sets A' and B' such that $A' \subseteq A$, $B' \subseteq B$, $A' \cap B' = \varnothing$, and $A' \cup B' = A \cup B$.

3.2.5. Let $\mathscr{C}$ be closed under recursive operations, with the pde-selector property. Assume A and B are $\mathscr{C}_{pd}$-enumerable sets. Prove that there are $\mathscr{C}_{pd}$-enumerable sets A' and B' such that $A' \subseteq A$, $B' \subseteq B$, $A' \cap B' = \varnothing$, and $A' \cup B' = A \cup B$.

3.2.6. Let $\mathscr{C}$ be closed under enumeration operations, and let h be a k-ary $\mathscr{C}$-computable function. Prove that h is $\mathscr{C}_d$-enumerable if and only if D_h is $\mathscr{C}_d$-enumerable.

3.2.7. Let $\mathscr{C}$ be closed under enumeration operations. Prove that there is a class $\mathscr{C}'$ such that:

(a) $\mathscr{C}'$ is closed under recursive operations.

(b) $\mathscr{C}'$ is bounded by RC.

(c) $\mathscr{C} = \mathscr{C}'_{pd\#}$.

3.2.8. Let $\mathscr{C}$ be e-total and bounded by $\mathsf{RC}(c)$, where c is a $\mathscr{C}$-computable function. Prove that $\mathscr{C} = \mathsf{RC}(c)$.

3.2.9. Let $\mathscr{C}$ be a class of functions closed under recursive operations. Assume A and B are sets such that $A \cap B = \varnothing$, both A and B are $\mathscr{C}_d$-enumerable, and $A \cup B$ is $\mathscr{C}$-decidable. Prove that both A and B are $\mathscr{C}$-decidable.

3.2.10. Let $\mathscr{C}$ be closed under enumeration operations. Assume A and B are sets such that $A \cap B = \varnothing$, both A and B are partially $\mathscr{C}$-decidable, and $A \cup B$ is $\mathscr{C}$-decidable. Prove that both A and B are $\mathscr{C}$-decidable.

3.2.11. Let $\mathscr{C}$ be a class closed under elementary operations such that $\mathscr{C}_{d\#} \subseteq \mathscr{C}$. Assume A and B are $\mathscr{C}_d$-enumerable sets such that $A \cup B = \mathbb{N}$ and $A \cap B \neq \varnothing$. Prove that there is a total unary $\mathscr{C}$-computable function h such that

$$x \in A \equiv h(x) \in A \cap B.$$

3.2.12. Let $\mathscr{C}$ be a class closed under recursive operations, and let A and B be $\mathscr{C}_{pd}$-enumerable sets such that $A \cup B = \mathbb{N}$ and $A \cap B \neq \varnothing$. Assume that $\mathscr{C}$ has the pd-selector property. Prove that there is a total unary $\mathscr{C}$-computable function h such that

$$x \in A \equiv h(x) \in A \cap B.$$

3.2.13. Prove $\mathsf{RC}_d = \mathsf{EL}_{de} \cap \mathsf{EL}_{du} = \mathsf{PR}_{de} \cap \mathsf{PR}_{du}$.

Notes

The notion of e-total class is a formalization of a classical result in the theory of recursive functions, namely, Kleene's normal form theorem (see Kleene [9]). This result, together with reflexivity (to be discussed in Chapter 4) and closure, provides the foundations for most of the material in this area. In fact, it is fair to say that recursive function theory is usually the theory of e-total reflexive structures. This direction is well covered in the literature, and it is avoided to some extent in this work, which is oriented to more general structures.

In the general case, given a reflexive structure $\mathscr{C}$, we must distinguish between $\mathscr{C}_{de}$ and $\mathscr{C}_{pd}$ (and sometimes between $\mathscr{C}_{pd}$ and $\mathscr{C}_{pde}$). The case where $\mathscr{C} = \mathscr{C}_{pd\#}$ appears to be important. Proofs of this relation usually require a selector property. In Chapter 5 we give important examples of reflexive structures where $\mathscr{C} = \mathscr{C}_{pd\#}$ but $\mathscr{C}$ is not e-total.

§3. Induction

Induction can be considered a special form of recursion, restricted to predicates and sets. The mathematical foundations are similar, but the implementation is different in several ways. Recursion, as discussed in Chapter 2, is essentially a deterministic procedure, and if at some place of the evaluation a value becomes undefined, then the whole evaluation is undefined (for that particular input). On the other hand, induction involves total functions, and predicates, which are technically also total functions. But this is not the real difference, because in induction predicates are treated as partial functions, that is, via the partial characteristic function. A value F (= false) during an inductive evaluation is similar to a value undefined in a recursive evaluation, a situation that is made explicit by the particular partial ordering imposed on predicates (which in fact is the normal ordering on functions but applied to partial characteristic functions). The real difference between recursion and induction is that in the latter the occurrence of an undefined value ($\doteq$ **false**) by no means blocks the evaluation, which may still be defined.

Roughly, we may describe induction as a recursive procedure to evaluate a partial characteristic function ψ_P of a predicate P, which is nondeterministic, in the sense that the evaluation may proceed along different lines, and it is not affected by the fact that one of these lines produces undefined values. This effect is obtained by the introduction of disjunction, a crucial element in this type of specification. A more detailed discussion of nondeterministic evaluation is given in the next section.

EXAMPLE 3.3.1. Consider the following inductive specification of a unary predicate P, by the following three clauses:

(a) If $x = 0$ then $P(x)$.
(b) If $P(x + x)$ then $P(x)$.
(c) If $P(x \times x)$ then $P(x)$.

We consider these clauses as rules to evaluate the function ψ_P. From the first clause it is clear that $\psi_P(0) \simeq 0$. If $x \neq 0$ we must start evaluations of $\psi_P(x + x)$ and $\psi_P(x \times x)$. These two evaluations are independent one from the other and it is sufficient that one of the two succeeds with value 0, which implies $\psi_P(x) \simeq 0$. By independent we understand that the alternative of either computation has no effect whatsoever on the other. In this sense we say that the evaluation is nondeterministic. On the other hand, it is obvious that in this example neither of the two evaluations can succeed.

We discuss first the mathematical foundations, which are similar to the theory of recursion. Here we consider a k-ary predicate P to be a set, that is, a subset of $\mathbb{N}^k$. The basic ordering between predicates is the usual set inclusion. Recall that $P_1 \subseteq_k P_2$ means that both P_1 and P_2 are k-ary predicates, and $P_2(\mathbf{x}) \equiv \mathsf{T}$ whenever $P_1(\mathbf{x}) \equiv \mathsf{T}$.

A k-ary *predicate transformation* is an operation that can be applied to a k-ary predicate, and the result is also a k-ary predicate. If F is a k-ary predicate transformation and P is a k-ary predicate, then $F(P)$ denotes the result of applying the transformation F to the predicate P. For example, we may define a k-ary predicate transformation Z_k such that for any k-ary predicate P we have $\mathsf{Z}_k(P) = \bar{P}$.

Let F be a k-ary predicate transformation such that for a given k-ary predicate P we have $F(P) = P'$. It follows that for any $(\mathbf{x}) \in \mathbb{N}^k$ the following relation is valid:

$$F(P)(\mathbf{x}) \equiv P'(\mathbf{x}).$$

We can use this notation to specify F. For example, a specification of a binary predicate transformation F can be given by the expression

$$F(P)(x_1, x_2) \equiv P(\mathsf{exp}(2, x_1), \mathsf{exp}(3, x_2)).$$

We say that a k-ary predicate transformation F is *monotonic* if whenever $P_1 \subseteq_k P_2$, then $F(P_1) \subseteq_k F(P_2)$. A k-ary predicate P is *closed under* F if $F(P) \subseteq_k P$. If $F(P) = P$ we say that P is a *fixed point* of F.

Theorem 3.3.1. *Let F be a monotonic predicate transformation. There is a k-ary predicate P that is a fixed point of F, and furthermore whenever P' is a k-ary predicate closed under F then $P \subseteq_k P'$.*

PROOF. We give two proofs of this classical result. First, we introduce the predicate P by the condition

$$P(\mathbf{x}) \equiv \forall P'\text{: } \textit{if } F(P') \subseteq_k P' \textit{ then } P'(\mathbf{x}).$$

In a more explicit set-theoretical terminology, P is the set intersection of all predicates that are closed under F, a collection that is nonempty, since $\mathbb{N}^k$ is trivially closed under F. From this definition it follows that whenever P' is closed under F then $P \subseteq_k P'$; hence $F(P) \subseteq_k F(P') \subseteq_k P'$, and this implies that $F(P) \subseteq_k P$. Using again monotonicity we get $F(F(P)) \subseteq_k F(P)$, which means that $F(P)$ is closed under F; hence $P \subseteq_k F(P)$, so $F(P) = P$.

The second proof generates P using transfinite recursion on the countable ordinals, where an ordinal is countable if the set of all preceding ordinals is countable in the usual set-theoretical sense. For every countable ordinal ν we define a k-ary predicate P_ν. Assume that P_μ has been defined for all ordinals $\mu < \nu$, in such a way that $P_\mu \subseteq_k F(P_\mu)$. We put $P_{<\nu} = \bigcup_{\mu<\nu} P_\mu$, and $P_\nu = F(P_{<\nu})$. If $\mu < \nu$, it follows that $P_\mu \subseteq_k F(P_{<\nu}) = P_\nu$, $P_{<\nu} \subseteq_k P_\nu$, and $P_\nu \subseteq_k F(P_\nu)$. Since $\mathbb{N}^k$ is countable, but the set of all countable ordinals is not countable, it follows that there is a least countable ordinal ν such that $P_\nu = P_{\nu+1}$, but

$P_{\nu+1} = F(P_\nu)$, so P_ν is a fixed point of F. If P' is another predicate closed under F, it follows by transfinite induction that $P_\mu \subseteq_k P'$ for all countable ordinals; hence $P_\nu \subseteq_k P'$. □

We move now to the implementation of inductive specifications. Since the application of Theorem 3.3.1 requires only monotonicity, there are many operations that can be used, not all of them finitary. The formalization in this section involves only finitary operations. A more complicated formalization will be described later in this work.

The basic idea here is similar to the one used to formalize recursion, but instead of basic terms we use E-*terms*, which are boolean terms, and take value T or F. The initial symbols are numerical variables, numerals, function symbols, and predicate symbols. The construction rules are as follows:

E 1: If Q is a predicate symbol of arity n, $n \geq 0$, and $U_1, \ldots, U_n$ are basic terms, then $Q(U_1, \ldots, U_n)$ is an E-term.

E 2: If U_1 and U_2 are E-terms, then $(U_1) \vee (U_2)$ and $(U_1) \wedge (U_2)$ are also E-terms.

E 3: If U is an E-term, V is a basic term, and y is a variable not occurring in V, then $\forall y < V : U$ and $\exists y < V : U$ are also E-terms.

E 4: If U is an E-term and y is a variable, then $\exists y : U$ is also an E-term.

The semantics for E-terms is self-explanatory, and there is no need to enter into many details. We only mention that function symbols must be interpreted as total functions, and predicate symbols as predicates, in either case of the proper arity. Given such an interpretation, the value of an E-term U depends only on the values of the free variables occurring in U, where, as usual, a variable y in U is *free* if it does not occur in the scope of a construction $\forall y < V$, $\exists y < V$ or $\exists y$, according to rules E 3 and E 4. The occurrence of y in one such scope is said to be *bound*, and the evaluation of U does not require that values be given to bound occurrences of variables.

If U is an E-term with a fixed interpretation for the function and predicate symbols, and a value is assigned to each variable occurring free in U, then U evaluates to one of the boolean values T or F. Note that U is always defined, since we restrict the interpretation of the function symbols to total functions. We can use U to specify a new predicate P whose value is determined by U. More precisely, if $\mathbf{x}$ is a list of variables containing all variables that occur free in U, we write

$$P(\mathbf{x}) \equiv U.$$

We say that P is *explicitly specified* from U. But this is a minor application. The main purpose of introducing E-terms is to give a formalization of inductive specification.

Let $U_1, \ldots, U_m$ be given E-terms with a fixed interpretation for the function and predicate symbols, but allowing for a k-ary predicate symbol P that is uninterpreted, which may occur in some or all of the given E-terms. Let **x** be a list of variables containing all the variables occurring free in the terms $U_1, \ldots, U_m$. An inductive specification consists of a sequence of m statements of the form:

(1) If U_1 then $P(\mathbf{x})$.
(2) If U_2 then $P(\mathbf{x})$.
⋮
(m) If U_m then $P(\mathbf{x})$.

Note that, since the symbol P is uninterpreted, these statements have no well-defined meaning. On the other hand, if P is given some interpretation, then it is possible that some of the statements are valid, where a statement is valid if it is true for all $(\mathbf{x}) \in \mathbb{N}^k$. For example, if P is given the interpretation $\mathbb{N}^k$, it is clear that all the statements are valid.

If the predicate symbol P is interpreted as a predicate that makes all the statements valid, we say that the predicate is a *solution* of the inductive specification. It follows that $\mathbb{N}^k$ is always a solution of any inductive specification with k variables.

The above format for inductive specifications, in the form of m statements, is useful in applications and will be used in examples. For the general discussion it is more convenient to reduce it to one with only one statement. For example, we may write

$$\text{If } U_1 \vee U_2 \vee \cdots \vee U_m \text{ then } P(\mathbf{x}),$$

which is equivalent to the original specification, in the sense that they have exactly the same solutions.

The notion of solution is not sufficient to interpret an inductive specification, because in general there are many solutions. In order to identify a *minimal solution* we introduce predicate transformations and apply Theorem 3.3.1. We assume the inductive transformation contains only one statement, so it is of the form

$$\text{If } U \text{ then } P(\mathbf{x}),$$

where P is the inductive k-ary symbol. We associate with this specification a k-ary predicate transformation F such that

$$F(P)(\mathbf{x}) \equiv U,$$

where P is now a symbol that denotes an arbitrary k-ary predicate. There are some obvious connections between the inductive specification and the predicate transformation F. First, a k-ary predicate P is a solution of the inductive specification if and only if P is closed under F. Second, the predicate transformation F is monotonic, a fact that can easily be proved by induction in the construction of U. It follows that F has a minimal fixed point, which is also a minimal solution of the inductive specification. Such a minimal fixed point is, by definition, the predicate specified by the inductive specification. The minimal fixed point P is a solution, which means that all statements are valid, and it is minimal, so if P' is another solution then $P \subseteq_k P'$.

EXAMPLE 3.3.2. Let g be a binary function, not necessarily total. We give an inductive specification for a binary predicate P_g. The specification has two statements, as follows:

$g1$: If $\exists w(\mathsf{exp}(2, w) = y \wedge (x = w \vee \mathsf{P}_g(x, w)))$ then $\mathsf{P}_g(x, y)$.
$g2$: If $\exists w \exists z \exists v(\mathsf{exp}(3, w + 1) = y \wedge \mathsf{G}_g(w, z, v) \wedge \mathsf{P}_g(x, v))$ then $\mathsf{P}_g(x, y)$.

For example, from $g1$ we get $\mathsf{P}_g(0, 1)$, $\mathsf{P}_g(1,2)$, $\mathsf{P}_g(3, 8)$, and in general $\mathsf{P}_g(x, \mathsf{exp}(2, x))$ for any number x. From $g1$ we also get $\mathsf{P}_g(0, 2)$ and $\mathsf{P}_g(3, \mathsf{exp}(2, 8))$. If g is a binary function such that there is a z and $g(0, z) \simeq 8$, then $\mathsf{P}_g(3, 3)$ holds.

We have proved that if P is the minimal solution of an inductive specification of the form given above, then P is a fixed point of the predicate transformation F such that $F(P)(\mathbf{x}) \equiv U$. This means that the relation $P(\mathbf{x}) \equiv U$ is valid. When we use this implication from left to right, that is, to infer U from $P(\mathbf{x})$, we say that U holds by *inversion* of the specification. For example, if $\mathsf{P}_g(x, y)$ holds, where P_g is the predicate of Example 3.3.2, we know that either the condition of $g1$ holds or the condition of $g2$ holds. If $y = \mathsf{exp}(2, w)$ we infer that $x = w$ or $\mathsf{P}_g(x, w)$. On the other hand, if $y = \mathsf{exp}(3, w + 1)$ we infer that there are z and v such that $g(w, z) \simeq v$, and $\mathsf{P}_g(x, v)$ holds.

EXAMPLE 3.3.3. Let A be a set and $f_1, \ldots, f_m$, $m \geq 1$, be m unary functions. A set B is specified by the following statements:

(0) If $y \in A$ then $y \in B$.
(1) If $\exists x(x \in B \wedge \mathsf{G}_{f_1}(x, y))$ then $y \in B$.
⋮
(m) If $\exists x(x \in B \wedge \mathsf{G}_{f_m}(x, y))$ then $y \in B$.

We say that B is the *minimal extension of A closed under the functions $f_1, \ldots,$ f_m*. Note the procedure of using nontotal functions via their graph predicates.

For the time being we are considering inductive specifications in which the statements are restricted to E-terms. This restriction is not essential, and it is assumed here in order to fit the closure properties of classes of the form $\mathscr{C}_{\mathsf{de}}$ or $\mathscr{C}_{\mathsf{pde}}$. Clearly, we might allow terms in which the universal quantifier is used, since this operation is monotonic. In fact, an extension of this kind will be introduced later. Still, note that the restriction to E-terms has the great advantage that such terms are in some sense *finitary*; that is, the evaluation requires only a finite number of values of the functions and predicates involved, as long as the evaluation gives value T. For example, if U is the E-term $\exists y P(f(y) + f(y + 1))$ and $U \equiv \mathsf{T}$, the evaluation requires two values of f, namely, $f(y)$ and $f(y + 1)$ for some y, and one value of P, namely, $P(y') \equiv \mathsf{T}$, where $y' = f(y) + f(y+1)$. On the other hand, the evaluation $U \equiv \mathsf{F}$ requires an infinite number of values of f and P.

The preceding remarks suggest the following definition: If P is a k-ary predicate we say that the number u *partially represents* P if whenever $y < \ell u$ and $(u)_y \neq 0$ then $P([y]_1, \ldots, [y]_k) \equiv \mathsf{T}$. More formally, we introduce a unary predicate pr_P such that

$$\mathsf{pr}_P(u) \equiv \forall y < \ell u : (u)_y = 0 \vee P([y]_1, \ldots, [y]_k).$$

Let F be a k-ary predicate transformation. A *finitary specification* for F is a $(k+1)$-ary predicate Q such that

$$F(P)(\mathbf{x}) \equiv \exists u : \mathsf{pr}_P(u) \wedge Q(\mathbf{x}, u).$$

We show next that a finitary specification for a predicate transformation F provides an effective procedure to evaluate the minimal fixed point of F.

Theorem 3.3.2. *Let $\mathscr{P}$ be a class of predicates closed under substitution with elementary functions, conjunction, disjunction, bounded quantification, and existential unbounded quantification. Assume $\mathsf{EL_d} \subseteq \mathscr{P}$. If F is a k-ary predicate transformation, and there is a finitary specification for F that belongs to $\mathscr{P}$, then the minimal fixed point of F also belongs to $\mathscr{P}$.*

PROOF. Assume Q is a finitary specification for the k-ary predicate transformation F, and Q is in the class $\mathscr{P}$. A finite sequence of numbers $(u_1, \ldots, u_m)$, $m \geq 1$, is said to be a *derivation sequence* if for each $i = 1, \ldots, m$, the following condition is satisfied: if $y < \ell u_i$ and $(u_i)_y \neq 0$, there is j, $1 \leq j < i$, such that $Q([y]_1, \ldots, [y]_k, u_j)$ holds. Clearly, the concatenation of several derivation sequences is a derivation sequence, and a nonempty initial segment of a derivation sequence is also a derivation sequence. We introduce a k-ary predicate P such that

$$P(\mathbf{x}) \equiv \text{there is a derivation sequence } (u_1, \ldots, u_m) \text{ such that } (u_m)_{\langle \mathbf{x} \rangle} \neq 0.$$

We shall prove that P is the minimal fixed point of F. To show that $F(P) \subseteq_k P$ assume that $F(P)(\mathbf{x})$ holds. Hence there is a number u such that $\mathsf{pr}_P(u)$ and $Q(\mathbf{x}, u)$. This means that whenever $y < \ell u$ and $(u)_y \neq 0$, then $P([y]_1, \ldots, [y]_k)$; hence there is a derivation sequence $(u_1, \ldots, u_m)$, and $(u_m)_{y'} \neq 0$, where $y' = \langle [y]_1, \ldots, [y]_k \rangle$. If we concatenate all the derivation sequences obtained in this way we get another derivation sequence. Finally, by adding: $u, \mathsf{p}_{\langle \mathbf{x} \rangle} \dot{-} 1$, we get another derivation; hence $P(\mathbf{x})$ holds.

Assume now that P' is another k-ary predicate such that $F(P') \subseteq_k P'$. We prove by induction on m that whenever $(u_1, \ldots, u_m)$ is a derivation sequence and $(u_m)_y \neq 0$, then $P'([y]_1, \ldots, [y]_k)$ holds. The case $m = 1$ is trivial. If $(u_1, \ldots, u_m, u_{m+1})$ is a derivation sequence then the property holds for $u_1, \ldots, u_m$. If $(u_{m+1})_y \neq 0$ then there is $j \leq m$ such that $Q([y]_1, \ldots, [y]_k, u_j)$ holds. By the induction hypothesis it follows that $\mathsf{pr}_{P'}(u_j)$, hence $F(P')([y]_1, \ldots, [y]_k)$, so $P'([y]_1, \ldots, [y]_k)$.

To complete the proof, we show that P is a predicate in $\mathscr{P}$. Note that a derivation sequence $(u_1, \ldots, u_m)$ can be encoded in the number $\langle u_1, \ldots, u_m \rangle$. With this in mind we introduce the 3-ary predicate D and the unary predicate D' such that

$$D(u, v, y) \equiv ([u]_{v+1})_y = 0 \vee \exists w < vQ([y]_1, \ldots, [y]_k, [u]_{w+1})$$
$$D'(u) = \forall v < \ell u \forall y < \ell [u]_{v+1} D(u, v, y).$$

It is clear that the predicates D and D' are in $\mathscr{P}$. Furthermore,

$$P(\mathbf{x}) \equiv \exists u \colon u \neq 0 \wedge D'(u) \wedge [\nabla u]_{\langle \mathbf{x} \rangle} \neq 0.$$

It follows that the predicate P is also in $\mathscr{P}$. □

Corollary 3.3.2.1. *Let $\mathscr{P}$ be a class of predicates closed under substitution with $\mathscr{C}$-computable functions, conjunction, disjunction, bounded quantification, and existential quantification, where $\mathscr{C}$ is a class of functions closed under elementary operations. Assume also that $\mathsf{EL_d} \subseteq \mathscr{P}$. Let the k-ary predicate P be the minimal solution of an inductive specification having the following form: If U then P(x), where U is an* E*-term, the predicates in U are in $\mathscr{P}$, and the functions in U are $\mathscr{C}$-computable. Then P is in $\mathscr{P}$.*

PROOF. Since the class $\mathscr{P}$ satisfies all closure properties of Theorem 3.3.2, we need only show that the associated predicate transformation has a finitary specification that is in $\mathscr{P}$. Let U' be the E-term obtained from U by replacing all occurrences of the form $P(U_1, \ldots, U_k)$ by $(u)_{\langle U_1, \ldots, U_k \rangle} \neq 0$. A finitary specification is given by

$$Q(\mathbf{x}, u) \equiv U',$$

which is in $\mathscr{P}$. □

The most important applications of Corollary 3.3.2.1 are classes of the form $\mathscr{C}_{\mathsf{de}}$, where $\mathscr{C}$ is closed under elementary operations, or $\mathscr{C}_{\mathsf{pde}}$, where $\mathscr{C}$ is closed under recursive operations. Note that in either case the classes are closed under substitution with $\mathscr{C}$-computable functions.

EXAMPLE 3.3.4. Consider the specification of the predicate P_g in Example 3.3.2, where g is a given binary function. If g is a recursive function it follows that P_g is a recursively enumerable predicate. If g is $\mathscr{C}_{\mathsf{pd}}$-enumerable, where $\mathscr{C}$ is closed under recursive operations, P_g is also $\mathscr{C}_{\mathsf{pd}}$-enumerable.

The most important feature of an inductive specification is the possibility of proving assertions by induction on the specification. For example, let P be a k-ary predicate, which is the minimal solution of an inductive specification of the form:

(1) If U_1 then $P(\mathbf{x})$.

$\vdots$

(m) If U_m then $P(\mathbf{x})$.

Let P' be another k-ary predicate. We want to prove that $P(\mathbf{x})$ implies $P'(\mathbf{x})$, that is, that $P \subseteq_k P'$. To apply induction we transform each term U_i, $i = 1, \ldots, m$, into another term U_i', which is the result of replacing P by $P \wedge P'$ in U_i, or more precisely, the result of replacing every occurrence of $P(V_1, \ldots, V_k)$ in U_i by $P(V_1, \ldots, V_k) \wedge P'(V_1, \ldots, V_k)$. We prove now that for each i the condition U_i' implies $P'(\mathbf{x})$. From this we infer that $P(\mathbf{x})$ implies $P'(\mathbf{x})$.

The validity of this procedure is immediate. The proof actually shows that $F(P \wedge P') \subseteq_k P'$, where F is the associated predicate transformation. We know also that $F(P \wedge P') \subseteq_k F(P) \subseteq_k P$; hence $F(P \wedge P') \subseteq_k P \wedge P'$. Since P is a minimal solution this means that $P \subseteq_k P \wedge P' \subseteq_k P'$.

It is convenient to organize a proof by induction as a proof by cases, so that each case corresponds to one of the statements in the induction. In the ith case we assume that U_i' holds and prove $P'(\mathbf{x})$. The assumption that U_i' holds is called the *induction hypothesis* for that case.

As an example of this technique we shall prove that the predicate P_g in Example 3.3.2 is transitive.

Theorem 3.3.3. *Let* P_g *be the predicate of Example* 3.3.2. *If* $\mathsf{P}_g(x, y)$ *and* $\mathsf{P}_g(u, x)$ *then* $\mathsf{P}_g(u, y)$.

PROOF. To use induction on the specification of P_g we introduce a predicate P' such that

$$P'(x, y) \equiv \forall u\colon \sim \mathsf{P}_g(u, x) \vee \mathsf{P}_g(u, y).$$

We use induction to prove that $\mathsf{P}_g(x, y)$ implies $P'(x, y)$. From the induction hypothesis for case $g1$ we get a number w such that

$$y = \mathsf{exp}(2, w) \wedge (x = w \vee (\mathsf{P}_g(x, w) \wedge P'(x, w)).$$

To prove $P'(x, y)$ assume u is such that $\mathsf{P}_g(u, x)$ holds. If $x = w$ we have $\mathsf{P}_g(u, w)$ hence $\mathsf{P}_g(u, y)$ by $g1$. Otherwise we have $P'(x, w)$ hence $\mathsf{P}_g(u, w)$, so again $\mathsf{P}_g(u, y)$ by $g1$.

The case $g2$ is similar. Here the induction hypothesis gives numbers w, z, and v such that

$$y = \mathsf{exp}(3, w + 1) \wedge \mathsf{G}_g(w, z, v) \wedge \mathsf{P}_g(x, v) \wedge P'(x, v).$$

Again we assume that $\mathsf{P}_g(u, x)$ and from $P'(x, v)$ it follows that $\mathsf{P}_g(u, v)$, hence $\mathsf{P}_g(u, y)$ follows by $g2$. □

Note that in this proof only parts of the induction hypothesis are used in each case. In particular, the assumptions that $\mathsf{P}_g(x, w)$ holds in case $g1$ and

$\mathsf{P}_g(x, v)$ holds in case $g2$ are not used at all. But this is not always the case and some proofs may require the use of this type of information.

Exercises

3.3.1. Let F be a k-ary predicate transformation. We define $\bar{F}$ as the k-ary predicate transformation such that $\bar{F}(P) = \overline{F(\bar{P})}$. Prove that if F is monotonic then $\bar{F}$ is also monotonic.

3.3.2. Let F be a k-ary monotonic transformation. Prove that there is a fixed point P of F such that whenever $P' \subseteq_k F(P')$, then $P' \subseteq_k P$.

3.3.3. Consider the binary predicate P' such that

$$P'(x, y) \equiv \exists w\colon y = \mathsf{exp}(2, w) \vee y = \mathsf{exp}(3, w + 1).$$

Prove that if g is an arbitrary binary function then P' is a solution of the inductive specification of Example 3.3.2. Give an example of a binary function g such that P' is not the minimal solution P_g.

3.3.4. Assume P_g is the minimal solution of the specification in Example 3.3.2 and $\mathsf{P}_g(x, y)$ holds. Prove the following by inversion:
(a) $\exists w\colon y = \mathsf{exp}(2, w) \vee y = \mathsf{exp}(3, w + 1)$.
(b) If $y = \mathsf{exp}(2, w)$, then $x = w \vee \mathsf{P}_g(x, w)$.
(c) If $y = \mathsf{exp}(3, w + 1)$, then $\exists z \exists v\colon \mathsf{G}_g(w, z, v) \wedge \mathsf{P}_g(x, v)$.

3.3.5. Let $\mathscr{C}$ be a class closed under elementary operations, and P a k-ary predicate. Prove the following:
(a) If P is $\mathscr{C}$-decidable, then pr_P is $\mathscr{C}$-decidable.
(b) If P is $\mathscr{C}_{\mathsf{d}}$-enumerable, then pr_P is $\mathscr{C}_{\mathsf{d}}$-enumerable.

3.3.6. Let $\mathscr{C}$ be a class of functions closed under recursive operations, and let P be a k-ary predicate. Prove the following:
(a) If P is partially $\mathscr{C}$-decidable, then pr_P is partially $\mathscr{C}$-decidable.
(b) If P is $\mathscr{C}_{\mathsf{pd}}$-enumerable, then pr_P is $\mathscr{C}_{\mathsf{pd}}$-enumerable.

3.3.7. Let F be a monotonic k-ary predicate transformation, and let P_0 be a fixed k-ary predicate. Prove that there is a k-ary predicate P such that:
(a) $P_0 \subseteq_k P$.
(b) P is closed under F.
(c) If $P_0 \subseteq_k P'$, and P' is closed under F, then $P \subseteq_k P'$.
(d) If $P_0 \subseteq_k F(P_0)$, then P is a fixed point of F.

3.3.8. If F_1 and F_2 are k-ary predicate transformations, the composition $F_1 \circ F_2 = F$ is defined by $F(P) = F_1(F_2(P))$. Prove the following:
(a) If F_1 and F_2 are monotonic transformations, then F is also a monotonic transformation.
(b) If F_1 and F_2 are monotonic, $F = F_1 \circ F_2 = F_2 \circ F_1$, and P is the minimal fixed point of F, then P is also a fixed point of F_1 and F_2.

3.3.9. Let g and g' be two binary functions such that $g \subseteq_2 g'$. Prove that $\mathsf{P}_g \subseteq_2 \mathsf{P}_{g'}$, where P_g and $\mathsf{P}_{g'}$ are given by the specification of Example 3.3.2.

3.3.10. Let P be a k-ary predicate that is the minimal solution of an inductive specification of the form: If U then $P(\mathbf{x})$. Assume P' is a k-ary predicate that satisfies the condition: If $P(\mathbf{x})$ and U' then $P'(\mathbf{x})$, where U' is the result of replacing all occurrences of P in U with $P \wedge P'$. Prove that $P \subseteq_k P'$.

Notes

This section provides an explicit treatment of induction, which is extended to nonfinitary induction in Chapter 5. In fact, induction pervades the whole approach to computability in this book, because minimal closure is a form of induction and so is of course recursion. Even reflexivity is derived by induction, as it is explained in Chapter 4.

A deeper treatment of induction is given by Hinman [8] and Fenstad [3], where more references are given. Our main interest here is the formalization of induction as a method of proof. This will be of great significance in connection with nonfinitary induction, where the interaction of quantifiers induces constructions of great complexity.

§4. Nondeterministic Computability

The best example of nondeterministic computation is provided by the inductive specification of a predicate P, which can be interpreted as the specification of the function ψ_P. This is not the only way in which induction takes the role of function specification. We shall show later that nontrivial functions can be specified via inductions in which the graph predicate is generated in a nondeterministic manner.

In this section we discuss nondeterministic computability relative to a given function c. Our purpose is to formulate a reasonable version of Church's thesis under some very general assumptions. Essentially, we assume that nondeterministic computations are organized in the form of a tree. Along each branch of the tree the computation is deterministic, and whenever a situation arises where an undefined value of c is required the computation is suspended, but only in that particular branch. All branches are independent one from the other, computations are assumed to be simultaneous, and a computation along one branch is pursued until an output is produced. This may fail because the computation is suspended in case an undefined value of c is required, or because the computation proceeds indefinitely without ever producing output. The computation as a whole halts if and only if one of the branches halts with output. We assume of course that the algorithm is single-valued, and the output is the same no matter which is the halting branch. More precisely, we assume that in case two different branches halt with output, this is the same in the two cases. This can be arranged in different ways. One way is to make sure that at most one branch ever halts with output. Another is to make the output a constant value.

Example 3.4.1. Let c be a 1-1, not necessarily total, unary function. We want to compute, relative to c, the inverse of c, that is, the function f such that $f(y) \simeq x$ if and only if $c(x) \simeq y$. An algorithm to compute $f(y)$ is given by the following program:

Set $x = 0$

[1]: *Branch to BR1 and BR2.*

BR1: *Call* $c(x)$. *If* $c(x) \simeq y$ *then halt computation with output* x. *If* $c(x) \not\simeq y$ *suspend computation.*

BR2: *Set* $x = x + 1$, *go to* [1].

The execution of this algorithm is self-explanatory. Computation along a branch *BR1* is suspended without output in two cases: one when $c(x)$ is undefined, and the other when it is defined with value different from y. The algorithm is single-valued because the function c is 1-1, and this implies that at most one branch halts with output.

Example 3.4.2. Let P be a k-ary predicate satisfying the relation

$$P(\mathbf{x}) \equiv \exists y Q(\mathbf{x}, y),$$

where Q is a given $(k+1)$-ary predicate. We want to compute ψ_P relative to Q. This can be interpreted in two ways: relative to χ_Q or relative to ψ_Q. A deterministic computation of ψ_P relative to χ_Q is given by the following expression:

$$\psi_P(\mathbf{x}) \simeq \mu y Q(\mathbf{x}, y) \times 0.$$

A nondeterministic algorithm to compute ψ_P relative to ψ_Q is given by the following program:

Set $y = 0$

[1]: *Branch to BR1 and BR2*

BR1: *Call* $\psi_Q(\mathbf{x}, y)$. *If* $\psi_Q(\mathbf{x}, y) \simeq 0$ *halt computation with output* 0.

BR2: *Set* $y = y + 1$, *go to* [1].

This algorithm has the same structure as the one given in the preceding example, that is, a nondeterministic search for a value satisfying some condition. But in this case single-valuedness is assured because the output is always 0, no matter which is the halting branch.

Example 3.4.3. The inductive specification of Example 3.3.2 provides a nondeterministic algorithm to evaluate the function ψ_{P_g} relative to g. The complete description of this algorithm is rather involved and will not be attempted here. It is a recursive evaluation, with the usual stack organization. A nondeterministic branching is required at the time of a recursive call, to determine which of the two statements is to be evaluated. Furthermore, both statements contain existential quantifiers but these can be evaluated by a deterministic search. Single-valuedness is assured because the output is always 0.

Church's thesis for nondeterministic relative computability. A function is nondeterministically computable relative to a function c if and only if it is partially recursively enumerable in c.

This means that $\mathsf{RC}_{\mathsf{pd}\#}(c)$ is the class of all functions that are nondeterministically computable relative to c. To validate this assertion we consider first the case in which a k-ary function h is partially recursively enumerable in c; hence

$$\mathsf{G}_h(\mathbf{x}, v) \equiv \exists y Q(\mathbf{x}, v, y),$$

where ψ_Q is recursive in c. The following nondeterministic program computes $h(\mathbf{x})$:

Set $y = 0$

[1]: *Branch to BR1 and BR2*

BR1: *Evaluate* $\psi_Q(\mathbf{x}, (y)_0, (y)_1)$. If $\psi_Q(\mathbf{x}, (y)_0, (y)_1) \simeq 0$ *halt with output* $(y)_0$.

BR2: *Set* $y = y + 1$, *go to* [1].

The computation described in *BR1* is deterministic, because ψ_Q is recursive in c, and may run indefinitely, or may be suspended without output, or may halt with output 0, in which case $Q(\mathbf{x}, (y)_0, (y)_1) \equiv \mathsf{T}$ and $h(\mathbf{x}) \simeq (y)_0$. It is quite possible that many branches halt for a given input in a program like this. The assumption about h makes sure that the output is the same no matter which is the halting branch.

To validate the converse relation we consider a k-ary function h that is nondeterministically computable relative to c, and we apply again the encoding technique. We immediately find a difficulty with the specification of the function tr, which is now many-valued due to the existence of branching instructions in the algorithm that computes h. To overcome this problem we change tr to a binary function tr′, with an extra parameter that determines the branch whenever a branching instruction is applied. For example, if all branches are of the form *BR1*, *BR2* (as in Example 3.4.1), we may agree that $z = 0$ means *BR1* and $z \neq 0$ means *BR2*. So now $\mathsf{tr}'(z, y) = y'$ means that y' is the next structure after y, with z determining the branch. Otherwise, tr′ is identical to tr, but is given by partial definition by cases and involves the function c. It follows that tr′ is recursive in c. Still, note that in the case that c is total tr′ is elementary in c.

Similar problems arise with the specification of the function ev, which are solved by introducing an extra parameter that encodes a sequence of branchings, rather than only one branching. Formally, we put

$$\mathsf{ev}(\mathbf{x}, z, 0) \simeq \text{the encoding of the initial structure}$$

$$\mathsf{ev}(\mathbf{x}, z, y + 1) \simeq \mathsf{tr}'((z)_y, \mathsf{ev}(\mathbf{x}, z, y)).$$

In this way a number z determines uniquely a branch in the computation, and every halting branch is determined by a number z. We assume again that there are elementary functions g_1 and g_2, which determine when a structure is halting and which is the output. In this way we get a $(k + 1)$-ary function f such that

$$f(\mathbf{x}, z) \simeq g_2(\mathsf{ev}(\mathbf{x}, z, \mu y(g_1(\mathsf{ev}(\mathbf{x}, z, y)) \simeq 0))),$$

which is recursive in c. Finally, we have

$$\mathsf{G}_h(\mathbf{x}, v) \equiv \exists z \mathsf{G}_f(\mathbf{x}, z, v),$$

so h is partially recursively enumerable in c.

Exercises

3.4.1. Let c be a unary function. The unary predicate pr_c is given by

$$\mathsf{pr}_c(u) \equiv \forall y < \ell u\colon (u)_y = 0 \vee c(y) \simeq (u)_y \dot{-} 1.$$

Let $\mathscr{C}$ be a class of functions. Prove the following:

(a) If $\mathscr{C}$ is closed under elementary operations, and c is $\mathscr{C}_{\mathsf{d}}$-enumerable, then pr_c is $\mathscr{C}_{\mathsf{d}}$-enumerable.

(b) If $\mathscr{C}$ is closed under recursive operations, and c is $\mathscr{C}_{\mathsf{pd}}$-enumerable, then pr_c is $\mathscr{C}_{\mathsf{pd}}$-enumerable.

3.4.2. Refine the analysis in Church's thesis and show that a k-ary function h is nondeterministically computable relative to a unary function c if and only if there is a recursive $(k + 2)$-ary function f such that

$$\mathsf{G}_h(\mathbf{x}, v) \equiv \exists u \exists z\colon \mathsf{pr}_c(u) \wedge f(\mathbf{x}, u, z) \simeq v.$$

3.4.3. Using the result in the preceding exercise, prove that a k-ary function h is nondeterministically computable relative to a unary function c if and only if there is an elementary $(k + 3)$-ary predicate Q such that

$$\mathsf{G}_h(\mathbf{x}, v) \equiv \exists u \exists y\colon \mathsf{pr}_c(u) \wedge Q(\mathbf{x}, v, u, y).$$

Notes

Nondeterministic computability relative to a function c is equivalent to single-valued enumeration reducibility relative to G_c in the sense of Rogers [20]. This relation is clear in one direction, and it will be proved to be a true equivalence in Chapter 4.

This type of computability is a kind of limit case, and to some extent it can be considered pathological. As far as we know it does not induce ordinary reflexivity properties, although some forms of weak reflexivity in terms of predicates are available.

Ordinary computability in the form of e-total classes implies that nondeterministic computability can be reduced to deterministic. This reduction is possible even in cases where e-totality does not hold and corresponds to the notion of closure under enumeration operations discussed in Section 2. Important examples of this situation are described in Chapter 5.

CHAPTER 4

Reflexive Structures

In this chapter we continue the study of classes closed under recursive operations, with emphasis on those classes of the form $\mathscr{C} = \mathsf{RC}(c)$, where c is some arbitrary function. There are many problems about these classes which cannot be solved with the techniques used in Chapter 2. For example, we do not know yet if $\mathscr{C}_{\mathsf{d}} = \mathscr{C}_{\mathsf{de}}$ or $\mathscr{C}_{\mathsf{d}} \neq \mathscr{C}_{\mathsf{de}}$. We shall see that both relations are possible, depending on the function c, but when c is a total function then $\mathscr{C}_{\mathsf{d}} \neq \mathscr{C}_{\mathsf{de}}$, which means that there are predicates that are $\mathscr{C}_{\mathsf{d}}$-enumerable (i.e., recursively enumerable in c) but are not recursive in c. Results of this type require a diagonalization technique, involving a kind of internal enumeration, or indexing for the class $\mathsf{RC}(c)$.

§1. Interpreters

We are interested here in classes of the form $\mathsf{RC}(\mathscr{C})$, where $\mathscr{C}$ is a finite class of functions. Our first result shows that this situation can be reduced to the case where $\mathscr{C}$ contains exactly one function.

Theorem 4.1.1. *Let $\mathscr{C}$ be a finite class of functions. There is a unary function c such that $\mathsf{RC}(\mathscr{C}) = \mathsf{RC}(c)$. Furthermore, if all functions in $\mathscr{C}$ are total, then c can be taken as a total function.*

Proof. If $\mathscr{C}$ is empty we take c to be some elementary function, for example, $c = \sigma$. Otherwise let $c_1, \ldots, c_m$ be all functions in $\mathscr{C}$, of arity $k_1, \ldots, k_m$, respectively, $m \geq 1$. We use partial definition by cases to introduce a unary function c as follows:

$$\begin{aligned} c(x) &\simeq c_1([x]_2, \ldots, [x]_{k_1+1}) \quad \text{if } [x]_1 = 0 \\ &\simeq c_2([x]_2, \ldots, [x]_{k_2+1}) \quad \text{if } [x]_1 = 1 \\ &\quad \vdots \qquad\qquad\qquad\qquad \vdots \\ &\simeq c_m([x]_2, \ldots, [x]_{k_m+1}) \quad \text{if } [x]_1 \geq m \dot{-} 1. \end{aligned}$$

From this specification it follows that c is recursive in $\mathscr{C}$; hence $\mathsf{RC}(c) \subseteq \mathsf{RC}(\mathscr{C})$. On the other hand, we have

$$c_i(\mathbf{x}) \simeq c(\langle i \dot{-} 1, \mathbf{x} \rangle)$$

for $i = 1, \ldots, m$; hence $\mathscr{C} \subseteq \mathsf{RC}(c)$, so $\mathsf{RC}(\mathscr{C}) \subseteq \mathsf{RC}(c)$. □

Let S be a binary function. For each $k \geq 0$ we specify a $(k + 1)$-ary function S^k as follows:

$$\mathsf{S}^0(z) \simeq z$$

$$\mathsf{S}^{k+1}(z, y, \mathbf{x}) \simeq \mathsf{S}^k(\mathsf{S}(z, y), \mathbf{x}).$$

Note that $\mathsf{S} = \mathsf{S}^1$. If $k \geq 0$, $m \geq 0$, $(\mathbf{x}) \in \mathbb{N}^k$, and $(\mathbf{y}) \in \mathbb{N}^m$, it is easy to prove by induction on k that

$$\mathsf{S}^{k+m}(z, \mathbf{x}, \mathbf{y}) \simeq \mathsf{S}^m(\mathsf{S}^k(z, \mathbf{x}), \mathbf{y}).$$

In particular, $\mathsf{S}^{k+1}(z, \mathbf{x}, y) \simeq \mathsf{S}(\mathsf{S}^k(z, \mathbf{x}), y)$.

Each one of the functions S^k is derived from the given function S by substitution. It follows that if $\mathscr{C}$ is class closed under basic operations, and S is $\mathscr{C}$-computable, then S^k is $\mathscr{C}$-computable for $k \geq 0$. Furthermore, if S is a total function, then S^k is also total for $k \geq 0$.

Let $\mathscr{C}$ be a class of functions. An *interpreter* for $\mathscr{C}$ is a pair (ϕ, S), where ϕ is a unary function and S is a total binary function, and if a k-ary function h is $\mathscr{C}$-computable then there is a number z such that

$$h(\mathbf{x}) \simeq \phi(\mathsf{S}^k(\mathsf{z}, \mathbf{x})).$$

In dealing with interpreters it is convenient to introduce an alternative notation. For example, if (ϕ, S) is an interpreter for the class $\mathscr{C}$, we specify a $(k + 1)$-ary function ϕ^k for each $k \geq 0$, such that

$$\phi^k(z, \mathbf{x}) \simeq \phi(\mathsf{S}^k(z, \mathbf{x})).$$

It follows that whenever h is a k-ary $\mathscr{C}$-computable function there is a number z such that

$$h(\mathbf{x}) \simeq \phi^k(\mathsf{z}, \mathbf{x}).$$

A number z satisfying this relation is said to be a (ϕ, S)-*index for h*. We shall see later that a function in $\mathscr{C}$ may have many indexes, in fact an infinite number of indexes.

The functions ϕ^k are related by the following transition equations:

$$\begin{aligned}\phi^{k+1}(z, y, \mathbf{x}) &\simeq \phi(\mathsf{S}^{k+1}(z, y, \mathbf{x}))\\ &\simeq \phi(\mathsf{S}^k(\mathsf{S}(z, y), \mathbf{x}))\\ &\simeq \phi^k(\mathsf{S}(z, y), \mathbf{x}).\end{aligned}$$

In general, if $k \geq 0$ and $m \geq 0$, we have

$$\phi^{k+m}(z, \mathbf{x}, \mathbf{y}) \simeq \phi^m(\mathsf{S}^k(z, \mathbf{x}), \mathbf{y}).$$

We conclude that if h is a $(k + m)$-ary $\mathscr{C}$-computable function then

$$h(\mathbf{x}, \mathbf{y}) \simeq \phi^m(g(\mathbf{x}), \mathbf{y}),$$

where g is a k-ary total function such that $g(\mathbf{x}) = \mathsf{S}^k(\mathbf{z}, \mathbf{x})$ and $\mathbf{z}$ is a (ϕ, S)-index for h. This relation will be referred to as the *abstraction property* for the interpreter (ϕ, S).

The use of interpreters is a traditional technique in computability theory, although in general they are introduced via the functions ϕ^k, while S is an elementary function that relates the functions ϕ^{k+1} and ϕ^k in the manner described above. Our terminology is taken from computer science, where an interpreter is a function that takes a program (which we may assume is encoded in a number z) and some input $(\mathbf{x}) \in \mathbb{N}^k$ and carries out the computation determined by the program z with such an input. Implicit in this definition is a function that reads the input $(\mathbf{x})$ and makes it available to the interpreter. Hence, if (ϕ, S) is an interpreter for a class $\mathscr{C}$ of functions, we say that ϕ is the *evaluation* function and S is the *input* function.

Note that up to this point we have not assumed that whenever (ϕ, S) is an interpreter for a class $\mathscr{C}$ then ϕ is $\mathscr{C}$-computable. In fact, there are situations where such a restriction is impossible. For example, there is no interpreter (ϕ, S) for the class PR, where ϕ is primitive recursive, although there is one where ϕ is a recursive function. Still, we must exclude these situations from our discussion.

An interpreter (ϕ, S) for a class $\mathscr{C}$ is *internal* if both functions ϕ and S are $\mathscr{C}$-computable.

A *functional reflexive structure* (FRS) is a class $\mathscr{C}$ of functions which satisfies the following conditions:

(i) $\mathscr{C}$ is closed under basic operations.
(ii) The functions σ, pd, and cd are $\mathscr{C}$-computable.
(iii) $\mathscr{C}$ has an internal interpreter.

Note that from the assumptions that $\mathscr{C}$ is closed under basic operations and the function cd is $\mathscr{C}$-computable, it follows that the functions sg and csg are $\mathscr{C}$-computable.

Theorem 4.1.2. *Let $\mathscr{C}$ be a* FRS *with internal interpreter* (ϕ, S). *Then* ϕ *is a nontotal function, and the functions* UD_k, $k \geq 0$, *are* $\mathscr{C}$*-computable.*

Proof. Let f be the unary function such that

$$f(x) \simeq \phi^1(x, x) + 1.$$

It follows that f is $\mathscr{C}$-computable. If z is a (ϕ, S)-index for f we have

$$f(\mathsf{z}) \simeq \phi^1(\mathsf{z}, \mathsf{z}) \simeq \phi^1(\mathsf{z}, \mathsf{z}) + 1;$$

hence $\phi^1(\mathsf{z}, \mathsf{z})$ is not defined. If $\mathsf{S}(\mathsf{z}, \mathsf{z}) = \mathsf{m}$, this means that $\phi(\mathsf{m})$ is not defined. Hence

$$\mathsf{UD}_k(\mathbf{x}) \simeq \phi(\mathsf{C}^k_{\mathsf{m}}(\mathbf{x})),$$

so UD_k is $\mathscr{C}$-computable. □

Theorem 4.1.3. If $\mathscr{C}$ is a FRS then $\mathscr{C}$ is closed under partial definition by cases.

Proof. We use induction in m = the number of cases in the specification. If $m = 1$ we have

$$h(\mathbf{x}) \simeq f(\mathbf{x}) \quad \text{if } g(\mathbf{x}) \simeq 0,$$

where f and g are given k-ary $\mathscr{C}$-computable functions. Let (ϕ, S) be an internal interpreter for $\mathscr{C}$, z_1 a (ϕ, S)-index for f, and z_2 a (ϕ, S)-index for UD_k. It follows that

$$h(\mathbf{x}) \simeq \phi^k(\mathsf{cd}(\mathsf{z}_1, \mathsf{z}_2, g(\mathbf{x})), \mathbf{x});$$

hence h is $\mathscr{C}$-computable. Assume the property is true for m cases, and consider a specification with $m + 1$ cases,

$$\begin{aligned} h(\mathbf{x}) &\simeq f_1(\mathbf{x}) && \text{if } g_1(\mathbf{x}) \simeq 0 \\ &\simeq f_2(\mathbf{x}) && \text{if } g_2(\mathbf{x}) \simeq 0 \\ &\ \ \vdots && \quad \vdots \\ &\simeq f_m(\mathbf{x}) && \text{if } g_m(\mathbf{x}) \simeq 0 \\ &\simeq f_{m+1}(\mathbf{x}) && \text{if } g_{m+1}(\mathbf{x}) \simeq 0. \end{aligned}$$

Introduce a k-ary function h' using only m cases in the form

$$\begin{aligned} h'(\mathbf{x}) &\simeq f_2(\mathbf{x}) && \text{if } g_2(\mathbf{x}) \simeq 0 \\ &\ \ \vdots && \quad \vdots \\ &\simeq f_{m+1}(\mathbf{x}) && \text{if } g_{m+1}(\mathbf{x}) \simeq 0. \end{aligned}$$

From the induction hypothesis it follows that h' is $\mathscr{C}$-computable, so let z_1 be a (ϕ, S)-index for f_1 and z_2 a (ϕ, S)-index for h'. It follows that

$$h(\mathbf{x}) \simeq \phi^k(\mathsf{cd}(\mathbf{z}_1, \mathbf{z}_2, g_1(\mathbf{x})), \mathbf{x});$$

hence h is $\mathscr{C}$-computable. □

We do not yet have examples of functional reflexive structures, but eventually we shall prove that all classes of the form $\mathsf{RC}(c)$, where c is an arbitrary function, satisfy the conditions in the definition. For the time being we note that whenever $\mathscr{C}$ is a FRS with internal interpreter (ϕ, S), then $\mathscr{C} \subseteq \mathsf{EL}(\{\phi, \mathsf{S}\}) \subseteq \mathsf{RC}(\{\phi, \mathsf{S}\})$. In this section we prove that the converse relation $\mathsf{RC}(\{\phi, \mathsf{S}\}) \subseteq \mathscr{C}$ also holds.

Example 4.1.1. Let $\mathscr{C}$ be a FRS with internal interpreter (ϕ, S), and for a fixed $k \geq 0$, the identity function I_1^{k+1} is $\mathscr{C}$-computable; hence

$$\mathsf{I}_1^{k+1}(z, \mathbf{x}) \simeq z.$$

From the abstraction property referred to above it follows that there is a unary total $\mathscr{C}$-computable function g such that

$$\mathsf{I}_1^{k+1}(z, \mathbf{x}) \simeq \phi^k(g(z), \mathbf{x}).$$

This means that for every z the value $g(z)$ is a (ϕ, S)-index for the constant function C_z^k. Such a function depends, of course, on the number k.

Example 4.1.2. Let $\mathscr{C}$ be a FRS with internal interpreter (ϕ, S), and for a fixed $k \geq 0$, let d be the $(k+2)$-ary function such that

$$d(z_1, z_2, \mathbf{x}) \simeq \phi^{k+1}(z_1, \mathbf{x}, \phi^k(z_2, \mathbf{x})).$$

By the abstraction property we get a total $\mathscr{C}$-computable function g such that

$$d(z_1, z_2, \mathbf{x}) \simeq \phi^k(g(z_1, z_2), \mathbf{x}).$$

It follows that if $\mathbf{z}_1$ is a (ϕ, S)-index for a $(k+1)$-ary function f_1, and $\mathbf{z}_2$ is a (ϕ, S)-index for a k-ary function f_2, then $g(\mathbf{z}_1, \mathbf{z}_2)$ is a (ϕ, S)-index for the k-ary function h such that

$$h(\mathbf{x}) \simeq f_1(\mathbf{x}, f_2(\mathbf{x})).$$

Note again that the function g in this example depends on the particular number k.

Theorem 4.1.4. *Let $\mathscr{C}$ be a FRS with internal interpreter (ϕ, S), and let h be a $(k+1)$-ary $\mathscr{C}$-computable function. There is a number $\mathbf{z}$ such that*

$$h(\mathbf{z}, \mathbf{x}) \simeq \phi^k(\mathbf{z}, \mathbf{x}).$$

Proof. Let $\mathbf{z}'$ be a (ϕ, S)-index for the $\mathscr{C}$-computable function h' such that

$$h'(y, \mathbf{x}) \simeq h(\mathsf{S}(y, y), \mathbf{x}).$$

We take $\mathbf{z} = \mathsf{S}(\mathbf{z}', \mathbf{z}')$ and compute

$$\begin{aligned} h(\mathsf{z}, \mathbf{x}) &\simeq h(\mathsf{S}(\mathsf{z}', \mathsf{z}'), \mathbf{x}) \\ &\simeq h'(\mathsf{z}', \mathbf{x}) \\ &\simeq \phi^{k+1}(\mathsf{z}', \mathsf{z}', \mathbf{x}) \quad (\mathsf{z}' \text{ is } (\phi, \mathsf{S})\text{-index for } h') \\ &\simeq \phi^{k}(\mathsf{S}(\mathsf{z}', \mathsf{z}'), \mathbf{x}) \quad (\text{abstraction}) \\ &\simeq \phi^{k}(\mathsf{z}, \mathbf{x}). \end{aligned}$$ □

In applications, Theorem 4.1.4 will be referred to as the *recursion theorem* (RT).

EXAMPLE 4.1.3. Let $\mathscr{C}$ be a FRS with internal interpreter (ϕ, S), and apply RT to the $(k + 1)$-ary function I_1^{k+1}, $k \geq 0$. It follows that there is a number z such that

$$\phi^k(\mathsf{z}, \mathbf{x}) \simeq \mathsf{z}.$$

This means that the number z is a (ϕ, S)-index for the function $\mathsf{C}_{\mathsf{z}}^k$.

Corollary 4.1.4.1. *Let $\mathscr{C}$ be a* FRS *with internal interpreter (ϕ, S), and let g be a unary $\mathscr{C}$-computable function. If $k \geq 0$, there is a number* z *such that*

$$\phi^k(\mathsf{z}, \mathbf{x}) \simeq \phi^k(g(\mathsf{z}), \mathbf{x}).$$

PROOF. Apply the RT to the function $h(y, \mathbf{x}) \simeq \phi^k(g(y), \mathbf{x})$. □

Note that in Corollary 4.1.4.1 the function g is not necessarily total, and $g(\mathsf{z})$ may be undefined, in which case the number z is a (ϕ, S)-index for the function UD_k. If $g(\mathsf{z})$ is defined, then z and $g(\mathsf{z})$ are (ϕ, S)-indexes for the same k-ary $\mathscr{C}$-computable function. In applications Corollary 4.1.4.1 will be referred to as the *fixed point theorem* (FPT).

EXAMPLE 4.1.4. If we apply the FPT with the function σ we get a number z such that z and z + 1 are (ϕ, S)-indexes of the same k-ary function.

Theorem 4.1.5. *If $\mathscr{C}$ is a* FRS *then $\mathscr{C}$ is closed under recursive operations.*

PROOF. We know that $\mathscr{C}$ contains σ and it is closed under basic operations and under partial definition by cases. It remains only to show that $\mathscr{C}$ is closed under primitive recursion and unbounded minimalization with functions.

To prove $\mathscr{C}$ is closed under primitive recursion consider the following specification of the $(k + 1)$-ary function h such that

$$h(\mathbf{x}, 0) \simeq f(\mathbf{x})$$

$$h(\mathbf{x}, y + 1) \simeq g(\mathbf{x}, y, h(\mathbf{x}, y)),$$

where f and g are $\mathscr{C}$-computable. Let (ϕ, S) be an internal interpreter for $\mathscr{C}$, and let h' be the $(k+2)$-ary function such that

$$\begin{aligned} h'(z, \mathbf{x}, y) &\simeq f(\mathbf{x}) && \text{if } y \simeq 0 \\ &\simeq g(\mathbf{x}, \mathsf{pd}(y), \phi^{k+1}(z, \mathbf{x}, \mathsf{pd}(y))) && \text{if } \mathsf{csg}(y) \simeq 0. \end{aligned}$$

Since this is a specification by partial definition by cases, it follows that h' is $\mathscr{C}$-computable. Let z be the number given by the RT such that

$$h'(\mathrm{z}, \mathbf{x}, y) \simeq \phi^{k+1}(\mathrm{z}, \mathbf{x}, y).$$

It follows immediately that $h(\mathbf{x}, y) \simeq h'(\mathrm{z}, \mathbf{x}, y)$; hence h is $\mathscr{C}$-computable.

To prove closure under unbounded minimalization with functions, we consider a specification of a k-ary function h such that

$$h(\mathbf{x}) \simeq \mu y(f(\mathbf{x}, y) \simeq 0),$$

where f is a $(k+1)$-ary $\mathscr{C}$-computable function. The proof is based in the reduction described in the proof of Theorem 2.3.1 so we introduce a $(k+1)$-ary function g by the following recursive specification:

$$\begin{aligned} g(\mathbf{x}, v) &\simeq 0 && \text{if } f(\mathbf{x}, v) \simeq 0 \\ &\simeq g(\mathbf{x}, v+1) + 1 && \text{if } \mathsf{csg}(f(\mathbf{x}, v)) \simeq 0. \end{aligned}$$

In order to apply the RT we introduce a $(k+2)$-ary function g' such that

$$\begin{aligned} g'(z, \mathbf{x}, v) &\simeq 0 && \text{if } f(\mathbf{x}, v) \simeq 0 \\ &\simeq \phi^{k+1}(z, \mathbf{x}, v+1) + 1 && \text{if } \mathsf{csg}(f(\mathbf{x}, v)) \simeq 0. \end{aligned}$$

It follows that g' is $\mathscr{C}$-computable, so let z be the number given by the RT such that

$$g'(\mathrm{z}, \mathbf{x}, v) \simeq \phi^{k+1}(\mathrm{z}, \mathbf{x}, v).$$

It follows that $g(\mathbf{x}, v) \simeq g'(\mathrm{z}, \mathbf{x}, v)$ is the unique solution of the above recursive specification; hence g is $\mathscr{C}$-computable. Furthermore, $h(\mathbf{x}) \simeq g(\mathbf{x}, 0)$, so h is also $\mathscr{C}$-computable. □

Corollary 4.1.5.1. *If $\mathscr{C}$ is a* FRS *with internal interpreter* (ϕ, S), *then* $\mathscr{C} = \mathsf{RC}(\{\phi, \mathsf{S}\})$.

PROOF. We have already remarked that $\mathscr{C} \subseteq \mathsf{RC}(\{\phi, \mathsf{S}\})$, and from Theorem 4.1.5 it follows that $\mathsf{RC}(\{\phi, \mathsf{S}\}) \subseteq \mathscr{C}$. □

We shall show in the next section that any FRS $\mathscr{C}$ has an internal interpreter (ϕ, S), where S satisfies special conditions. Some of them are identified here, and a few consequences are derived.

A total k-ary function f, $k \geq 1$, is *strict* if it is 1-1, and furthermore the conditions $x_i < f(x_1, \ldots, x_k)$ are satisfied for $i = 1, \ldots, k$. For example, the function σ is strict, but the identity functions are not strict.

An interpreter (ϕ, S) is *strict* if the input function S is strict. In this case all functions S^k, $k \geq 1$, are strict, but S^0 is not strict. Note also that in all nontrivial applications of the abstraction property (i.e., with $k \geq 1$), the k-ary function g satisfying the condition

$$h(\mathbf{x}, \mathbf{y}) \simeq \phi^m(g(\mathbf{x}), \mathbf{y})$$

is strict.

EXAMPLE 4.1.5. Let $\mathscr{C}$ be a FRS and assume (ϕ, S) is a strict internal interpreter for $\mathscr{C}$. If we apply the abstraction property to the $(1 + k)$-ary function ϕ^k we obtain a strict unary $\mathscr{C}$-computable function g such that

$$\phi^k(z, \mathbf{x}) \simeq \phi^k(g(z), \mathbf{x}).$$

This means that z and $g(z)$ are (ϕ, S)-indexes of the same k-ary function. Hence if z is a (ϕ, S)-index for the k-ary function h, the function g generates an infinite sequence $\mathsf{z} < g(\mathsf{z}) < g(g(\mathsf{z})) < \cdots$, in which all the elements are (ϕ, S)-indexes for h.

EXAMPLE 4.1.6. Let $\mathscr{C}$ be a FRS with internal interpreter (ϕ, S), and consider the set K_ϕ, such that

$$x \in \mathsf{K}_\phi \equiv \phi^1(x, x)\downarrow.$$

It is clear that K_ϕ is partially $\mathscr{C}$-decidable. In fact, $\psi_{\mathsf{K}_\phi}(x) \simeq \phi^1(x, x) \times 0$. On the other hand, the complement $\overline{\mathsf{K}}_\phi$ is not partially $\mathscr{C}$-decidable. For if $\overline{\mathsf{K}}_\phi$ is partially $\mathscr{C}$-decidable then there is a (ϕ, S)-index z for $\psi_{\overline{\mathsf{K}}_\phi}$; hence

$$\mathsf{z} \in \mathsf{K}_\phi \equiv \phi^1(\mathsf{z}, \mathsf{z})\downarrow \equiv \mathsf{z} \in \overline{\mathsf{K}}_\phi,$$

which is a contradiction. It follows from this that K_ϕ is not $\mathscr{C}$-decidable.

EXAMPLE 4.1.7. Let $\mathscr{C}$ be a FRS with internal interpreter (ϕ, S), and let K_ϕ be the set introduced in the preceding example. We introduce sets A and B such that

$$y \in A \equiv y \text{ is a } (\phi, \mathsf{S})\text{-index for the function } \mathsf{C}_0^1$$

$$y \in B \equiv y \text{ is a } (\phi, \mathsf{S})\text{-index for the function } \mathsf{UD}_1.$$

Let h be the binary $\mathscr{C}$-computable function such that

$$\begin{aligned} h(y, x) &\simeq \psi_{\mathsf{K}_\phi}(y) \\ &\simeq \phi^1(g(y), x), \end{aligned}$$

where g is the total unary $\mathscr{C}$-computable function given by the abstraction property. It follows that

$$\begin{aligned} y \in \mathsf{K}_\phi &\equiv g(y) \in A \\ &\equiv g(y) \in \overline{B}; \end{aligned}$$

hence neither A nor B is $\mathscr{C}$-decidable.

Let $\mathscr{C}$ be a FRS with internal interpreter (ϕ, S). We introduce a set W_ϕ such that

$$z \in \mathsf{W}_\phi \equiv \phi(z)\downarrow \equiv z \in \mathsf{D}_\phi.$$

The set W_ϕ is clearly partially $\mathscr{C}$-decidable, that is, an element of $\mathscr{C}_{\mathrm{pd}}$. In fact, W_ϕ is a kind of universal element in $\mathscr{C}_{\mathrm{pd}}$, as described in the next theorem.

Theorem 4.1.6. *Let $\mathscr{C}$ be a FRS with internal interpreter (ϕ, S). A k-ary predicate P is partially $\mathscr{C}$-decidable if and only if there is a number* $\mathbf{z}$ *such that*

$$P(\mathbf{x}) \equiv \mathsf{S}^k(\mathbf{z}, \mathbf{x}) \in \mathsf{W}_\phi.$$

PROOF. The condition is clearly sufficient. To prove it is necessary assume P is a partially $\mathscr{C}$-decidable k-ary predicate; hence the function ψ_P is $\mathscr{C}$-computable, and there is a number $\mathbf{z}$ such that

$$\psi_P(\mathbf{x}) \simeq \phi(\mathsf{S}^k(\mathbf{z}, \mathbf{x})).$$

It follows that

$$P(\mathbf{x}) \equiv \mathsf{S}^k(\mathbf{z}, \mathbf{x}) \in \mathsf{W}_\phi.$$ □

Theorem 4.1.7. *Let $\mathscr{C}$ be a FRS with internal interpreter (ϕ, S) and let h be a k-ary $\mathscr{C}$-computable function. If B is the set of all (ϕ, S)-indexes for h, then B is not $\mathscr{C}$-decidable.*

PROOF. We assume B is decidable, take $y \in B$ and $y' \notin B$, and introduce using total definition by cases a total unary $\mathscr{C}$-computable function f such that

$$\begin{aligned} f(x) &= y' \quad \text{if } x \in B \\ &= y \quad \text{otherwise.} \end{aligned}$$

From the FPT (Corollary 4.1.4.1) it follows that there is a number $\mathbf{z}$ such that

$$\phi^k(\mathbf{z}, \mathbf{x}) \simeq \phi^k(f(\mathbf{z}), \mathbf{x});$$

that is, $\mathbf{z}$ and $f(\mathbf{z})$ are (ϕ, S)-indexes for the same k-ary $\mathscr{C}$-computable function. But this is a contradiction because $\mathbf{z} \in B$ if and only if $f(\mathbf{z}) \notin B$. □

Corollary 4.1.7.1. *Let $\mathscr{C}$ be a FRS with internal interpreter (ϕ, S), and let B be the set of all (ϕ, S)-indexes for a k-ary $\mathscr{C}$-computable function h. Then B is infinite.*

PROOF. Immediate from Theorem 4.1.7 noting that if B is finite then it is $\mathscr{C}$-decidable. □

EXERCISES

4.1.1. Prove that if $\mathscr{C}$ is a FRS then $\mathscr{C}$ is countable.

4.1.2. Let $\mathscr{C}$ be a FRS with internal interpreter (ϕ, S). Prove the following:
(a) There is a number $\mathbf{z}$ that is a (ϕ, S)-index for the unary function $f(x) = \mathbf{z} + x$.

(b) There is a number z that is a (ϕ, S)-index for the characteristic function of the set $\{\mathsf{z}\}$.

(c) There is a number z that for all x satisfies the condition

$$\phi^1(\mathsf{z}, x) \simeq \phi^1(x, \mathsf{z}).$$

4.1.3. Let $\mathscr{C}$ be a FRS with internal interpreter (ϕ, S), and let $k \geq 0$. Prove the following:

(a) There is a total $\mathscr{C}$-computable unary function g_1 such that if z is a (ϕ, S)-index for a $\mathscr{C}$-computable k-ary function h, then $g_1(z)$ is a (ϕ, S)-index for the partial characteristic function of the predicate D_h (= domain of h).

(b) There is a total $\mathscr{C}$-computable unary function g_2 such that if z is a (ϕ, S)-index for a $\mathscr{C}$-computable k-ary function h, then $g_2(z)$ is a (ϕ, S)-index for the partial characteristic function of the predicate G_h.

4.1.4. Let $\mathscr{C}$ be a FRS with internal interpreter (ϕ, S), and let $k \geq 0$. Prove the following:

(a) There is a total $\mathscr{C}$-computable unary function g_1 such that if z is a (ϕ, S)-index for a $\mathscr{C}$-computable $(k + 1)$-ary function f, then $g_1(z)$ is a (ϕ, S)-index for the k-ary function h such that

$$h(\mathbf{x}) \simeq \mu y(f(\mathbf{x}, y) \simeq 0).$$

(b) There is a total $\mathscr{C}$-computable binary function g_2 such that if z_1 is a (ϕ, S)-index for a $\mathscr{C}$-computable k-ary function f, and z_2 is a (ϕ, S)-index for a $\mathscr{C}$-computable $(k + 2)$-ary function g, then $g_2(z_1, z_2)$ is a (ϕ, S)-index for the $(k + 1)$-ary function h specified by primitive recursion from f and g.

4.1.5. Let $\mathscr{C}$ be a FRS with strict internal interpreter (ϕ, S), and let h be a $(k + 1)$-ary $\mathscr{C}$-computable function. Prove there is a strict $\mathscr{C}$-computable function g such that

$$h(g(v), \mathbf{x}) \simeq \phi^k(g(v), \mathbf{x}).$$

4.1.6. Consider the sets A and B introduced in Example 4.1.7, in relation with a FRS $\mathscr{C}$. Prove that neither $\bar{A}$ nor B is partially $\mathscr{C}$-decidable and that $\bar{B}$ is $\mathscr{C}_{\mathsf{pd}}$-enumerable.

4.1.7. Let $\mathscr{C}$ be a FRS with internal interpreter (ϕ, S), and let K_ϕ be the set introduced in Example 4.1.6. Prove the following:

(a) A set A is partially $\mathscr{C}$-decidable if and only if there is a total $\mathscr{C}$-computable function g such that

$$y \in A \equiv g(y) \in \mathsf{K}_\phi.$$

(b) A k-ary predicate P is partially $\mathscr{C}$-decidable if and only if there is a total $\mathscr{C}$-computable function g such that

$$P(\mathbf{x}) \equiv g(\mathbf{x}) \in \mathsf{K}_\phi.$$

(c) If the input function S is strict (elementary), the function g in (a) and (b) can be taken as strict (elementary).

4.1.8. Let $\mathscr{C}$ be a FRS with internal interpreter (ϕ, S). Prove that the set $\overline{\mathsf{W}_\phi}$ is not partially $\mathscr{C}$-decidable and that the W_ϕ is not $\mathscr{C}$-decidable.

4.1.9. Let $\mathscr{C}_0, \mathscr{C}_1, \ldots, \mathscr{C}_n, \ldots$ be an infinite sequence of FRSs such that $\mathscr{C}_n \subset \mathscr{C}_{n+1}$ for all $n \geq 0$, and let $\mathscr{C}'$ be the union of all classes in the sequence, that is,

$$\mathscr{C}' = \mathscr{C}_0 \cup \mathscr{C}_1 \cup \cdots \cup \mathscr{C}_n \cup \cdots.$$

Prove that $\mathscr{C}'$ is closed under recursive operations but is not a FRS.

4.1.10. Let $\mathscr{C}$ be a FRS with internal interpreter (ϕ, S). Consider the partially $\mathscr{C}$-decidable predicates K_ϕ^0 and K_ϕ^1 such that

$$x \in \mathsf{K}_\phi^0 \equiv \phi^1(x, x) \simeq 0$$

$$x \in \mathsf{K}_\phi^1 \equiv \phi^1(x, x) \simeq 1.$$

Prove that if A is a $\mathscr{C}$-decidable set then either $\mathsf{K}_\phi^0 \not\subseteq A$ or $\mathsf{K}_\phi^1 \not\subseteq \bar{A}$.

4.1.11. Let $\mathscr{C}$ be a FRS. Prove that if $\mathscr{C}$ is e-total then $\mathscr{C}_{\mathsf{de}} \not\subseteq \mathscr{C}_{\mathsf{d}}$.

4.1.12. Define an interpreter (ϕ, S) for the class PR, where ϕ is a total recursive function and S is an elementary function. Prove that the unary predicate P such that

$$P(x) \equiv \phi(\mathsf{S}(x, x)) = 1$$

is not primitive recursive.

4.1.13. Prove that $\mathsf{PR}_\mathsf{d} \subseteq \mathsf{RC}_\mathsf{d} = \mathsf{PR}_{\mathsf{de}} \cap \mathsf{PR}_{\mathsf{du}}$.

4.1.14. Let $\mathscr{C}$ be a FRS with internal interpreter (ϕ, S), and consider sets C and D such that

$$y \in C \equiv y \text{ is } (\phi, \mathsf{S})\text{-index for a total function}$$

$$y \in C \equiv y \text{ is } (\phi, \mathsf{S})\text{-index for a finite function.}$$

Prove that neither $\bar{C}$ nor $\bar{D}$ is partially $\mathscr{C}$-decidable.

Notes

Interpreters are a traditional element of computability theory and usually are introduced via some form of machine computation and gödelization. The characterization of this process as reflexivity was proposed in Wagner [31]. The underlying idea is more general and applies, for example, to the semantics of the lambda-calculus. Structures of this form has been called reflexive domains, because they require a partial order structure (see Sanchis [22, 23]).

The idea of taking reflexivity as a primitive construction and deriving recursive operations from there is due to Kleene [12]. This approach has had a considerable influence on the development of abstract computability and generalized recursion theory (see Moschovakis [15], Friedman [5], and Fenstad [3]) and has proved to be useful in the development of ordinary recursive function theory (see Hinman [8] and Tourlakis [29]).

§2. A Universal Interpreter

In this section we prove that whenever $\mathscr{C} = \mathsf{RC}(\mathscr{C}')$, and $\mathscr{C}'$ is finite, then $\mathscr{C}$ is a FRS. From Theorem 4.1.1 it follows that we need to consider only the case where $\mathscr{C} = \mathsf{RC}(c)$ and c is a unary function. The interpreter for $\mathsf{RC}(c)$ is derived by recursion in c, which is actually uniform in c; hence in some sense it is universal, that is, valid for any c. Later we shall go a step further and introduce a truly universal recursive function, from which interpreters for $\mathsf{RC}(c)$ can be obtained by specialization.

Recursion in c is used to derive a binary function δ_c, from which an internal interpreter $(\phi_c, \mathsf{S}_\delta)$ is obtained by substitution. Note that the input function S_δ is independent of c; in fact, it is the elementary strict function given by

$$\mathsf{S}_\delta(z, y) = (2 \times \mathsf{exp}(3, z) \times \mathsf{exp}(5, y)) \dot{-} 1.$$

The function δ_c is given by the following recursive equations:

$$
\begin{array}{lll}
(0) & \delta_c(z, x) \simeq \delta_c([x]_1, [x]_2) & \text{if } (z)_0 = 0 \\
(1) & \simeq \delta_c((z)_1, \langle (z)_2 \rangle \square x) & \text{if } (z)_0 = 1 \\
(2) & \simeq \mathsf{S}_\delta([x]_1, [x]_2) & \text{if } (z)_0 = 2 \\
(3) & \simeq (z)_1 & \text{if } (z)_0 = 3 \\
(4) & \simeq [x]_1 & \text{if } (z)_0 = 4 \\
(5) & \simeq [x]_1 + 1 & \text{if } (z)_0 = 5 \\
(6) & \simeq [x]_1 \dot{-} 1 & \text{if } (z)_0 = 6 \\
(7) & \simeq \mathsf{cd}([x]_1, [x]_2, [x]_3) & \text{if } (z)_0 = 7 \\
(8) & \simeq \delta_c((z)_1, x\S) & \text{if } (z)_0 = 8 \\
(9) & \simeq \delta_c((z)_1, \S x) & \text{if } (z)_0 = 9 \\
(10) & \simeq \delta_c((z)_1, x \square \langle \delta_c((z)_2, x) \rangle) & \text{if } (z)_0 = 10 \\
(11) & \simeq c([x]_1) & \text{if } (z)_0 = 11.
\end{array}
$$

This recursion involves only elementary functions and the function c, so it follows that δ_c is recursive in c. To define the interpreter we consider the function $\phi_c(z) \simeq \delta_c(z, 0)$. Let $\mathscr{C}_\delta$ be the class that contains all k-ary functions h, $k \geq 0$, such that there is a number z and

$$h(\mathbf{x}) \simeq \delta_c(\mathsf{z}, \langle \mathbf{x} \rangle).$$

Theorem 4.2.1. *The class* $\mathscr{C}_\delta$ *is a* FRS *with internal interpreter* $(\phi_c, \mathsf{S}_\delta)$, and $\mathsf{RC}(c) = \mathscr{C}_\delta$.

PROOF. We prove first that $\mathscr{C}_\delta$ is a FRS with internal interpreter $(\phi_c, \mathsf{S}_\delta)$. Since $\phi_c(z) \simeq \delta_c(0, \langle z \rangle)$ it follows that ϕ_c is in $\mathscr{C}_\delta$, and S_δ is in $\mathscr{C}_\delta$ by equation (2); that is, the number 3, for example, satisfies the relation

$$\mathsf{S}_\delta(z, y) \simeq \delta_c(3, \langle z, y \rangle).$$

Using equation (10) we can prove that $\mathscr{C}_\delta$ is closed under partial substitution, because if h is a k-ary function such that

$$h(\mathbf{x}) \simeq f(\mathbf{x}, g(\mathbf{x})),$$

where f and g are in $\mathscr{C}_\delta$, then there are numbers z_1 and z_2 such that

$$f(\mathbf{x}, y) \simeq \delta_c(\mathsf{z}_1, \langle \mathbf{x}, y \rangle)$$
$$g(\mathbf{x}) \simeq \delta_c(\mathsf{z}_2, \langle \mathbf{x} \rangle).$$

If we take $\mathsf{z} = (\mathsf{exp}(2, 10) \times \mathsf{exp}(3, \mathsf{z}_1) \times \mathsf{exp}(5, \mathsf{z}_2)) \dot{-} 1$ we have

$$\begin{aligned} \delta_c(\mathsf{z}, \langle \mathbf{x} \rangle) &\simeq \delta_c(\mathsf{z}_1, \langle \mathbf{x} \rangle \,\square\, \langle \delta_c(\mathsf{z}_2, \langle \mathbf{x} \rangle) \rangle) \\ &\simeq \delta_c(\mathsf{z}_1, \langle \mathbf{x}, g(\mathbf{x}) \rangle) \\ &\simeq f(\mathbf{x}, g(\mathbf{x})) \\ &\simeq h(\mathbf{x}). \end{aligned}$$

By a similar argument, using equations (8) and (9) we can prove that $\mathscr{C}_\delta$ is closed under adjoining of variables, and hence that it is closed under full substitution. Finally, $\mathscr{C}_\delta$ contains all constant functions by equation (3), and all identity functions by equation (4) and adjoining of variables. It follows that $\mathscr{C}_\delta$ is closed under basic operations.

From equations (5), (6), (7) it follows that $\mathscr{C}_\delta$ contains the functions σ, pd, and cd. It remains to show that $(\phi_c, \mathsf{S}_\delta)$ is an interpreter for $\mathscr{C}_\delta$. First note that it is easy to prove by induction on k that

$$\delta_c(\mathsf{S}_\delta^k(z, \mathbf{x}), 0) \simeq \delta_c(z, \langle \mathbf{x} \rangle).$$

Now assume that h is a k-ary function in $\mathscr{C}_\delta$; hence there is a number z such that

$$h(\mathbf{x}) \simeq \delta_c(\mathsf{z}, \langle \mathbf{x} \rangle).$$

It follows that

$$\begin{aligned} \phi_c(\mathsf{S}_\delta^k(\mathsf{z}, \mathbf{x})) &\simeq \delta_c(\mathsf{S}_\delta^k(\mathsf{z}, \mathbf{x}), 0) \\ &\simeq \delta_c(\mathsf{z}, \langle \mathbf{x} \rangle) \\ &\simeq h(\mathbf{x}), \end{aligned}$$

and this means that z is a $(\phi, \mathsf{S}_\delta)$-index for the function h.

From the definition of $\mathscr{C}_\delta$, and noting that δ_c is recursive in c, it follows that $\mathscr{C}_\delta \subseteq \mathsf{RC}(c)$. On the other hand, $\mathscr{C}_\delta$ is closed under recursive operations and constains c (equation (11)), so it follows that $\mathsf{RC}(c) \subseteq \mathscr{C}_\delta$. We have proved that $\mathsf{RC}(c) = \mathscr{C}_\delta$, hence $\mathsf{RC}(c)$ is a FRS with internal interpreter $(\phi_c, \mathsf{S}_\delta)$. □

Now we turn our attention to the evaluation of the function δ_c, which is determined by the recursive specification given above in the manner explained in Chapter 2. We recall that the evaluation is given by a deterministic stack algorithm, which is encoded via several numerical functions, one of them being the transition function tr, which involves elementary operations and partial definition by cases. We shall show that partial definition by cases is not necessary in the case of the function δ_c, which can be encoded using primitive recursive operations and unbounded minimalization with functions.

Theorem 4.2.2. *Let $\mathscr{C}$ be a class of functions closed under primitive recursive operations, and let h be a k-ary function given by a partial definition by cases of the form*

$$\begin{aligned} h(\mathbf{x}) &\simeq f_1(\mathbf{x}) \quad \textit{if } P(\mathbf{x}) \\ &\simeq f_2(\mathbf{x}) \quad \textit{otherwise}, \end{aligned}$$

where P is a $\mathscr{C}$-decidable predicate, f_1 is a total $\mathscr{C}$-computable function, and f_2 is a $\mathscr{C}$-computable function. Then h is also $\mathscr{C}$-computable.

PROOF. Let h' be the $(k+1)$-ary function introduced using basic primitive recursion in the form

$$\begin{aligned} h'(\mathbf{x}, 0) &\simeq f_1(\mathbf{x}) \\ h'(\mathbf{x}, y+1) &\simeq f_2(\mathbf{x}). \end{aligned}$$

Clearly h' is $\mathscr{C}$-computable, and since f_1 is total it follows that

$$h(\mathbf{x}) \simeq h'(\mathbf{x}, \chi_P(\mathbf{x}));$$

hence h is $\mathscr{C}$-computable □

Corollary 4.2.2.1. *Let $\mathscr{C}$ be a class of functions closed under primitive recursive operations, and let h be a k-ary function given by a partial definition by cases of the form*

$$\begin{aligned} h(\mathbf{x}) &\simeq f_1(\mathbf{x}) \qquad \textit{if } P_1(\mathbf{x}) \\ &\ \ \vdots \qquad\qquad\quad \vdots \\ &\simeq f_m(\mathbf{x}) \qquad \textit{if } P_m(\mathbf{x}) \\ &\simeq f_{m+1}(\mathbf{x}) \quad \textit{otherwise}, \end{aligned}$$

where $m \geq 1$, $P_1, \ldots, P_m$ are $\mathscr{C}$-decidable predicates, $f_1, \ldots, f_m$ are total $\mathscr{C}$-computable functions, and f_{m+1} is a $\mathscr{C}$-computable function. Then h is also $\mathscr{C}$-computable.

PROOF. By induction on m = the number of cases. If $m = 1$ we apply Theorem 4.2.2. If $m = m_1 + 1$ we write the specification in the form

$$\begin{aligned} h(\mathbf{x}) &\simeq f_1(\mathbf{x}) \quad \text{if } P_1(\mathbf{x}) \\ &\simeq h'(\mathbf{x}) \quad \text{otherwise}, \end{aligned}$$

where h' is obtained using partial definition with m_1 cases. By the induction hypothesis it follows that h' is $\mathscr{C}$-computable; hence by Theorem 4.2.2 it follows that h is $\mathscr{C}$-computable. □

Theorem 4.2.3. *Let c be a $\mathscr{C}$-computable function, where $\mathscr{C}$ is a class of functions closed under primitive recursive operations and unbounded minimalization with functions. Then* $\mathsf{RC}(c) \subseteq \mathscr{C}$.

PROOF. First we assume that the function c is unary and note that it is sufficient to show that the function δ_c is $\mathscr{C}$-computable. We refer again to the evaluation algorithm for δ_c and note that the specification of the function tr can be put in the form considered in Corollary 4.2.2.1. It follows that tr is $\mathscr{C}$-computable; hence δ_c is $\mathscr{C}$-computable. If c is not unary we use Theorem 4.1.1. □

Corollary 4.2.3.1. *Let c be an arbitrary function. Then* $\mathsf{RC}(c)$ *is the smallest class that contains c and is closed under primitive recursive operations and unbounded minimalization with functions.*

PROOF. Immediate from the definition and Theorem 4.2.3. □

The specification of the function δ_c is uniform in c, and this suggests that we replace c in the recursion by an extra argument in the function being specified, which becomes independent of c. Now in general it is not possible to replace a function, which is an infinite object, by a number, which is finite, but in this case it is possible because the evaluation of $\delta_c(z, x)$ involves only a finite number of values of c.

We use here a modification of the technique applied in Chapter 3 to encode finite information about predicates. We say that a number u *partially represents* a function c, if whenever $(u)_y \neq 0$ then $c(y) \simeq (u)_y \dot{-} 1$. Formally, we introduce a unary predicate pr_c such that

$$\mathsf{pr}_c(u) \equiv \forall y < \ell u\colon (u)_y = 0 \vee \mathsf{G}_c(y, (u)_y \dot{-} 1).$$

To manipulate this representation we need several functions and predicates, which in general can be taken to be elementary. For example, note the two elementary predicates cn and ex such that

$$\mathsf{cn}(u, u') \equiv \forall y < \ell u\colon (u)_y = 0 \vee (u')_y = 0 \vee (u)_y = (u')_y$$

$$\mathsf{ex}(u, u') \equiv \forall y < \ell u\colon (u)_y = 0 \vee (u)_y = (u')_y.$$

Clearly, cn is an equivalence relation such that whenever $\mathsf{pr}_c(u)$ and $\mathsf{pr}_c(u')$ then $\mathsf{cn}(u, u')$. On the other hand, $\mathsf{ex}(u, u')$ implies $\mathsf{cn}(u, u')$. Furthermore, if $\mathsf{pr}_c(u')$ and $\mathsf{ex}(u, u')$ then $\mathsf{pr}_c(u)$.

In applications of this representation we need a decoding function, which can be applied to a number u to get values of a function c whenever u partially represents c. We put

$$\mathsf{dc}(u, y) \simeq \mu v\colon (u)_y = v + 1,$$

which is a nonelementary recursive function.

The idea is to replace the binary function δ_c by a function δ that is independent of c (in fact it is a recursive function), has an extra argument u, and works exactly as $\delta_c(z, x)$ whenever u partially represents c. Hence, δ is a 3-ary function, and $\delta(z, x, u)$ is given by a recursive specification derived from the specification for δ_c. For example, equations (0) and (1) become

$$(0') \quad \delta(z, x, u) \simeq \delta([x]_1, [x]_2, u) \qquad \text{if } (z)_0 = 0$$

$$(1') \quad \simeq \delta((z)_1, \langle (z)_2 \rangle \,\square\, x, u) \quad \text{if } (z)_0 = 1.$$

The same transformation applies to equations (2), ..., (10), which become (2′), ..., (10′). The only change here is to replace every occurrence of δ_c by δ with the extra argument u. On the other hand, equation (11) becomes

$$(11') \quad \delta(z, x, u) \simeq \mathsf{dc}(u, [x]_1) \quad \text{if } (z)_0 \geq 11.$$

Since the recursive specification for δ involves only recursive functions, it follows that δ is a recursive function. We know that $\mathsf{RC} = \mathsf{EL}_{\mathsf{d}\#}$ (Theorem 3.2.4 with $\mathscr{C} = \varnothing$); hence there is a 5-ary elementary predicate $\mathfrak{H}$ such that

$$\mathsf{G}_\delta(z, x, u, v) \equiv \exists y \mathfrak{H}(z, x, u, v, y).$$

Theorem 4.2.4. *Let c be an arbitrary unary function. Then*

(i) $\delta_c(z, x) \simeq v \equiv \exists u\colon \mathsf{pr}_c(u) \wedge \delta(z, x, u) \simeq v.$

(ii) *A k-ary function h is recursive in c if and only if there is a number z such that*

$$h(\mathbf{x}) \simeq v \equiv \exists u\colon \mathsf{pr}_c(u) \wedge \delta(z, \langle \mathbf{x} \rangle, u) \simeq v.$$

(iii) *A k-ary predicate P is partially recursive in c if and only if there is a number z such that*

$$P(\mathbf{x}) \equiv \exists u \exists y\colon \mathsf{pr}_c(y) \wedge \mathfrak{H}(z, \langle \mathbf{x} \rangle, u, 0, y).$$

(iv) *A k-ary function h is recursive if and only if there is a number z such that*

$$h(\mathbf{x}) \simeq \delta(z, \langle \mathbf{x} \rangle, 0).$$

(v) *A k-ary predicate P is recursively enumerable (= partially recursive) if and only if there is a number z such that*

$$P(\mathbf{x}) \equiv \exists y \mathfrak{H}(z, \langle \mathbf{x} \rangle, 0, 0, y)$$

Proof. In order to prove (i) we assume first that $\delta_c(z, x) \simeq v$, which means that the evaluation algorithm halts with output v. The evaluation involves only a finite number of values of c, which can be encoded in a number u; hence the evaluation of $\delta(z, x, u)$ also halts with output v. Conversely, if $\delta(z, x, u) \simeq v$, where $\mathsf{pr}_c(u)$ holds, the evaluation algorithm involves steps of the form $\mathsf{dc}(u, y) \simeq v'$, where in each case we have $c(y) \simeq v'$; hence the evaluation of $\delta_c(z, x)$ also halts with output v.

Now (ii) follows from (i) and Theorem 4.2.1. To prove (iii) note that a predicate P is partially recursive in c if and only if ψ_P is recursive in c. If P is partially recursive in c, then there is a number $\mathbf{z}$ such that

$$P(\mathbf{x}) \equiv \exists u\colon \mathsf{pr}_c(u) \wedge \delta(\mathbf{z}, \langle \mathbf{x} \rangle, u) \simeq 0$$
$$\equiv \exists u \exists y\colon \mathsf{pr}_c(u) \wedge \mathfrak{H}(\mathbf{z}, \langle \mathbf{x} \rangle, u, 0, y).$$

To prove the converse we assume there is a number z that satisfies the condition and prove that P is partially recursive in c. Note that although $\mathsf{RC}_{\mathsf{pd}}(c)$ is closed under conjunction, in general it is not closed under existential quantification. We get around this difficulty by noting that

$$\psi_P(\mathbf{x}) \simeq \mu y(\delta_c(\mathbf{z}, \langle \mathbf{x} \rangle) + y \simeq 0).$$

Parts (iv) and (v) follow from (ii) and (iii) noting that $\mathsf{RC} = \mathsf{RC}(\mathsf{UD}_1)$, and $\mathsf{pr}_{\mathsf{UD}_1}(u)$ holds if and only if $u = 0$. □

If c is a total function the preceding results can be improved, because in this case we can encode all the values of c up to some value y. Formally, we put

$$\bar{c}(y) = \left(\prod_{v<y} \mathsf{exp}(\mathsf{p}_v, c(v) + 1) \right) \dot{-} 1$$

and it follows that $\mathsf{pr}_c(\bar{c}(y))$ holds for all y. Furthermore, $\bar{c}(0) = 0$, and whenever $y > 0$ then

$$\bar{c}(y) = \langle c(0), \ldots, c(y \dot{-} 1) \rangle;$$

hence $\ell(\bar{c}(y)) = y$, $[\bar{c}(y)]_1 = c(0), \ldots, [\bar{c}(y)]_y = c(y \dot{-} 1)$.

Theorem 4.2.5. *Let c be an arbitrary total unary function. Then*

(i) $\delta_c(z, x) \simeq v \equiv \exists w \delta(z, x, \bar{c}(w)) \simeq v$.

(ii) *A k-ary function h is recursive in c if and only if there is a number z such that*

$$h(\mathbf{x}) \simeq v \equiv \exists w \delta(z, \langle \mathbf{x} \rangle, \bar{c}(w)) \simeq v.$$

(iii) *A k-ary predicate P is recursively enumerable in c if and only if there is a number z such that*

$$P(\mathbf{x}) \equiv \exists w \exists y \mathfrak{H}(z, \langle \mathbf{x} \rangle, \bar{c}(w), 0, y).$$

Proof. Immediate from Theorem 4.2.4, noting that whenever $\delta(z, x, u) \simeq v$, and $\mathsf{ex}(u, u')$ then $\delta(z, x, u') \simeq v$, a property that can be derived easily from the specification of the function δ. □

The formulation of part (iii) in the preceding theorem can be given a more compact form by contracting the two existential quantifiers. We introduce a unary elementary function f such that

$$f(y) = \left(\prod_{v<(\ell y)_0} \mathsf{exp}(\mathsf{p}_v, (y)_v) \right) \dot{-} 1,$$

and it follows that $f(\bar{c}(y)) = \bar{c}((y)_0)$. Now we introduce a 3-ary elementary predicate $\mathfrak{T}$ such that

$$\mathfrak{T}(z, x, y) \equiv \mathfrak{H}(z, x, f(y), 0, (\ell y)_1).$$

Corollary 4.2.5.1. *Let c be an arbitrary total unary function. A k-ary predicate P is recursively enumerable in c if and only if there is a number z such that*

$$P(\mathbf{x}) \equiv \exists y \mathfrak{T}(z, \langle \mathbf{x} \rangle, \bar{c}(y)).$$

PROOF. The condition is clearly sufficient. To prove it is necessary we assume P is recursively enumerable in c; hence there are numbers w and y such that condition (iii) of Theorem 4.2.5 is satisfied. Take y' such that

$$y' = (\mathsf{exp}(2, w) \times \mathsf{exp}(3, y)) \dot{-} 1,$$

and it follows that $\mathfrak{T}(z, \langle \mathbf{x} \rangle, \bar{c}(y'))$ holds. □

Let $\mathscr{C}$ be a class of functions. We recall that $\mathscr{C}_{\mathsf{pdd}}$ is the class that contains the characteristic functions of the predicates, which are partially $\mathscr{C}$-decidable, and $\mathscr{C}_{\mathsf{pdpd}}$ is the class that contains the partial characteristic functions of the same predicates. Clearly, we have $\mathscr{C}_{\mathsf{pdpd}} \subseteq \mathscr{C}$, but the inclusion $\mathscr{C}_{\mathsf{pdd}} \subseteq \mathscr{C}$ fails in general.

We set $\mathscr{C}_{\mathsf{j}} = \mathsf{RC}(\mathscr{C}_{\mathsf{pdd}})$ and $\mathscr{C}_{\mathsf{r}} = \mathsf{RC}(\mathscr{C}_{\mathsf{pdpd}})$. The class $\mathscr{C}_{\mathsf{j}}$ is called the *jump* of $\mathscr{C}$, and the class $\mathscr{C}_{\mathsf{r}}$ is called the *retraction* of $\mathscr{C}$. Note that $\mathscr{C}_{\mathsf{j}}$ is always an e-total class. On the other hand, $\mathscr{C}_{\mathsf{r}}$ is e-total if and only if $\mathscr{C}_{\mathsf{pd}} \subseteq \mathsf{RC}_{\mathsf{de}}$.

Theorem 4.2.6. *Let $\mathscr{C}$ be a class of functions. Then*

(i) $\mathscr{C}_{\mathsf{pd}} \subseteq \mathscr{C}_{\mathsf{jd}}$ *and* $\mathscr{C}_{\mathsf{pd}} \subseteq \mathscr{C}_{\mathsf{rpd}}$.
(ii) $\mathscr{C}_{\mathsf{j}}$ *and* $\mathscr{C}_{\mathsf{r}}$ *are closed under recursive operations.*
(iii) *If* $\mathscr{C}_{\mathsf{pd}} \subseteq \mathscr{C}'_{\mathsf{d}}$, *where* $\mathscr{C}'$ *is closed under recursive operations, then* $\mathscr{C}_{\mathsf{j}} \subseteq \mathscr{C}'$.
(iv) *If* $\mathscr{C}_{\mathsf{pd}} \subseteq \mathscr{C}'_{\mathsf{pd}}$, *where* $\mathscr{C}'$ *is closed under recursive operations, then* $\mathscr{C}_{\mathsf{r}} \subseteq \mathscr{C}'$.

PROOF. Parts (i) and (ii) are trivial from the definitions. To prove (iii) note that $\mathscr{C}_{\mathsf{pdd}} \subseteq \mathscr{C}'_{\mathsf{dd}} \subseteq \mathscr{C}'$; hence $\mathsf{RC}(\mathscr{C}_{\mathsf{pdd}}) \subseteq \mathscr{C}'$ since $\mathscr{C}'$ is closed under recursive operations. A similar argument applies to (iv). □

Corollary 4.2.6.1. *Let $\mathscr{C}$ and $\mathscr{C}'$ be classes such that $\mathscr{C}_{\mathsf{pd}} \subseteq \mathscr{C}'_{\mathsf{pd}}$. Then $\mathscr{C}_{\mathsf{j}} \subseteq \mathscr{C}'_{\mathsf{j}}$ and $\mathscr{C}_{\mathsf{r}} \subseteq \mathscr{C}'_{\mathsf{r}}$.*

PROOF. Immediate from Theorem 4.2.6 (i), (iii), and (iv). □

Theorem 4.2.7. *Let $\mathscr{C}$ be a class closed under recursive operations. Then*

(i) $\mathscr{C}_{\mathsf{r}} \subseteq \mathscr{C} \subseteq \mathscr{C}_{\mathsf{pd\#}} \subseteq \mathscr{C}_{\mathsf{j}}$.
(ii) $\mathscr{C}_{\mathsf{pd}} = \mathscr{C}_{\mathsf{rpd}}$, $\mathscr{C}_{\mathsf{pde}} = \mathscr{C}_{\mathsf{rpde}}$, *and* $\mathscr{C}_{\mathsf{pd\#}} = \mathscr{C}_{\mathsf{rpd\#}}$.

PROOF. The inclusion $\mathscr{C}_r \subseteq \mathscr{C}$ follows from Theorem 4.2.6 (iv) (with $\mathscr{C}' = \mathscr{C}$). On the other hand,

$$\mathscr{C} \subseteq \mathscr{C}_{\mathsf{pd}\#} \subseteq \mathscr{C}_{\mathsf{jd}\#} = \mathscr{C}_{\mathsf{j}},$$

since $\mathscr{C}_{\mathsf{j}}$ is e-total.

The equality $\mathscr{C}_{\mathsf{pd}} = \mathscr{C}_{\mathsf{rpd}}$ follows from (i) and Theorem 4.2.6 (i). The rest of part (ii) is immediate. □

Corollary 4.2.7.1. *Let $\mathscr{C}$ and $\mathscr{C}'$ be classes closed under recursive operations. The following conditions are equivalent:*

(i) $\mathscr{C}_r \subseteq \mathscr{C}'$.
(ii) $\mathscr{C}_{\mathsf{rpd}} \subseteq \mathscr{C}'_{\mathsf{pd}}$.
(iii) $\mathscr{C}_{\mathsf{pd}} \subseteq \mathscr{C}'_{\mathsf{pd}}$.
(iv) $\mathscr{C}_{\mathsf{pd}} \subseteq \mathscr{C}'_{\mathsf{rpd}}$.
(v) $\mathscr{C}_r \subseteq \mathscr{C}'_r$.

PROOF. The implications from (i) to (ii), (ii) to (iii), and (iii) to (iv) are trivial using Theorem 4.2.7 (ii). The implication from (iv) to (v) follows from Theorem 4.2.6 (iv). The implication from (v) to (i) follows from Theorem 4.2.7 (i). □

EXAMPLE 4.2.1. If $\mathscr{C}$ is a class closed under recursive operations it follows that $\mathsf{RC} \subseteq \mathscr{C}$; hence $\mathsf{RC} \subseteq \mathsf{RC}_{\mathsf{j}} \subseteq \mathscr{C}_{\mathsf{j}}$. On the other hand, $\mathsf{RC} \neq \mathsf{RC}_{\mathsf{j}}$ because $\mathsf{RC}_{\mathsf{de}} \subseteq \mathsf{RC}_{\mathsf{jd}}$ and $\mathsf{RC}_{\mathsf{de}}$ is not closed under negation (see Example 4.1.6).

Theorem 4.2.8. *Let $\mathscr{C}$ be a class of functions, and assume there is a set K that is partially $\mathscr{C}$-decidable, and furthermore that whenever P is a k-ary partially $\mathscr{C}$-decidable predicate there is a total k-ary recursive function f such that*

$$P(\mathbf{x}) \equiv f(\mathbf{x}) \in K.$$

Then $\mathscr{C}_{\mathsf{j}} = \mathsf{RC}(\chi_K)$ and $\mathscr{C}_r = \mathsf{RC}(\psi_K)$.

PROOF. From the assumption it follows that

$$\chi_P(\mathbf{x}) = \chi_K(f(\mathbf{x}))$$
$$\psi_P(\mathbf{x}) \simeq \psi_K(f(\mathbf{x}));$$

hence $\mathscr{C}_{\mathsf{pdd}} \subseteq \mathsf{RC}(\chi_K)$ and $\mathscr{C}_{\mathsf{pdpd}} \subseteq \mathsf{RC}(\psi_K)$. Furthermore, if $\mathscr{C}'$ is closed under recursive operations and $\mathscr{C}_{\mathsf{pdd}} \subseteq \mathscr{C}'_{\mathsf{d}}$, it follows that χ_K is $\mathscr{C}'$-computable and hence that $\mathsf{RC}(\chi_K) \subseteq \mathscr{C}'$. From Theorem 4.2.6 it follows that $\mathscr{C}_{\mathsf{j}} = \mathsf{RC}(\chi_K)$. By a similar argument, using also Theorem 4.2.6, it follows that $\mathscr{C}_r = \mathsf{RC}(\psi_K)$. □

Corollary 4.2.8.1. *If $\mathscr{C}$ is a* FRS *then* $\mathscr{C}_{\mathsf{j}}$ and $\mathscr{C}_r$ are also FRSs.

PROOF. Immediate from Theorem 4.2.8 and Theorem 4.1.6. □

From the preceding results it follows that whenever c is a unary function then there is a unary function c' such that $\mathsf{RC}_{\mathsf{j}}(c) = \mathsf{RC}(c')$. The function c' is by no means unique and can be obtained in different ways. The method of Theorem 4.2.8 shows that we can always take c' to be the characteristic function of some set partially recursive in c. We proceed now to give a more uniform construction of c' which works for any function c and depends on the universal function δ.

If c is a unary function then J_c is the set such that

$$z \in \mathsf{J}_c \equiv \delta_c(z, \langle z \rangle)\downarrow.$$

It is clear that J_c is partially recursive in c, because $\psi_{\mathsf{J}_c}(z) \simeq \delta_c(z, \langle z \rangle) \times 0$. On the other hand, the set J_c satisfies the conditions of Theorem 4.2.8. To show that this is the case, assume P is a k-ary predicate partially recursive in c, and $(\phi_c, \mathsf{S}_\delta)$ is the universal interpreter for $\mathsf{RC}(c)$. Note that $\phi_c^1(z, z) \simeq \delta_c(z, \langle z \rangle)$. Let h be the $k + 1$-ary function such that

$$\begin{aligned} h(\mathbf{x}, y) &\simeq \psi_P(\mathbf{x}) \\ &\simeq \phi_c^1(g(\mathbf{x}), y), \end{aligned}$$

where the function h is recursive in c and g is the total recursive function given by the abstraction property. It follows from this that

$$\begin{aligned} P(\mathbf{x}) \equiv h(\mathbf{x}, g(\mathbf{x}))\downarrow &\equiv \phi_c^1(g(\mathbf{x}), g(\mathbf{x}))\downarrow \\ &\equiv g(\mathbf{x}) \in \mathsf{J}_c. \end{aligned}$$

We set $c_{\mathsf{j}} = \chi(\mathsf{J}_c)$ and $c_{\mathsf{r}} = \psi(\mathsf{J}_c)$ and it follows that $\mathsf{RC}_j(c) = \mathsf{RC}(c_{\mathsf{j}})$ and $\mathsf{RC}_{\mathsf{r}}(c) = \mathsf{RC}(c_{\mathsf{r}})$.

It is useful here to recall the characterization of the function δ_c in terms of the recursive function δ. In relation with the set J_c we get

$$z \in \mathsf{J}_c \equiv \exists u\colon \mathsf{pr}_c(u) \wedge \delta(z, \langle z \rangle, u)\downarrow.$$

When the function c is total we can express this relation in the sharper form

$$z \in \mathsf{J}_c \equiv \exists w \delta(z, \langle z \rangle, \bar{c}(w))\downarrow.$$

EXAMPLE 4.2.2. If $\mathscr{C}$ is closed under recursive operations and $\mathscr{C}_{\mathsf{pd}}$ is closed under existential quantification (e.g., if $\mathscr{C} = \mathsf{RC}$), then $\mathscr{C}_{\mathsf{j}}$ has special closure properties. For example, if Q is a $(k + 1)$-ary predicate that is partially $\mathscr{C}$-decidable, then we can introduce a k-ary function h such that

$$\begin{aligned} h(\mathbf{x}) &\simeq \mu y Q(\mathbf{x}, y) && \text{if } \exists y Q(\mathbf{x}, y) \\ &\simeq 0 && \text{otherwise.} \end{aligned}$$

Since Q and the existential quantification of Q are both $\mathscr{C}_{\mathsf{j}}$-decidable, it follows that the function h is $\mathscr{C}_{\mathsf{j}}$-computable. □

Theorem 4.2.9. *Let c be a total unary function. There is a total unary function c' such that*

$$\mathsf{RC}(\mathsf{RC}_\mathsf{j} \cup \{c'\}) = \mathsf{RC}(c'_\mathsf{j}) = \mathsf{RC}(\mathsf{RC}_\mathsf{j} \cup \{c\}).$$

PROOF. We introduce two predicates R and Q such that

$$R(v) \equiv \forall y < \ell v\colon (v)_y \neq 0$$

$$Q(z, u, v) \equiv \delta(z, \langle z \rangle, v)\downarrow \wedge R(v) \wedge \mathsf{ex}(u, v).$$

It is clear that R is elementary and Q is recursively enumerable; hence both are RC_j-decidable. Using primitive recursion we specify a unary function g such that

$$g(0) \simeq 0$$

$$g(z + 1) \simeq \mu v Q(z, g(z) \,\Box\, \langle c(z) \rangle, v) \quad \text{if } \exists v Q(z, g(z) \,\Box\, \langle c(z) \rangle, v)$$

$$\simeq g(z) \,\Box\, \langle c(z) \rangle \quad \text{otherwise.}$$

We set $c'(z) = (g(z + 1))_z \dot{-} 1 = [g(z + 1)]_{z+1}$, and it follows that g is recursive in $\mathsf{RC}_\mathsf{j} \cup \{c\}$, c' is recursive in g, and hence c' is recursive in $\mathsf{RC}_\mathsf{j} \cup \{c\}$. This means that $\mathsf{RC}(\mathsf{RC}_\mathsf{j} \cup \{c'\}) \subseteq \mathsf{RC}(\mathsf{RC}_\mathsf{j} \cup \{c\})$.

To prove the converse of the preceding inclusion we note that $c(z) = c'(\ell g(z))$, and using this relation we can rewrite the specification of the function g by replacing all occurrences of $c(z)$. This gives another primitive recursive specification of g involving c' rather than c. It follows that g is recursive in $\mathsf{RC}_\mathsf{j} \cup \{c'\}$, and since c is recursive in g it follows that $RC(RC_\mathsf{j} \cup \{c\}) \subseteq \mathsf{RC}(\mathsf{RC}_\mathsf{j} \cup \{c'\})$.

To complete the proof we need to prove only that

$$\mathsf{RC}(\mathsf{RC}_\mathsf{j} \cup \{c'\}) \subseteq \mathsf{RC}(c'_\mathsf{j}) \subseteq \mathsf{RC}(\mathsf{RC}_\mathsf{j} \cup \{c\}).$$

The first inclusion is trivial. Regarding the second we note the relation

$$z \in \mathsf{J}_{c'} \equiv \exists w \delta(z, \langle z \rangle, \bar{c}'(w))\downarrow$$

$$\equiv \delta(z, \langle z \rangle, g(z + 1))\downarrow,$$

which means that c'_j is recursive in $\mathsf{RC}_\mathsf{j} \cup \{g\}$. Since g is recursive in $\mathsf{RC}_\mathsf{j} \cup \{c\}$, we have $\mathsf{RC}(c'_\mathsf{j}) \subseteq \mathsf{RC}(\mathsf{RC}_\mathsf{j} \cup \{c\})$. □

Corollary 4.2.9.1. *Let $\mathscr{C}$ be a* FRS *that is* e-*total and* $\mathsf{RC}_\mathsf{j} \subseteq \mathscr{C}$. *There is a* FRS $\mathscr{C}'$ *such that:*

(i) $\mathscr{C}'$ *is* e-*total.*
(ii) $\mathscr{C}'_\mathsf{j} = \mathscr{C}$.
(iii) $\mathsf{RC}_\mathsf{j} \nsubseteq \mathscr{C}'$.

PROOF. We can put $\mathscr{C} = \mathsf{RC}(c)$, where c is a unary total function. By Theorem 4.2.9 there is a total c' such that if $\mathscr{C}' = \mathsf{RC}(c')$ then

$$\mathsf{RC}(\mathsf{RC}_\mathsf{j} \cup \mathscr{C}') = \mathscr{C}'_\mathsf{j} = \mathsf{RC}(\mathsf{RC}_\mathsf{j} \cup \mathscr{C}) = \mathscr{C}.$$

If $\mathsf{RC}_\mathsf{j} \subseteq \mathscr{C}'$ we have $\mathscr{C}' = \mathscr{C}'_\mathsf{j}$, which is a contradiction. □

Theorem 4.2.10. *Let $\mathscr{C}$ be a* FRS. *If $\mathscr{C}$ is* e-*total then* $\mathscr{C}_\mathsf{j} = \mathscr{C}_{\mathsf{du\#}}$.

PROOF. In one direction we compute

$$\mathscr{C}_{\mathsf{de}} = \mathscr{C}_{\mathsf{pd}} \subseteq \mathscr{C}_{\mathsf{jd}}$$

$$\mathscr{C}_{\mathsf{du}} \subseteq \mathscr{C}_{\mathsf{jd}}$$

$$\mathscr{C}_{\mathsf{du\#}} \subseteq \mathscr{C}_{\mathsf{jd\#}} = \mathscr{C}_\mathsf{j}.$$

To prove the converse note that $\mathscr{C}_{\mathsf{du\#}} = \mathsf{RC}(\mathscr{C}_{\mathsf{dud}})$ by Theorem 3.1.6 (iv); hence it is sufficient to show that whenever c is a unary $\mathscr{C}$-computable function then c_j is recursive in $\mathscr{C}_{\mathsf{dud}}$. Since $\mathscr{C}$ is e-total it follows that J_c is in $\mathscr{C}_{\mathsf{de}}$; hence $\bar{J_c}$ is in $\mathscr{C}_{\mathsf{du}}$ and c_j is recursive in $\mathscr{C}_{\mathsf{dud}}$. □

Now we study the jump operator from another point of view. We recall that δ is a nontotal universal recursive function, and there is an elementary predicate $\mathfrak{H}$ such that

$$\delta(z, x, u) \simeq v \equiv \exists y \mathfrak{H}(z, x, u, v, y).$$

From this expression we derive a total 4-ary elementary function $\delta^\dagger$ such that

$$\delta^\dagger(z, x, u, w) \simeq \mu v < w \exists y < w \mathfrak{H}(z, x, u, v, y)$$

It is clear that the function $\delta^\dagger$ satisfies the condition $\delta^\dagger(z, x, u, w) \le w$.

Note that if $\delta(z, x, u) \simeq v$ it follows from the analysis above that there is a number y such that $\mathfrak{H}(z, x, u, v, y)$. If we take $w = \max\{v, y\} + 1$ it follows that whenever $w' \ge w$ then $\delta^\dagger(z, x, u, w') \simeq v$.

On the other hand, if $\delta(z, x, u)$ is undefined it is clear that the relation $\delta^\dagger(z, x, u, w) = w$ holds for all w.

We interpret the extra argument w in the function $\delta^\dagger$ as a counter in the evaluation of $\delta(z, x, u)$. If $w = 0$ we have the start of the evaluation and $\delta^\dagger(z, x, u, 0) = 0$. As long as the evaluation is undefined we have $\delta^\dagger(z, x, u, w) = w$. If $\delta(z, x, u)$ is defined there is a least value w such that $\delta^\dagger(z, x, u, w) = v < w$, and this means that $\delta(z, x, u) \simeq v$. From now on, that is, as long as $w' \ge w$, we have $\delta^\dagger(z, x, u, w') = v$.

The relation between δ and $\delta^\dagger$ described in the preceding paragraph can be used to obtain an expression for δ in terms of $\delta^\dagger$:

$$\delta(z, x, u) \simeq \delta^\dagger(z, x, u, \mu w\colon \delta^\dagger(z, x, u, w) < w).$$

In applications it is convenient to have a more general notion of counter function, which is given in the next definition.

Let h be a k-ary function. A *counter expansion* of h is a total $(k + 1)$-ary function f such that if $h(\mathbf{x}) \simeq v$ then there is a number w such that whenever

$w' \geq w$ then $f(\mathbf{x}, w') = v$. For example, if h is a total function we can take $f(\mathbf{x}, w) = h(\mathbf{x})$, which is a counter expansion of h elementary in h.

Theorem 4.2.11. *Let $\mathscr{C}$ be a class of functions closed under elementary operations, and let $\mathscr{C}'$ be the class of all functions with counter expansion in $\mathscr{C}$. Then $\mathscr{C}'$ is closed under recursive operations.*

Proof. It is clear that $\mathscr{C}'$ contains all the elementary functions, so we need only prove that $\mathscr{C}'$ is closed under full substitution, primitive recursion, unbounded minimalization with functions, and partial definition by cases.

Closure under full substitution is straightforward, for if h is a k-ary function such that

$$h(\mathbf{x}) \simeq f(g_1(\mathbf{x}), \ldots, g_m(\mathbf{x}))$$

and f', g_1', ..., g_m' are $\mathscr{C}$-computable counter expansions of f, g_1, ..., g_m respectively, we introduce

$$h'(\mathbf{x}, w) = f'(g_1'(\mathbf{x}, w), \ldots, g_m'(\mathbf{x}, w), w)$$

and it follows that h' is a $\mathscr{C}$-computable counter expansion of h.

Closure under primitive recursion is slightly more complicated. Consider a specification of a $(k+1)$-ary function h such that

$$h(\mathbf{x}, 0) \simeq f(\mathbf{x})$$

$$h(\mathbf{x}, y+1) \simeq g(\mathbf{x}, y, h(\mathbf{x}, y)),$$

and let f', g' be $\mathscr{C}$-computable counter expansions of f and g. We introduce a $\mathscr{C}$-decidable $(k+3)$-ary predicate Q such that

$$Q(\mathbf{x}, y, w, z) \equiv (z)_0 = f'(\mathbf{x}, w) \wedge \forall v < y\colon (z)_{v+1} = g'(\mathbf{x}, v, (z)_v, w).$$

and a total $(k+2)$-ary function h' such that

$$h'(\mathbf{x}, y, w) = (\mu z < wQ(\mathbf{x}, y, w, z))_y$$

It follows that h' is a $\mathscr{C}$-computable counter expansion of h.

The treatment of unbounded minimalization is similar, using bounded minimalization with the counter expansion and the counter argument as the upper bound. Finally, partial definition by cases reduces to total definition by cases with the counter expansions. □

Theorem 4.2.12. *If c is a total unary function then c_j has a counter expansion that is elementary in c.*

Proof. We recall that c_j can be defined in the form

$$c_j(z) = 0 \quad \text{if } \exists w \delta(z, \langle z \rangle, \bar{c}(w))\downarrow$$

$$= 1 \quad \text{otherwise.}$$

We introduce a binary total function f such that

$$\begin{aligned} f(z, w) &= 0 \quad \text{if } \delta^\dagger(z, \langle z \rangle, \bar{c}(w), w) < w \\ &= 1 \quad \text{otherwise.} \end{aligned}$$

This is a total definition by cases involving elementary functions and c, so f is elementary in c. Clearly, f is a counter expansion of c_{j}. □

Corollary 4.2.12.1. *Let c be a total unary function and h a k-ary function recursive in c_{j}. Then h has a counter expansion elementary in c.*

PROOF. Immediate from Theorem 4.2.11 and Theorem 4.2.12. □

Corollary 4.2.12.2. *Let c be a total unary function and h a total k-ary function. The following conditions are equivalent:*

(i) *h has a counter expansion elementary in c.*
(ii) *h has a counter expansion recursive in c.*
(iii) *h is in the class $\mathsf{RC}_{\mathsf{du}\#}(c)$.*
(iv) *h is recursive in c_{j}.*

PROOF. The implication from (i) to (ii) is trivial. Assume (ii) holds, so f is a counter expansion of h which is recursive in c. Since h is total it follows that

$$\mathsf{G}_h(\mathbf{x}, v) \equiv \exists w \forall w' \colon w' \leq w \vee f(\mathbf{x}, w') = v;$$

hence h is in $\mathsf{RC}_{\mathsf{du}\#}(c)$. The implication from (iii) to (iv) follows from the equality $\mathsf{RC}(c_{\mathsf{j}}) = \mathsf{RC}_{\mathsf{du}\#}(c)$ (Theorem 4.2.10). Finally, the implication from (iv) to (i) follows from Corollary 4.2.12.1. □

EXERCISES

4.2.1. Prove that if $\mathscr{C}$ is a FRS then $\mathscr{C}$ has an internal interpreter where the input function is S_δ.

4.2.2. Let ϕ be the unary function such that $\phi(z) \simeq \delta(z, 0, 0)$. Prove that $(\phi, \mathsf{S}_\delta)$ is an internal interpreter for RC.

4.2.3. Let $\mathscr{C}$ be a class of functions closed under elementary operations, and let $\mathsf{RC} \subseteq \mathscr{C}$. Prove that the following conditions are equivalent:
(a) $\mathscr{C}$ is a FRS.
(b) There is a binary $\mathscr{C}$-computable function ϕ' such that whenever h is a unary $\mathscr{C}$-computable function then there is a number z such that

$$h(x) \simeq \phi'(\mathsf{z}, x).$$

(c) There is a unary $\mathscr{C}$-computable function ϕ such that whenever h is a k-ary $\mathscr{C}$-computable function then there is a k-ary elementary function f such that

$$h(\mathbf{x}) \simeq \phi(f(\mathbf{x})).$$

(d) There is a unary $\mathscr{C}$-computable function ϕ such that whenever h is a k-ary $\mathscr{C}$-computable function then there is a k-ary recursive function f such that

$$h(\mathbf{x}) \simeq \phi(f(\mathbf{x})).$$

4.2.4. Let $\mathscr{C}$ be a FRS. Prove that the following conditions are equivalent:
(a) $\mathscr{C}$ is e-total.
(b) There is a 3-ary predicate Q such that

$$\mathscr{C} = \mathsf{RC}(\chi_Q) = \mathsf{RC}_{\mathsf{pd}\#}(\psi_Q)$$

(c) There is a set A such that

$$\mathscr{C} = \mathsf{RC}(\chi_A) = \mathsf{RC}_{\mathsf{pd}\#}(\psi_A).$$

(d) There is a total unary function c such that $\mathscr{C} = \mathsf{RC}(c)$.

4.2.5. Prove that if $\delta(z, x, u) \simeq v$ and $\mathsf{ex}(u, u')$ then $\delta(z, x, u') \simeq v$.

4.2.6. Let c be a unary function, and assume $\mathsf{RC}(c)$ is closed under enumeration operations. Let P be a k-ary predicate. Prove that the following conditions are equivalent:
(a) P is recursive in c.
(b) There are k-ary elementary functions f and f' such that

$$\chi_P = c_{\mathsf{j}} \circ f \qquad \chi_{\bar{P}} = c_{\mathsf{j}} \circ f'.$$

(c) P and $\bar{P}$ are partially recursive in c.

4.2.7. Prove that if $\mathscr{C}$ is a FRS then $\mathscr{C}_{\mathsf{rr}} = \mathscr{C}_{\mathsf{r}}$ and $\mathscr{C}_{\mathsf{rj}} = \mathscr{C}_{\mathsf{j}}$.

4.2.8. Let $\mathscr{C}$ and $\mathscr{C}'$ be FRSs. Prove the following:
(a) If $\mathscr{C}'$ is closed under enumeration operations, then $\mathscr{C} \subseteq \mathscr{C}'$ if and only if $\mathscr{C}_{\mathsf{r}} \subseteq \mathscr{C}'_{\mathsf{r}}$.
(b) If both $\mathscr{C}$ and $\mathscr{C}'$ are closed under enumeration operations, then $\mathscr{C} = \mathscr{C}'$ if and only if $\mathscr{C}_{\mathsf{r}} = \mathscr{C}'_{\mathsf{r}}$.

4.2.9. Let $\mathscr{C}$ and $\mathscr{C}'$ be FRSs such that $\mathscr{C}_{\mathsf{r}} \subseteq \mathscr{C}'$ and $\mathscr{C}'$ is closed under enumeration operations. Prove that $\mathscr{C} \subseteq \mathscr{C}'$.

4.2.10. Let $\mathscr{C}$ be a FRS. Prove there is a total FRS $\mathscr{C}'$ such that

$$\mathsf{RC}(\mathsf{RC}_{\mathsf{j}} \cup \mathscr{C}') = \mathscr{C}'_{\mathsf{j}} = \mathscr{C}_{\mathsf{jj}}$$

and $\mathscr{C}' \nsubseteq \mathscr{C}_{\mathsf{j}}$, $\mathscr{C}_{\mathsf{j}} \nsubseteq \mathscr{C}'$.

4.2.11. Let $\mathscr{C}$ be a FRS with internal interpreter (ϕ, S). Prove that $\mathscr{C}_{\mathsf{j}} = \mathsf{RC}(\{\chi_\phi, \mathsf{S}\})$, where χ_ϕ denotes the characteristic function of the graph predicate G_ϕ.

Notes

The idea of a universal interpreter relative to a given function c can be traced to the notion of uniformity in Kleene [9], and a more formal treatment was given in Davis [2], although in both cases the relativization is taken to a total function c. The generalization to nontotal functions has been proposed in several places and was formalized in Sasso [24].

The reduction of the general interpreter δ_c to the recursive function δ, as given in Theorem 4.2.4 (i), shows that the whole theory can be reduced to ordinary recursive functions, and in fact it is sufficient to have an interpreter for the class RC. This approach has been taken by Rogers [20], where general reduction to arbitrary functions is handled via enumeration reducibility. The problem is that the deterministic element is lost, because enumeration reducibility is essentially nondeterministic.

The results about the jump operator are taken from Friedberg [4] and Shoenfield [25].

§3. Two Constructions

A typical application of the universal interpreter is the construction of functions and predicates satisfying an infinite number of conditions. This is a crucial technique in computability theory, and we give two examples of moderate complexity. But first it is convenient to introduce some special notation.

Let P be a k-ary predicate. An occurrence of the symbol P in a context that is meaningful only for functions is understood as an occurrence of χ_P. An occurrence of $P|$ in a similar context is understood as an occurrence of ψ_P.

As a consequence we have that $\mathsf{RC}(P) = \mathsf{RC}(\chi_P)$ and $\mathsf{RC}(P|) = \mathsf{RC}(\psi_P)$, and similarly with other operators. To say that a function h is recursive in P is equivalent to saying that h is recursive in χ_P, but h is recursive in $P|$ is equivalent to h is recursive in ψ_P.

Note that P is recursive in P for any predicate P, but P is not necessarily recursive in $P|$. On the other hand, P is always partially recursive in $P|$ and also in P.

Most of the time this notation is used with sets. Note that if A and B are sets such that $A \subseteq B$ then $A| \subseteq_1 B|$, where the latter is equivalent to $\psi_A \subseteq_1 \psi_B$.

In particular, if A is a set then δ_A is the function δ_{χ_A}, and $\delta_{A|}$ is the function δ_{ψ_A}.

Theorem 4.3.1. *Let B be a fixed set. Then*

(i) *If $\delta_{B|}(z, x) \simeq v$ then there is a finite set $C \subseteq B$ such that $\delta_{C|}(z, x) \simeq v$.*
(ii) *If $\delta_{B|}(z, x) \simeq v$ and $B \subseteq A$ then $\delta_{A|}(z, x) \simeq v$.*
(iii) *If for some number z the binary predicate P such that*

$$P(x, v) \equiv \exists y(\delta_{B|}(z, \langle x, v, y\rangle) \simeq 0)$$

is not single-valued, then there is a finite set $C \subseteq B$ such that the binary predicate P' such that

$$P'(x, v) \equiv \exists y(\delta_{C|}(z, \langle x, v, y\rangle) \simeq 0)$$

is not single-valued.

PROOF. Parts (i) and (ii) follow easily from the recursive evaluation of $\delta_{B|}(z, x) \simeq v$. To prove (iii) assume P is not single-valued; hence there are numbers x, v, and v' such that $P(x, v)$ and $P(x, v')$ both hold and $v \neq v'$. This means that there are numbers y and y' such that

$$\delta_{B|}(z, \langle x, v, y\rangle) \simeq 0$$

$$\delta_{B|}(z, \langle x, v', y'\rangle) \simeq 0;$$

hence using (i) and (ii) we get a finite set $C \subseteq B$ such that

$$\delta_{C|}(z, \langle x, v, y\rangle) \simeq 0$$

$$\delta_{C|}(z, \langle x, v', y'\rangle) \simeq 0$$

and this means that P' is not single-valued. □

The first construction is related to the following situation. If B is a set we know that $\psi_B \subseteq_1 \mathsf{C}_0^1$ = the unary constant 0 function, and this implies that $\mathsf{RC}(B|)$ is bounded by RC. It follows that if $\mathsf{RC}(B|)$ is e-total then $\mathsf{RC}(B|) = \mathsf{RC}$; hence B is recursively enumerable.

EXAMPLE 4.3.1. Let B be a set that is not recursively enumerable, but $\bar{B}$ is recursively enumerable (see Example 4.1.6). It follows that $\mathsf{RC}(B|)$ is not e-total. On the other hand, $\mathsf{RC}_{\mathsf{pd}\#}(B|)$ is e-total; in fact, $\mathsf{RC}_{\mathsf{pd}\#}(B|) = \mathsf{RC}(B)$. Clearly, we have $\mathsf{RC}(B|) \subseteq \mathsf{RC}(B)$; hence $\mathsf{RC}_{\mathsf{pd}\#}(B|) \subseteq \mathsf{RC}_{\mathsf{pd}\#}(B) = \mathsf{RC}(B)$. In the other direction, note that both B and $\bar{B}$ are partially recursive in $B|$; hence χ_B is in $\mathsf{RC}_{\mathsf{pd}\#}(B|)$, so $\mathsf{RC}(B) \subseteq \mathsf{RC}_{\mathsf{pd}\#}(B|)$.

Theorem 4.3.2. *Let $R_0, R_1, \ldots, R_n, R_{n+1}, \ldots$ be an infinite sequence of sets such that R_n is infinite for every $n \geq 0$, and let (C_0, D_0) be a pair of finite sets such that $C_0 \cap D_0 = \varnothing$ (i.e., C_0 and D_0 are disjoint). There is an infinite set B such that:*

(i) $\mathsf{RC}_{\mathsf{pd}\#}(B|)$ *is bounded by* RC.
(ii) $C_0 \subseteq B$ *and* $B \cap D_0 = \varnothing$.
(iii) $R_n \nsubseteq B$ *for all* $n \geq 0$.

PROOF. We generate an infinite sequence of pairs (C_0, D_0), ..., (C_n, D_n), ... such that for each $n \geq 0$, C_n, D_n are finite sets, $C_n \subseteq C_{n+1}$, $D_n \subsetneq D_{n+1}$, and $C_n \cap D_n = \varnothing$. The initial pair (C_0, D_0) is given. We assume (C_n, D_n) has been defined, take $k = \max(C_n \cup D_n) + 1$, and k' the least element in R_n such that $k < k'$ (such an element exists because R_n is infinite). We put $D_{n+1} = D_n \cup \{k'\}$. To define C_{n+1} we consider two cases:

(a) The predicate P such that

$$P(x, v) \equiv \exists y(\delta_{\bar{D}_{n+1}|}(n, \langle x, v, y\rangle) \simeq 0)$$

is single-valued. We put $C_{n+1} = C_n \cup \{k\}$. Note that $C_{n+1} \subseteq \bar{D}_{n+1}$; hence the predicate P' such that

$$P'(x, v) \equiv \exists y(\delta_{C_{n+1}|}(n, \langle x, v, y\rangle) \simeq 0)$$

is also single-valued.

(b) The predicate P above is not single-valued. In this case there is a finite set $C \subseteq \bar{D}_{n+1}$ and the predicate P'' such that

$$P''(x, v) \equiv \exists y(\delta_{C|}(z, \langle x, v, y\rangle) \simeq 0)$$

is not single-valued. We put $C_{n+1} = C_n \cup C \cup \{k\}$. Note that in this case the predicate P' above is not single-valued.

We have completed the definition of the pair (C_{n+1}, D_{n+1}). Note that $k \notin C_n$ but $k \in C_{n+1}$. Now we put

$$B = C_0 \cup C_1 \cup C_2 \cup \cdots \cup C_n \cup C_{n+1} \cup \cdots.$$

It follows that B is infinite, $B \cap D_n = \varnothing$, and $R_n \nsubseteq B$ for all $n \geq 0$. To prove that $\mathsf{RC}_{\mathsf{pd}\#}(B|)$ is bounded by RC, we take a unary function h that is partially recursively enumerable in $B|$, hence there is a 3-ary predicate Q such that

$$h(x) \simeq v \equiv \exists y Q(x, v, y),$$

and ψ_Q is recursive in $B|$, hence there is a number n such that

$$\psi_Q(x, v, y) \simeq \delta_{B|}(n, \langle x, v, y\rangle),$$

and the predicate P' such that

$$P'(x, v) \equiv \exists y(\delta_{B|}(n, \langle x, v, y\rangle) \simeq 0)$$

is single valued. This means that in the definition of the pair (C_{n+1}, D_{n+1}) case (a) applies and we can introduce a unary function h' such that

$$h'(x) \simeq v \equiv \exists y(\delta_{\bar{D}_{n+1}|}(n, \langle x, v, y\rangle) \simeq 0).$$

This function is clearly recursive, and since $B \subseteq \bar{D}_{n+1}$ it follows that $h \subseteq_1 h'$. This is sufficient if the function h is unary. If it is k-ary and $k > 1$ we introduce a unary function h_1 such that

$$h_1(x) \simeq h([x]_1, \ldots, [x]_k),$$

and since h_1 is also recursive in $B|$ we can find a unary recursive function h_1' such that $h_1 \subseteq_1 h_1'$. Finally, we take h' such that

$$h'(\mathbf{x}) \simeq h_1'(\langle \mathbf{x} \rangle),$$

and it follows that $h \subseteq_k h'$. □

Corollary 4.3.2.1. *Let $\mathscr{C}$ be a* FRS. *There is a set B such that*:

(i) *B is infinite and $\mathsf{RC}_{\mathsf{pd}\#}(B|)$ is bounded by* RC.
(ii) *If A is an infinite set partially $\mathscr{C}$-decidable then $A \nsubseteq B$.*
(iii) *$\mathsf{RC}_{\mathsf{pd}\#}(B|)$ is not* e-*total.*

PROOF. We apply Theorem 4.3.2 with R_0, R_1, R_2, ... an enumeration of all infinite sets that are partially $\mathscr{C}$-decidable. Parts (i) and (ii) are immediate. Furthermore, if $\mathsf{RC}_{\mathsf{pd}\#}(B|)$ is e-total then $\mathsf{RC}_{\mathsf{pd}\#}(B|) = \mathsf{RC}$; hence B is recursively enumerable, contradicting (ii). □

Corollary 4.3.2.2. *The collection of all infinite sets B such that $\mathsf{RC}_{\mathsf{pd}\#}(B|)$ is bounded by* RC *is noncountable.*

PROOF. Given an enumeration of all sets in such a collection we can use Theorem 4.3.2 to construct a set B with the same properties which is not in the enumeration. □

For the second construction we need an operation on sets. First, we introduce the unary elementary function dt such that

$$\mathsf{dt}(x) = \mu y \leq x \colon x = 2 \times y \vee x = 2 \times y + 1.$$

If A and B are sets then $A \oplus B$ is the set such that

$$x \in A \oplus B \equiv x = 2 \times \mathsf{dt}(x) \wedge \mathsf{dt}(x) \in A$$
$$\vee\ (x = 2 \times \mathsf{dt}(x) + 1 \wedge \mathsf{dt}(x) \in B).$$

It follows that

$$x \in A \equiv 2 \times x \in A \oplus B$$
$$x \in B \equiv 2 \times x + 1 \in A \oplus B.$$

As a consequence we have the following relations:

$$\mathsf{RC}(\{A, B\}) = \mathsf{RC}(A \oplus B)$$
$$\mathsf{RC}(\{A|, B|\}) = \mathsf{RC}((A \oplus B)|).$$

Theorem 4.3.3. *Let B be a set such that $\delta_{B|}(z, x) \simeq v$. There is a finite set C such that $\delta_{C|}(z, x) \simeq v$, and if A is a set such that $C \nsubseteq A$ then $\delta_{A|}(z, x)\uparrow$.*

PROOF. We put in C all those elements of B which are required by the evaluation $\delta_{B|}(z, x) \simeq v$. It follows that C is finite and $\delta_{C|}(z, x) \simeq v$. If one element of C is missing in the set A then $\delta_{A|}(z, x)\uparrow$. □

Theorem 4.3.4. *Let R_0, R_1, ..., R_n, R_{n+1}, ... be an infinite sequence of sets such that R_n is infinite for every $n \geq 0$, and let (A_0, B_0, C_0, D_0) be a quadruple where A_0, B_0, C_0, and D_0 are finite sets, $A_0 \cap B_0 = \varnothing$, $A_0 \cap C_0 = \varnothing$, and*

$B_0 \cap D_0 = \varnothing$. *There are sets* A *and* B *such that:*

(i) $A \cap B = \varnothing$, $A_0 \subseteq A$, $B_0 \subseteq B$, $A \cap C_0 = \varnothing$, *and* $B \cap D_0 = \varnothing$.
(ii) $A \cup B$ *is not partially recursive in* $(A \oplus B)|$.
(iii) $R_n \nsubseteq A$ *and* $R_n \nsubseteq B$ *for all* $n \geq 0$.

Proof. We generate an infinite sequence of quadruples (A_0, B_0, C_0, D_0), $\ldots$, (A_n, B_n, C_n, D_n), $\ldots$, where A_n, B_n, C_n, D_n are finite sets, $A_n \cap B_n = \varnothing$, $A_n \cap C_n = \varnothing$, $B_n \cap D_n = \varnothing$, $A_n \subseteq A_{n+1}$, $B_n \subseteq B_{n+1}$, $C_n \subseteq C_{n+1}$, and $D_n \subseteq D_{n+1}$. The initial quadruple (A_0, B_0, C_0, D_0) is given. We assume the quadruple (A_n, B_n, C_n, D_n) has been defined, and x_0 is the least element in R_n such that $\max(A_n \cup B_n \cup C_n \cup D_n) < x_0$. We put $x_1 = x_0 + 1$, $C_n' = C_n \cup \{x_0\}$, and $D_n' = D_n \cup \{x_0\}$. We consider two cases:

(a) There is no set C such that $\delta_{C|}(n, \langle x_1 \rangle) \simeq 0$. We put $A_{n+1} = A_n \cup \{x_1\}$, $B_{n+1} = B_n$, $C_{n+1} = C_n'$, and $D_{n+1} = D_n'$.
(b) There is a set C such that $\delta_{C|}(n, \langle x_1 \rangle) \simeq 0$, and we take C the minimal given by Theorem 4.3.3. Here we consider two subcases.
(b1) $C \subseteq A_n \oplus B_n$. We put $A_{n+1} = A_n$, $B_{n+1} = B_n$, $C_{n+1} = C_n' \cup \{x_1\}$, and $D_{n+1} = D_n' \cup \{x_1\}$.
(b2) $C \nsubseteq A_n \oplus B_n$, and let $y \in C$, $y \notin A_n \oplus B_n$. If $y = 2 \times z$ we put $A_{n+1} = A_n$, $B_{n+1} = B_n \cup \{x_1\}$, $C_{n+1} = C_n' \cup \{z\}$, and $D_{n+1} = D_n'$. If $y = 2 \times z + 1$ we put $A_{n+1} = A_n \cup \{x_1\}$, $B_{n+1} = B_n$, $C_{n+1} = C_n'$, and $D_{n+1} = D_n' \cup \{z\}$.

This completes the definition of the quadruple $(A_{n+1}, B_{n+1}, C_{n+1}, D_{n+1})$. We put

$$A = A_0 \cup A_1 \cup A_2 \cup \cdots \cup A_n \cup \cdots$$
$$B = B_0 \cup B_1 \cup B_2 \cup \cdots \cup B_n \cup \cdots.$$

It is clear that $A \cap B = \varnothing$, $A_0 \subseteq A$, $B_0 \subseteq B$, $A \cap C_n = \varnothing$, $B \cap D_n = \varnothing$, $R_n \nsubseteq A$, and $R_n \nsubseteq B$ holds for all $n \geq 0$. We must prove that $A \cup B$ is not partially recursive in $(A \oplus B)|$, so assume that this is the case and there is a number n such that

$$\psi_{A \cup B}(x) \simeq \delta_{(A \oplus B)|}(n, \langle x \rangle)$$

holds for all x, and let x_1 be the value considered in the definition of $(A_{n+1}, B_{n+1}, C_{n+1}, D_{n+1})$. If $x_1 \in A \cup B$, then $\psi_{A \cup B}(x_1) \simeq 0$, and one of the cases (a) or (b2) applies, which is a contradiction because in case (a) $\delta_{(A \oplus B)|}(n, \langle x_1 \rangle)$ is not defined with value 0, and in case (b2) it is undefined. If $x_1 \notin A \cup B$ then $\psi_{A \cup B}(x_1)$ is undefined but case (b1) applies and $\delta_{(A \oplus B)|}(n, \langle x_1 \rangle) \simeq 0$. So in both cases we have a contradiction and $\psi_{A \cup B}$ is not recursive in $(A \oplus B)|$. □

Corollary 4.3.4.1. *There is a* FRS $\mathscr{C}$ *such that* $\mathscr{C}_{\mathrm{pd}}$ *is not closed under disjunction.*

Proof. Take $\mathscr{C} = \mathrm{RC}(\psi_{A \oplus B})$, where A and B are sets given by Theorem 4.3.4 (e.g., with $R_n = \mathbb{N}$, $A_0 = B_0 = C_0 = D_0 = \varnothing$). Now both A and B are partially $\mathscr{C}$-decidable, but $A \cup B$ is not partially $\mathscr{C}$-decidable. □

EXERCISES

4.3.1. Prove there is a FRS $\mathscr{C}$, where $\mathscr{C}_{\mathsf{pd}}$ is not closed under existential quantification.

4.3.2. Let A and B be sets. Prove the following:
(a) $\mathsf{RC}(\{A, B\}) = \mathsf{RC}(A \oplus B)$.
(b) $\mathsf{RC}(\{A|, B|\}) = \mathsf{RC}((A \oplus B)|)$.

4.3.3. Let A and B be sets. Prove the following
(a) $\mathsf{RC}_{\mathsf{pd}\#}((A \cup B)|) \subseteq \mathsf{RC}_{\mathsf{pd}\#}((A \oplus B)|)$.
(b) $\mathsf{RC}(A \cup B) \subseteq \mathsf{RC}(A \oplus B)$.
(c) If $A \cap B = \varnothing$ and A and B are recursively enumerable in $A \cup B$, then $\mathsf{RC}(A \oplus B) \subseteq \mathsf{RC}(A \cup B)$.

4.3.4. Prove that there is a binary recursive function δ' such that whenever A is a set then $\delta_{A|} \subseteq_2 \delta'$.

4.3.5. Let c be a function and A a set. Prove the following:
(a) $\mathsf{RC}(\{c, A|\}) \subseteq \mathsf{BD}(\mathsf{RC}(c))$.
(b) If h is a k-ary function recursive in $\{c, A|\}$ then h is recursive in $\{c, \mathsf{D}_h|\}$.

4.3.6. Let $\mathscr{C}$ be a class of functions, c a function, and A a set. Assume $\mathscr{C} \subseteq \mathsf{RC}(c) \subseteq \mathsf{RC}(\mathscr{C} \cup \{A|\})$. Prove that $\mathsf{RC}(c) = \mathsf{RC}(\mathscr{C} \cup \{\mathsf{D}_c|\})$.

4.3.7. Let A be a set. Prove that the following conditions are equivalent:
(a) A is recursively enumerable.
(b) A is recursively enumerable in $A|$.
(c) $\mathsf{RC}(A|)$ is e-total.
(d) $\mathsf{RC}(A|) = \mathsf{RC}$.

4.3.8. Let $\mathscr{C}$ be a FRS closed under enumeration operations. Prove that there is a set A such that $\mathscr{C} = \mathsf{RC}_{\mathsf{pd}\#}(A|)$.

Notes

The idea of using a universal interpreter to generate sets and functions satisfying an infinite number of conditions is widely used in recursive function theory. In most applications it is combined with the normal form theorem (Kleene [9]), which produces a very powerful technique. We have chosen two examples where the normal form theorem is not required. The first is taken from Medvedev (see reference in Rogers [20]), and the second from Sasso [24].

§4. The Recursion Theorem

The most important property of a FRS is the recursion theorem (RT), which was introduced as Theorem 4.1.4, although the fixed point theorem (FPT) given as Corollary 4.1.4.1 is sometimes called the recursion theorem by some authors. The RT was used to prove the fundamental Theorem 4.1.5 and in general can be used to solve recursive specifications. For us such applications are not very important, because closure under recursive specifications follows from closure under recursive operations. However, there are other applications that are recursive in a more general sense but do not reduce to recursion in the sense described in Chapter 2.

EXAMPLE 4.4.1. Let $\mathscr{C}$ be a FRS with internal interpreter (ϕ, S), and let h be the binary $\mathscr{C}$-computable function such that

$$h(y, x) \simeq \phi^1(x, \phi^1(x, y)).$$

From the RT we find there is a number z such that

$$\begin{aligned} h(\mathbf{z}, x) &\simeq \phi^1(x, \phi^1(x, \mathbf{z})) \\ &\simeq \phi^1(\mathbf{z}, x) \end{aligned}$$

holds for all x. Essentially, we prove there is a unary $\mathscr{C}$-computable function f (i.e., $f(x) \simeq h(\mathbf{z}, x)$) with (ϕ, S)-index z which satisfies the relation

$$f(x) \simeq \phi^1(x, \phi^1(x, \mathbf{z})).$$

This relation is recursive in the sense that the right side refers to something (i.e., the index z) that depends on the function on the left side.

In the next theorem we prove two extensions of the RT, which assert the existence of functions rather than numbers. These functions are total, because they are given by the abstraction property. Furthermore, when the given interpreter is strict, then the functions are also strict, and they are elementary in case the input function is elementary.

Theorem 4.4.1. *Let $\mathscr{C}$ be a* FRS *with internal interpreter* (ϕ, S). *Then*

(i) *If f is an $(m + 1 + k)$-ary $\mathscr{C}$-computable function, $m, k \geq 0$, then there is a total $\mathscr{C}$-computable m-ary function g, such that*

$$f(\mathbf{y}, g(\mathbf{y}), \mathbf{x}) \simeq \phi^k(g(\mathbf{y}), \mathbf{x}).$$

(ii) *If f is an $(1 + m + k)$-ary $\mathscr{C}$-computable function, $m, k \geq 0$, there is a total $\mathscr{C}$-computable $(m + 1)$-ary function g, such that*

$$f(g(\mathbf{y}, v), \mathbf{y}, \mathbf{x}) \simeq \phi^k(g(\mathbf{y}, v), \mathbf{x}).$$

PROOF. To prove (i) let g' be the total m-ary $\mathscr{C}$-computable function given by the abstraction property such that

$$f(\mathbf{y}, \mathsf{S}(z, z), \mathbf{x}) \simeq \phi^{k+1}(g'(\mathbf{y}), z, \mathbf{x}),$$

and take $g(\mathbf{y}) = \mathsf{S}(g'(\mathbf{y}), g'(\mathbf{y}))$. Then

$$\begin{aligned} \phi^k(g(\mathbf{y}), \mathbf{x}) &\simeq \phi^{k+1}(g'(\mathbf{y}), g'(\mathbf{y}), \mathbf{x}) \\ &\simeq f(\mathbf{y}, \mathsf{S}(g'(\mathbf{y}), g'(\mathbf{y})), \mathbf{x}) \\ &\simeq f(\mathbf{y}, g(\mathbf{y}), \mathbf{x}). \end{aligned}$$

To prove (ii) apply (i) to the function f' such that

$$f'(\mathbf{y}, v, z, \mathbf{x}) \simeq f(z, \mathbf{y}, \mathbf{x}). \qquad \square$$

EXAMPLE 4.4.2. If we apply part (i) of Theorem 4.4.1 to the 3-ary function f such that

$$f(y, z, x) \simeq \phi^1(x, z),$$

we get a unary total $\mathscr{C}$-computable function g such that

$$\begin{aligned}\phi^1(g(y), x) &\simeq f(y, g(y), x)\\ &\simeq \phi^1(x, g(y))\end{aligned}$$

holds for all x and y.

Let $\mathscr{C}$ be a FRS with internal interpreter (ϕ, S). If $\mathscr{C}'$ is a subset of $\mathscr{C}$ containing only unary functions we introduce a set $[\mathscr{C}']_{(\phi,\mathsf{S})}$ such that

$$z \in [\mathscr{C}']_{(\phi,\mathsf{S})} \equiv z \text{ is a } (\phi, \mathsf{S})\text{-index for some function in } \mathscr{C}'.$$

As usual, if c is a unary function we write $[c]_{(\phi,\mathsf{S})}$, which is the set that contains all the (ϕ, S)-indexes for the function c.

Theorem 4.4.2. *Let $\mathscr{C}$ be a* FRS *with internal interpreter (ϕ, S), and let f be a unary $\mathscr{C}$-computable function such that $\mathsf{UD}_1 \neq f$. If B is a set such that $[\mathsf{UD}_1]_{(\phi,\mathsf{S})} \subseteq B$ and $[f]_{(\phi,\mathsf{S})} \subseteq \bar{B}$, then B is not partially $\mathscr{C}$-decidable.*

PROOF. To get a contradiction we assume that B is partially $\mathscr{C}$-decidable. Let h be the 3-ary $\mathscr{C}$-computable function such that

$$h(y, z, x) \simeq \phi^1(y, x) + \psi_B(z).$$

From Theorem 4.4.1 (i) it follows that there is a unary total $\mathscr{C}$-computable function g such that

$$\begin{aligned}h(y, g(y), x) &\simeq \phi^1(y, x) + \psi_B(g(y))\\ &\simeq \phi^1(g(y), x).\end{aligned}$$

If $y \in \mathbb{N}$ then $g(y) \in B$. For if $g(y) \in \bar{B}$ then $g(y)$ is (ϕ, S)-index for UD_1, and $g(y) \in B$ follows from the assumptions. This means that y and $g(y)$ are (ϕ, S)-indexes for the same unary $\mathscr{C}$-computable function. In particular, if y is (ϕ, S)-index for the function f, we have $g(y) \in [f]_{(\phi,\mathsf{S})} \subseteq \bar{B}$ and this contradicts $g(y) \in B$. $\square$

Corollary 4.4.2.1. *Let $\mathscr{C}$ be a* FRS *with internal interpreter (ϕ, S), and let $\mathscr{C}' \subseteq \mathscr{C}$ be a class of unary functions such that UD_1 is $\mathscr{C}'$-computable. If $[\mathscr{C}']_{(\phi,\mathsf{S})}$ is partially $\mathscr{C}$-decidable then $[\mathscr{C}']_{(\phi,\mathsf{S})} = \mathbb{N}$.*

PROOF. We prove that $\mathscr{C}'$ contains all the unary $\mathscr{C}$-computable functions. To get a contradiction assume f is a unary function that is $\mathscr{C}$-computable but not $\mathscr{C}'$-computable. If we set $B = [\mathscr{C}']_{(\phi,\mathsf{S})}$ we have $[\mathsf{UD}_1]_{(\phi,\mathsf{S})} \subseteq B$ and $[f]_{(\phi,\mathsf{S})} \subseteq \bar{B}$, so from Theorem 4.4.2 it follows that B is not partially $\mathscr{C}$-decidable. □

EXAMPLE 4.4.3. Let $\mathscr{C}$ be a FRS with internal interpreter (ϕ, S) and $\mathscr{C}'$ the class of all finite unary functions. Since $\mathscr{C}'$ contains UD_1 and $A = [\mathscr{C}']_{(\phi,\mathsf{S})} \neq \mathbb{N}$, it follows that A is not partially $\mathscr{C}$-decidable, and hence $\bar{A}$ is not $\mathscr{C}$-decidable.

EXAMPLE 4.4.4. Let $\mathscr{C}'$ be a class that contains a unary $\mathscr{C}$-computable function f if and only if $f(0)\uparrow$. It follows that $A = [\mathscr{C}']_{(\phi,\mathsf{S})}$ is not partially $\mathscr{C}$-decidable, but $\bar{A}$ is partially $\mathscr{C}$-decidable, for

$$\psi_{\bar{A}}(z) \simeq \phi^1(z,0) \times 0.$$

EXAMPLE 4.4.5. If we take $\mathscr{C}' = \{\mathsf{UD}_1\}$ it follows that $A = [\mathscr{C}']_{(\phi,\mathsf{S})}$ is not partially $\mathscr{C}$-decidable, but $\bar{A}$ is $\mathscr{C}_{\mathsf{pd}}$-enumerable, for

$$z \in \bar{A} \equiv \exists y \phi^1(z, y)\downarrow;$$

hence if $\mathscr{C}$ is closed under enumeration operations it follows that $\bar{A}$ is partially $\mathscr{C}$-decidable.

Theorem 4.4.3. *Let $\mathscr{C}$ be a FRS closed under enumeration operations with internal interpreter (ϕ, S). If A is a partially $\mathscr{C}$-decidable set, f is a unary $\mathscr{C}$-computable function such that $[f]_{(\phi,\mathsf{S})} \subseteq A$, and f' is a unary $\mathscr{C}$-computable function such that $f \subseteq_1 f'$, then A contains at least one (ϕ, S)-index for the function f'.*

PROOF. Let P be the binary predicate such that

$$P(x,z) \equiv x \in \mathsf{D}_f \vee z \in A.$$

Since $\mathscr{C}$ is closed under enumeration operations it follows that P is partially $\mathscr{C}$-decidable. Let h be the binary $\mathscr{C}$-computable function such that

$$h(z,x) \simeq f'(x) + \psi_P(x,z).$$

By the RT there is a number z such that

$$\phi^1(\mathsf{z},x) \simeq f'(x) + \psi_P(x,\mathsf{z}).$$

It follows that $\mathsf{z} \in A$, for otherwise z is a (ϕ, S)-index for the function f, which implies that $\mathsf{z} \in A$. If $\mathsf{z} \in A$ then $P(x,\mathsf{z})$ holds for all x, and this means that z is a (ϕ, S)-index for the function f'. □

Theorem 4.4.4. *Let $\mathscr{C}$ be a FRS with internal interpreter (ϕ, S). If A is a $\mathscr{C}_{\mathsf{d}}$-enumerable set, and f is a unary $\mathscr{C}$-computable function such that $[f]_{(\phi,\mathsf{S})} \subseteq A$, then there is a number $\mathsf{z} \in A$ such that z is a (ϕ, S)-index for a finite unary $\mathscr{C}$-computable function f' and $f' \subseteq_1 f$.*

PROOF. Since A is $\mathscr{C}_{\mathsf{d}}$-enumerable, there is a binary $\mathscr{C}$-decidable predicate Q such that

$$z \in A \equiv \exists y Q(z, y).$$

Let Q' be the binary $\mathscr{C}$-decidable predicate such that

$$Q'(z, x) \equiv \forall y < x \sim Q(z, y).$$

Let h be the binary $\mathscr{C}$-computable function such that

$$h(z, x) \simeq f(x) + \psi_{Q'}(z, x).$$

By the RT there is a number z such that

$$\phi^1(\mathsf{z}, x) \simeq h(\mathsf{z}, x)$$
$$\simeq f(x) + \psi_{Q'}(\mathsf{z}, x).$$

It follows that $\mathsf{z} \in A$, because if $\mathsf{z} \notin A$ then z is a (ϕ, S)-index for the function f, and hence $\mathsf{z} \in A$. If $\mathsf{z} \in A$ then z is a (ϕ, S)-index for some finite unary $\mathscr{C}$-computable function such that $f' \subseteq_1 f$. □

Corollary 4.4.4.1. *Let $\mathscr{C}$ be a* FRS *closed under enumeration operations with internal interpreter (ϕ, S), and let $\mathscr{C}'$ be a subclass of $\mathscr{C}$ containing only unary functions such that the set $[\mathscr{C}']_{(\phi, \mathsf{S})}$ is $\mathscr{C}_{\mathsf{d}}$-enumerable. If f is a unary $\mathscr{C}$-computable function then the following conditions are equivalent:*

(i) *f is $\mathscr{C}'$-computable.*
(ii) *There is a finite unary function $f' \subseteq_1 f$ such that $f' \in \mathscr{C}'$.*

PROOF. The implication from (i) to (ii) follows from Theorem 4.4.4, and the implication from (ii) to (i) follows from Theorem 4.4.3. □

Corollary 4.4.4.2. *Let $\mathscr{C}$ be an* e-total FRS *with internal interpreter (ϕ, S), and let $\mathscr{C}'$ be a subclass of $\mathscr{C}$ containing only unary functions, such that the set $[\mathscr{C}']_{(\phi, \mathsf{S})}$ is partially $\mathscr{C}$-decidable. If f is a unary $\mathscr{C}$-computable function then the following conditions are equivalent:*

(i) *f is $\mathscr{C}'$-computable.*
(ii) *There is a finite unary function $f' \subseteq_1 f$ such that $f' \in \mathscr{C}'$.*

PROOF. Immediate from Corollary 4.4.4.1, noting that since $\mathscr{C}$ is e-total it follows that $\mathscr{C}$ is closed under enumeration operations and $\mathscr{C}_{\mathsf{de}} = \mathscr{C}_{\mathsf{pd}}$. □

EXAMPLE 4.4.6. Let $\mathscr{C}$ be a FRS with internal interpreter (ϕ, S). Let $\mathscr{C}'$ be the class of all unary total $\mathscr{C}$-computable functions, and let $\mathscr{C}''$ be the class of all unary nontotal $\mathscr{C}$-computable functions. From Theorem 4.4.4 it follows that $[\mathscr{C}']_{(\phi, \mathsf{S})}$ is not $\mathscr{C}_{\mathsf{d}}$-enumerable, and from Theorem 4.4.3 it follows that in case $\mathscr{C}$ is closed under enumeration operations then $[\mathscr{C}'']_{(\phi, \mathsf{S})}$ is not partially $\mathscr{C}$-decidable.

Let $\mathscr{C}$ be a FRS with internal interpreter (ϕ, S). A binary function po is said to be a ϕ-preorder if the following three conditions are satisfied:

PO 1: If $\mathsf{po}(z, z')\uparrow$, then $\phi(z)\uparrow$ and $\phi(z')\uparrow$.
PO 2: If $\mathsf{po}(z, z') \simeq 0$, then $\phi(z)\downarrow$ or $\phi(z')\uparrow$.
PO 3: If $\mathsf{po}(z, z') \neq 0$, then $\phi(z')\downarrow$.

A function satisfying these conditions can be defined in several ways, and in general it is not $\mathscr{C}$-computable.

Example 4.4.7. Let po be the total function such that

$$\begin{aligned} \mathsf{po}(z, z') &= 0 \quad \text{if } \phi(z)\downarrow \text{ or } \phi(z')\uparrow \\ &= 1 \quad \text{otherwise.} \end{aligned}$$

Condition PO 1 is trivial, since po is total, and conditions PO 2 and PO 3 are clear from the definition. This function is not $\mathscr{C}$-computable, because if m is a number such that $\phi(\mathsf{m})$ is undefined, then

$$\phi(z')\downarrow \equiv \mathsf{po}(\mathsf{m}, z') = 1.$$

Example 4.4.8. Let $\mathscr{C}$ be an e-total FRS with internal interpreter (ϕ, S). In this case there is a binary $\mathscr{C}$-decidable predicate Q such that

$$\phi(z)\downarrow \equiv \exists y Q(z, y).$$

We put

$$h(z, z') \simeq \mu y\colon Q(z, y) \vee Q(z', y)$$

$$\mathsf{po}(z, z') \simeq \chi_Q(z, h(z, z')).$$

It is easy to see that this is a $\mathscr{C}$-computable ϕ-preorder function, and condition PO 2 is satisfied in the form: if $\mathsf{po}(z, z') \simeq 0$ then $\phi(z)\downarrow$.

Theorem 4.4.5. *Let $\mathscr{C}$ be a* FRS *with internal interpreter (ϕ, S), and assume there is a ϕ-preorder function* po *that is $\mathscr{C}$-computable. If $k \geq 0$, there is a $(k + 1)$-ary $\mathscr{C}$-computable function h, such that whenever there is a y such that $\phi^{k+1}(z, \mathbf{x}, y)\downarrow$, then $\phi^{k+1}(z, \mathbf{x}, h(z, \mathbf{x}))\downarrow$.*

Proof. We introduce several auxiliary functions. The function f_1 is total unary, given by the abstraction property, such that

$$\phi^{k+1}(f_1(z), \mathbf{x}, y) \simeq \phi^{k+1}(z, \mathbf{x}, y + 1).$$

The functions f_2, f_3, and f_4 are given by substitution in the form

$$f_2(z, \mathbf{x}) = \mathsf{S}^{k+1}(z, \mathbf{x}, 0)$$

$$f_3(w, z, \mathbf{x}) = \mathsf{S}^{k+1}(w, f_1(z), \mathbf{x})$$

$$f_4(w, z, \mathbf{x}) \simeq \mathsf{S}^{k+1}(z, \mathbf{x}, \phi^{k+1}(w, f_1(z), \mathbf{x}) + 1).$$

These functions are $\mathscr{C}$-computable; f_2 and f_3 are total functions.

We use partial definition by cases to introduce a $(k + 2)$-ary function h' such that

$$\begin{aligned} h'(w,z,\mathbf{x}) &\simeq 0 && \text{if } \mathsf{po}(f_2(z,\mathbf{x}), f_3(w,z,\mathbf{x})) \simeq 0 \\ &\simeq 0 && \text{if } \mathsf{po}(f_2(z,\mathbf{x}), f_4(w,z,\mathbf{x})) \simeq 0 \\ &\simeq \phi^{k+1}(w, f_1(z), \mathbf{x}) + 1 && \text{if } \mathsf{po}(f_2(z,\mathbf{x}), f_4(w,z,\mathbf{x})) \not\simeq 0. \end{aligned}$$

It follows that h' is $\mathscr{C}$-computable, so let w be the number given by the RT such that

$$h'(\mathsf{w}, z, \mathbf{x}) \simeq \phi^{k+1}(\mathsf{w}, z, \mathbf{x}).$$

Consider the $(k+2)$-ary predicate P such that

$$P(z,\mathbf{x},y) \equiv \phi^{k+1}(z,\mathbf{x},y)\downarrow \wedge \forall v < y \phi^{k+1}(z,\mathbf{x},v)\uparrow.$$

We use the predicate P to formulate the following basic assertion: for all $(\mathbf{x}) \in \mathbb{N}^k$ and $z \in \mathbb{N}$, if $P(z,\mathbf{x},y)$ then there is a number v such that $h'(\mathsf{w},z,\mathbf{x}) \simeq v$, and $\phi^{k+1}(z,\mathbf{x},v)\downarrow$.

We prove the basic assertion by induction on y. Assume $y = 0$, and $P(z,\mathbf{x},0)$ holds. This means $\phi^{k+1}(z,\mathbf{x},0)\downarrow$, hence $\phi(f_2(z,\mathbf{x}))\downarrow$. From PO 1 it follows that

$$\mathsf{po}(f_2(z,\mathbf{x}), f_3(\mathsf{w},z,\mathbf{x})) \simeq t.$$

If $t = 0$ then $h'(\mathsf{w},z,\mathbf{x}) \simeq 0$, so we take $v = 0$. If $t \neq 0$ then from PO 3 it follows that $\phi(f_3(\mathsf{w},z,\mathbf{x}))\downarrow$; hence $\phi^{k+1}(\mathsf{w}, f_1(z), \mathbf{x}) \simeq u$, and $f_4(\mathsf{w},z,\mathbf{x})\downarrow$. We conclude that

$$\mathsf{po}(f_2(z,\mathbf{x}), f_4(\mathsf{w},z,\mathbf{x})) \simeq t'.$$

If $t' = 0$ we have again $h'(\mathsf{w},z,\mathbf{x}) \simeq 0$ and take $v = 0$. If $t' = 0$ then $\phi(f_4(\mathsf{w},z,\mathbf{x}))\downarrow$; hence $\phi^{k+1}(z,\mathbf{x},u+1)\downarrow$. Furthermore,

$$\begin{aligned} h'(\mathsf{w},z,\mathbf{x}) &\simeq \phi^{k+1}(\mathsf{w}, f_1(z), \mathbf{x}) + 1 \\ &\simeq u + 1; \end{aligned}$$

hence we take $v = u + 1$. This completes the case $y = 0$.

Now we assume the basic assertion is true for some y and prove it is true for $y + 1$, so we assume $P(z,\mathbf{x},y+1)$ holds. By definition this means that $P(f_1(z),\mathbf{x},y)$ holds; hence by the induction hypothesis there is v' such that $\phi^{k+1}(f_1(z),\mathbf{x},v')\downarrow$, and

$$\begin{aligned} \phi^{k+1}(f_1(z),\mathbf{x},v') &\simeq \phi^{k+1}(z,\mathbf{x},v'+1) \\ h'(\mathsf{w}, f_1(z), \mathbf{x}) &\simeq v'. \end{aligned}$$

We need only show that $h'(\mathsf{w},z,\mathbf{x}) \simeq v' + 1$, and take $v = v' + 1$. Note that $\phi(f_2(z,\mathbf{x}))\uparrow$, and furthermore

$$\phi^{k+1}(\mathsf{w}, f_1(z), \mathbf{x}) \simeq h'(\mathsf{w}, f_1(z), \mathbf{x}) \simeq v';$$

hence $\phi(f_3(\mathsf{w},z,\mathbf{x}))\downarrow$, and from PO 2 it follows that

$$\mathsf{po}(f_2(z,\mathbf{x}), f_3(\mathsf{w},z,\mathbf{x})) \not\simeq 0.$$

We have already shown that $\phi^{k+1}(z,\mathbf{x},v'+1)\downarrow$, hence $\phi(f_4(\mathsf{w},z,\mathbf{x}))\downarrow$. It follows that

$$\begin{aligned}\mathsf{po}(f_2(z,\mathbf{x}), f_4(\mathsf{w},z,\mathbf{x})) &\not\simeq 0\\ h'(\mathsf{w},z,\mathbf{x}) &\simeq \phi^{k+1}(\mathsf{w}, f_1(z), \mathbf{x}) + 1\\ &\simeq v' + 1.\end{aligned}$$

To complete the proof we take $h(z, \mathbf{x}) \simeq h'(\mathsf{w}, z, \mathbf{x})$, and from the basic assertion it follows that whenever $\phi^{k+1}(z, \mathbf{x}, y)\downarrow$ for some y, then $\phi^{k+1}(z, \mathbf{x}, h(z, \mathbf{x}))\downarrow$. □

Theorem 4.4.6. *Let $\mathscr{C}$ be a* FRS *with internal interpreter* (ϕ, S). *The following conditions are equivalent*:

(i) *$\mathscr{C}$ has the* pd-*selector property.*
(ii) *$\mathscr{C}$ has the* pde-*selector property.*
(iii) *There is a $\mathscr{C}$-computable ϕ-preorder function* po *that satisfies condition* PO 2 *in the form: if* $\mathsf{po}(z, z') \simeq 0$ *then* $\phi(z)\downarrow$.
(iv) *There is a $\mathscr{C}$-computable ϕ-preorder function.*

Proof. Assume (i) to prove (ii). Let P be a $(k + 1)$-ary $\mathscr{C}_{\mathsf{pd}}$-enumerable predicate. It follows that

$$P(\mathbf{x}, y) \equiv \exists v Q(\mathbf{x}, y, v),$$

where Q is a partially $\mathscr{C}$-decidable predicate. A $\mathscr{C}$-computable selector function for P is given by

$$f(\mathbf{x}) \simeq g(\sigma y Q(\mathbf{x}, (y)_0, (y)_1)),$$

and $g(y) = (y)_0$. Now assume (ii) and to prove (iii) let P be the 3-ary predicate such that

$$P(z, z', v) \equiv (\phi(z)\downarrow \wedge\ y = 0) \vee (\phi(z')\downarrow \wedge\ v = 1)$$

Then $\mathsf{po}(z, z') \equiv \sigma v P(z, z', v)$ is a $\mathscr{C}$-computable ϕ-preorder function satisfying the condition in part (iii). The implication from (iii) to (iv) is trivial.

Assume now (iv) to prove (i), so let P be a $(k + 1)$-ary partially $\mathscr{C}$-decidable predicate, let z be a (ϕ, S)-index for the function ψ_P, and let h be the $(k + 1)$-ary $\mathscr{C}$-computable function given by Theorem 4.4.5. It follows that

$$f(\mathbf{x}) \simeq h(z, \mathbf{x}) + \psi_P(\mathbf{x}, h(z, \mathbf{x}))$$

is a $\mathscr{C}$-computable selector function for the predicate P. □

An application of this result is given in Chapter 5, Section 3.

Exercises

4.4.1. Let $\mathscr{C}$ be a FRS with internal interpreter (ϕ, S), and let f be some $(m + 1)$-ary $\mathscr{C}$-computable function, $m \geq 0$. Prove there is a total m-ary $\mathscr{C}$-computable function g such that

$$f(\mathbf{y}, g(\mathbf{y})) \simeq \phi(g(\mathbf{y})).$$

4.4.2. Let $\mathscr{C}$ be a FRS with internal interpreter (ϕ, S), and let f be some unary $\mathscr{C}$-computable function. Prove there is a number z such that $f(\mathsf{z}) \simeq \phi(\mathsf{z})$.

4.4.3. Let $\mathscr{C}$ be a FRS with internal interpreter (ϕ, S). Prove the following:
(a) If h is a unary $\mathscr{C}$-computable function such that $\phi \subseteq_1 h$ then h is not total.
(b) The set D_ϕ is not $\mathscr{C}$-decidable.

4.4.4. Let $\mathscr{C}$ be a FRS with internal interpreter (ϕ, S), and let $\mathscr{C}'$ be a subclass of $\mathscr{C}$ that contains only unary functions. Assume that $\mathscr{C}'$ is nonempty, and $\mathsf{UD}_1 \notin \mathscr{C}'$. Prove that if W is a partially $\mathscr{C}$-decidable set then there is a total unary $\mathscr{C}$-computable function g such that

$$z \in W \equiv g(z) \in [\mathscr{C}']_{(\phi, \mathsf{S})}.$$

4.4.5. Let $\mathscr{C}$ be a FRS closed under enumeration operations with internal interpreter (ϕ, S). Assume there is a given set A, and f and f' are unary $\mathscr{C}$-computable functions such that $f \subseteq_1 f'$, $[f]_{(\phi,\mathsf{S})} \subseteq A$, and $[f']_{(\phi,\mathsf{S})} \subseteq \bar{A}$. Prove that if W is a partially $\mathscr{C}$-decidable set then there is a total unary $\mathscr{C}$-computable function g such that

$$z \in W \equiv g(z) \in \bar{A}.$$

4.4.6. Let $\mathscr{C}$ be a FRS with internal interpreter (ϕ, S). Assume there is a set A and a unary $\mathscr{C}$-computable function f such that $[f]_{(\phi,\mathsf{S})} \subseteq A$. Furthermore, if f' is a finite $\mathscr{C}$-computable function and $f' \subseteq_1 f$, then $[f']_{(\phi,\mathsf{S})} \subseteq \bar{A}$. Prove that if K is a $\mathscr{C}_\mathsf{d}$-enumerable set then there is a total unary $\mathscr{C}$-computable function g such that

$$z \in K \equiv g(z) \in \bar{A}.$$

4.4.7. Let $\mathscr{C}$ be a FRS with internal interpreter (ϕ, S) and A a partially $\mathscr{C}$-decidable set such that $[\mathsf{UD}_1]_{(\phi,\mathsf{S})} \subseteq A$. Assume f is a unary $\mathscr{C}$-computable function such that $f \neq \mathsf{UD}_1$ and $B = A \cap [f]_{(\phi,\mathsf{S})}$. Prove that $\bar{B}$ is not partially $\mathscr{C}$-decidable and B is infinite.

4.4.8. Let $\mathscr{C}$ be a FRS closed under enumeration operations with internal interpreter (ϕ, S). Assume f and f' are unary $\mathscr{C}$-computable functions such that $f \subseteq_1 f'$, and A is a set such that $[f]_{(\phi,\mathsf{S})} \subseteq A$ and $[f']_{(\phi,\mathsf{S})} \subseteq \bar{A}$. Prove that A is not partially $\mathscr{C}$-decidable.

4.4.9. Let $\mathscr{C}$ be a FRS closed under enumeration operations with internal interpreter (ϕ, S). Assume f and f' are unary $\mathscr{C}$-computable functions such that $f \subseteq_1 f'$, A is a partially $\mathscr{C}$-decidable set such that $[f]_{(\phi,\mathsf{S})} \subseteq A$, and $B = A \cap [f']_{(\phi,\mathsf{S})}$. Prove that $\bar{B}$ is not partially $\mathscr{C}$-decidable.

4.4.10. Let $\mathscr{C}$ be a FRS with internal interpreter (ϕ, S) and po a $\mathscr{C}$-computable ϕ-preorder. Prove that po is not total.

Notes

The recursion theorem is a fascinating feature peculiar to reflexive structures. It contains recursion, in the sense of Chapter 2, since it is used in the proof of closure under recursive operations. But it goes farther and involves a more general form of recursive relation. The proof of Theorem 4.4.5 (due to Gandy but here taken from Hinman [8]) is a remarkable example of this technique.

More information about the recursion theorem is given in Rogers [20] and Smullyan [26]. The discussion in Cutland [1] is illuminating. Other results in this section are derived from Rice [18].

§5. Relational Structures

Up to this point we have considered only functional reflexive structures, in which the elements are functions. Such structures induce in a natural way relational structures in which the elements are predicates. A general definition, with some examples and applications, is presented in this section.

Let $\mathscr{P}$ be a class of predicates. An *interpreter* for $\mathscr{P}$ is a pair (W, S), where W is a set, S is a total binary function, and whenever P is a k-ary predicate in $\mathscr{P}$, then there is a number z such that

$$P(\mathbf{x}) \equiv \mathsf{S}^k(\mathsf{z}, \mathbf{x}) \in \mathsf{W}.$$

A more flexible notation is obtained by introducing $(k+1)$-ary predicates W^k, $k \geq 0$, such that

$$\mathsf{W}^k(z, \mathbf{x}) \equiv \mathsf{S}^k(z, \mathbf{x}) \in \mathsf{W}.$$

It follows that whenever P is a k-ary predicate in $\mathscr{P}$, then there is a number z such that

$$P(\mathbf{x}) \equiv \mathsf{W}^k(\mathsf{z}, \mathbf{x}).$$

When the notation $\mathsf{W}^k(z, \mathbf{x})$ is used, usually the first argument will be written as a subscript, that is, in the form $\mathsf{W}^k_{\mathsf{z}}(\mathbf{x})$. Now, for every number z, $\mathsf{W}^k_{\mathsf{z}}$ denotes a k-ary predicate, and for every k-ary predicate P in $\mathscr{P}$ there is a number z such that $P = \mathsf{W}^k_{\mathsf{z}}$. We say that z is a (W, S)-*index* for P.

If $k = 1$ the superscript k is omitted, and $\mathsf{W}_z = \mathsf{W}^1_z$ is considered a set. Hence, whenever A is a set in $\mathscr{P}$, then there is a number z such that $A = \mathsf{W}_{\mathsf{z}}$.

We recall the relation $\mathsf{S}^{k+m}(z, \mathbf{x}, \mathbf{y}) = \mathsf{S}^m(\mathsf{S}^k(z, \mathbf{x}), \mathbf{y})$. It follows that whenever P is a $(k + m)$-ary predicate in $\mathscr{P}$, $k, m > 0$, then there is a number z such that

$$\begin{aligned} P(\mathbf{x}, \mathbf{y}) &\equiv \mathsf{W}^{k+m}(\mathsf{z}, \mathbf{x}, \mathbf{y}) \\ &\equiv \mathsf{W}^m(\mathsf{S}^k(\mathsf{z}, \mathbf{x}), \mathbf{y}) \\ &\equiv \mathsf{W}^m_{g(\mathbf{x})}(\mathbf{y}), \end{aligned}$$

where $g(\mathbf{x}) = \mathsf{S}^k(\mathsf{z}, \mathbf{x})$. We call this relation the *abstraction property* for the interpreter (W, S). Note that whenever S is strict (elementary) (recursive) then g is also strict (elementary) (recursive).

If (W, S) is an interpreter for the class $\mathscr{P}$, W is in $\mathscr{P}$, S is a recursive function, and $\mathscr{P}$ is closed under substitution with total recursive functions, we say that

(W, S) is an *internal interpreter* for $\mathscr{P}$. It follows that all functions S^k are recursive, all predicates W^k are in $\mathscr{P}$, and furthermore all predicates W_z^k are in $\mathscr{P}$. We conclude that whenever (W, S) is an internal interpreter for $\mathscr{P}$, then a k-ary predicate P is in $\mathscr{P}$ if and only if there is a number z such that $P = W_z^k$.

EXAMPLE 4.5.1. Let $\mathscr{C}$ be a FRS with internal interpreter (ϕ, S), where S is a recursive function. Then (W_ϕ, S) is an internal interpreter for $\mathscr{C}_{pd}$ (see Theorem 4.1.6). In particular, if $\mathscr{C} = RC$ and (ϕ, S) is any internal interpreter for RC, then (W_ϕ, S) is an internal interpreter for $RC_{de} = RC_{pd}$ = the class of all recursively enumerable predicates.

EXAMPLE 4.5.2. Let $\mathscr{P}$ be the class that contains all the k-ary false predicates, that is, predicates of the form $F_k(\mathbf{x}) \equiv F$ for all $(\mathbf{x}) \in \mathbb{N}^k$, $k \geq 0$. Then $(\varnothing, S)$, where S is an arbitrary total binary recursive function, is an internal interpreter for $\mathscr{P}$.

Theorem 4.5.1. *Let $\mathscr{P}$ be a class of predicates with internal interpreter* (W, S). *Then*

(i) *If P is a $(k + m + 1)$-ary predicate in $\mathscr{P}$, $k, m \geq 0$, there is a total m-ary recursive function g_1 such that*

$$P(\mathbf{x}, \mathbf{y}, g_1(\mathbf{y})) \equiv W^k(g_1(\mathbf{y}), \mathbf{x}).$$

(ii) *If P is an $(m + 1)$-ary predicate in $\mathscr{P}$, $m \geq 0$, there is a total m-ary recursive function g_2 such that*

$$P(\mathbf{y}, g_2(\mathbf{y})) \equiv g_2(\mathbf{y}) \in W.$$

PROOF. To prove (i) assume P is a $(k + m + 1)$-ary predicate in $\mathscr{P}$, and let f_1 be the total m-ary recursive function given by the abstraction property such that

$$P(\mathbf{x}, \mathbf{y}, S(z, z)) \equiv W^{k+1}(f_1(\mathbf{y}), z, \mathbf{x}).$$

We put $g_1(\mathbf{y}) = S(f_1(\mathbf{y}), f_1(\mathbf{y}))$, and compute

$$\begin{aligned} W^k(g_1(\mathbf{y}), \mathbf{x}) &\equiv W^{k+1}(f_1(\mathbf{y}), f_1(\mathbf{y}), \mathbf{x}) \\ &\equiv P(\mathbf{x}, \mathbf{y}, S(f_1(\mathbf{y}), f_1(\mathbf{y}))) \\ &\equiv P(\mathbf{x}, \mathbf{y}, g_1(\mathbf{y})). \end{aligned}$$

To prove part (ii) we apply part (i) with $k = 0$. □

In applications Theorem 4.5.1 will be referred to as the *recursion theorem* (RT). We derive immediately a few easy consequences.

Corollary 4.5.1.1. *Let $\mathscr{P}$ be a class of predicates with internal interpreter* (W, S). *Then*

(i) *If P is a* $(k+1)$*-ary predicate in* $\mathscr{P}$, $k \geq 0$, *there is a number* z *such that*

$$P(\mathbf{x}, \mathsf{z}) \equiv \mathsf{W}^k(\mathsf{z}, \mathbf{x}).$$

(ii) *If A is a set in* $\mathscr{P}$, *there is a number* z *such that*

$$\mathsf{z} \in A \equiv \mathsf{z} \in \mathsf{W}.$$

(iii) *The set* $\overline{\mathsf{W}}$ is not in $\mathscr{P}$.

Proof. Part (i) is immediate from Theorem 4.5.1 (i) with $m = 0$, and part (ii) follows from Theorem 4.5.1 (ii) also with $m = 0$. To prove (iii) note that if $\overline{\mathsf{W}}$ is in $\mathscr{P}$ then from (ii) it follows that there is a number z such that

$$\mathsf{z} \in \overline{\mathsf{W}} \equiv \mathsf{z} \in \mathsf{W},$$

which is obviously a contradiction. □

Up to this point substitution in predicates has been restricted to total functions, on the assumption that predicates are total operations, and substitution with nontotal functions would be inconsistent. There are ways to overcome this restriction, and the most natural is to assume that if a predicate expression is undefined it takes the value F. This was in fact the underlying interpretation for the theory of induction given in Chapter 3. We formalize this idea as a more general form of substitution.

Let $\mathscr{P}$ be a class of predicates and $\mathscr{C}$ a class of functions. We say that $\mathscr{P}$ is *closed under general substitution with* $\mathscr{C}$*-computable functions*, if whenever Q is an m-ary predicate in $\mathscr{P}$, $m \geq 1$, and $g_1, \ldots, g_m$ are k-ary $\mathscr{C}$-computable functions, then the k-ary predicate P such that

$$P(\mathbf{x}) \equiv \exists y_1 \cdots \exists y_m\colon g_1(\mathbf{x}) \simeq y_1 \wedge \cdots \wedge g_m(\mathbf{x}) \simeq y_m \wedge Q(y_1, \ldots, y_m)$$

is also in $\mathscr{P}$.

Example 4.5.3. Let $\mathscr{C}$ be a class of functions closed under basic operations. Then $\mathscr{C}_{\mathsf{pd}}$ is closed under general substitution with $\mathscr{C}$-computable functions. Note that $\mathscr{C}_{\mathsf{pd}}$ is not necessarily closed under existential quantification.

Theorem 4.5.2. *Assume* $(\mathsf{W}', \mathsf{S}')$ *is an interpreter for the class* $\mathscr{P}$, *where* W' *is in* $\mathscr{P}$, S' *is* $\mathscr{C}$*-computable,* $\mathsf{RC} \subseteq \mathscr{C}$, $\mathscr{C}$ *is closed under elementary operations, and* $\mathscr{P}$ *is closed under general substitution with* $\mathscr{C}$*-computable functions. If* (ϕ, S) *is an internal interpreter for* RC, *then* $\mathscr{P}$ *has an internal interpreter of the form* (W, S).

Proof. We introduce a unary function f and set W such that

$$f(x) \simeq \mathsf{S}'([\phi(x)]_1, \S\phi(x))$$

$$x \in \mathsf{W} \equiv \mathsf{E}v\colon f(x) \simeq v \wedge v \in \mathsf{W}'.$$

It is clear that f is $\mathscr{C}$-computable and W is in $\mathscr{P}$. To prove that (W, S) is an interpreter for $\mathscr{P}$ assume that P is a k-ary predicate in $\mathscr{P}$, and let P' be the unary predicate such that

$$\begin{aligned} P'(x) &\equiv P([x]_1, \dots, [x]_k) \\ &\equiv \mathsf{S}'(\mathsf{z}, x) \in \mathsf{W}', \end{aligned}$$

where z is a $(\mathsf{W}', \mathsf{S}')$-index for the predicate P'. Let z' be a (ϕ, S)-index for the k-ary elementary function $g(\mathbf{x}) \simeq \langle \mathsf{z}, \mathbf{x} \rangle$. It follows that

$$\begin{aligned} \mathsf{S}^k(\mathsf{z}', \mathbf{x}) \in \mathsf{W} &\equiv \exists v\colon f(\mathsf{S}^k(\mathsf{z}', \mathbf{x})) \simeq v \wedge v \in \mathsf{W}' \\ &\equiv \exists v\colon \mathsf{S}'([\phi(\mathsf{S}^k(\mathsf{z}', \mathbf{x}))]_1, \S\phi(\mathsf{S}^k(\mathsf{z}', \mathbf{x}))) \simeq v \wedge v \in \mathsf{W}' \\ &\equiv \mathsf{S}'(\mathsf{z}, \langle \mathbf{x} \rangle) \in \mathsf{W}' \\ &\equiv P'(\langle \mathbf{x} \rangle) \\ &\equiv P(\mathbf{x}). \end{aligned}$$

□

A class $\mathscr{P}$ of predicates is a *relational reflexive structure* (RRS) if the following conditions are satisfied:

(i) $\mathscr{P}$ is closed under general substitution with recursive functions, conjunction, and universal bounded quantification.
(ii) $\mathscr{P}$ has an internal interpreter.

EXAMPLE 4.5.4. Let $\mathscr{C}$ be a FRS with internal interpreter (ϕ, S), where S is a recursive function (e.g., $\mathsf{S} = \mathsf{S}_\delta$). Then $\mathscr{C}_{\mathsf{pd}}$ is a RRS with internal interpreter $(\mathsf{W}_\phi, \mathsf{S})$. Note that $\mathscr{C}_{\mathsf{pd}}$ is actually closed under general substitution with $\mathscr{C}$-computable functions.

Theorem 4.5.3. *Let $\mathscr{P}$ be a* RRS *with internal interpreter* (W, S). Then

(i) W *is nonempty.*
(ii) *If $\mathscr{C}$ is a class of functions closed under elementary operations, and $\mathscr{P}$ is closed under general substitution with $\mathscr{C}$-computable functions, then $\mathscr{C}_{\mathsf{pd}} \subseteq \mathscr{P}$.*
(iii) $\mathsf{RC}_{\mathsf{de}} \subseteq \mathscr{P}$.
(iv) *$\mathscr{P}_{\mathsf{e}}$ and $\mathscr{P}_{\mathsf{u}}$ are both* RRSs.
(v) *$\mathscr{P}_{\mathsf{e}}$ is closed under existential quantification and disjunction, and $\mathscr{P}_{\mathsf{u}}$ is closed under universal quantification.*

PROOF. To prove that W is nonempty it is sufficient to show that $\mathscr{P}$ contains at least one nonempty set. We take the set A such that

$$x \in A \equiv \forall y < x \mathsf{S}(y, y) \in \mathsf{W},$$

and it follows that $0 \in A$. To prove (ii) we consider a k-ary predicate Q that is partially $\mathscr{C}$-decidable and introduce the k-ary function f such that

$$f(\mathbf{x}) \simeq \psi_Q(\mathbf{x}) + \mathsf{m},$$

where $\mathsf{m} \in \mathsf{W}$. It follows that

$$Q(\mathbf{x}) \equiv \exists y\colon f(\mathbf{x}) \simeq y \wedge y \in \mathsf{W},$$

so Q is in $\mathscr{P}$. Part (iii) follows from (ii) by taking $\mathscr{C} = \mathsf{RC}$.

To prove (iv) it is sufficient to introduce internal interpreters for $\mathscr{P}_{\mathsf{e}}$ and $\mathscr{P}_{\mathsf{u}}$. Let W' and W'' be the sets such that

$$z \in \mathsf{W}' \equiv \exists y \mathsf{S}(z, y) \in \mathsf{W}$$

$$z \in \mathsf{W}'' \equiv \forall y \mathsf{S}(z, y) \in \mathsf{W}.$$

It follows that $(\mathsf{W}', \mathsf{S})$ is an internal interpreter for $\mathscr{P}_{\mathsf{e}}$, and $(\mathsf{W}'', \mathsf{S})$ is an internal interpreter for $\mathscr{P}_{\mathsf{u}}$.

From Theorem 3.1.2 (iii) it follows that $\mathscr{P}_{\mathsf{e}}$ is closed under existential quantification, and $\mathscr{P}_{\mathsf{u}}$ is closed under universal quantification. To prove that $\mathscr{P}_{\mathsf{e}}$ is closed under disjunction we consider k-ary predicates P_1 and P_2 in $\mathscr{P}$, with (W, S)-indexes z_1 and z_2. It follows that

$$P_1(\mathbf{x}) \vee P_2(\mathbf{x}) \equiv \exists y \mathsf{S}^k(\mathsf{cd}(\mathsf{z}_1, \mathsf{z}_2, y), \mathbf{x})) \in \mathsf{W};$$

hence $P_1 \vee P_2$ is in $\mathscr{P}$. □

Corollary 4.5.3.1. *Let $\mathscr{P}$ be a* RRS *and $\mathscr{C}$ a class of functions closed under recursive operations, where $\mathscr{P}$ is closed under general substitution with $\mathscr{C}$-computable functions. If P_1 is a k-ary $\mathscr{C}$-decidable predicate, and P_2 is a k-ary predicate in $\mathscr{P}$, then $P_1 \vee P_2$ is in $\mathscr{P}$.*

Proof. Let (W, S) be an internal interpreter for $\mathscr{P}$. From the inclusion $\mathscr{C}_{\mathsf{pd}} \subseteq \mathscr{P}$ it follows that P_1 is in $\mathscr{P}$. Let $\mathsf{z}_1, \mathsf{z}_2$ be (W, S)-indexes for P_1 and P_2, and let f be the k-ary total function such that

$$f(\mathbf{x}) = \mathsf{S}^k(\mathsf{cd}(\mathsf{z}_1, \mathsf{z}_2, \chi_{P_1}(\mathbf{x})), \mathbf{x}).$$

It follows that f is $\mathscr{C}$-computable, and

$$P_1(\mathbf{x}) \vee P_2(\mathbf{x}) \equiv f(\mathbf{x}) \in \mathsf{W};$$

hence $P_1 \vee P_2$ is in $\mathscr{P}$. □

Example 4.5.5. The preceding results have many applications. If $\mathscr{C}$ is a FRS we know that $\mathscr{C}_{\mathsf{pd}}$ is a RRS, not necessarily closed under existential quantification or disjunction. From Theorem 4.5.3 it follows that all classes $\mathscr{C}_{\mathsf{pde}}$, $\mathscr{C}_{\mathsf{pdu}}$, $\mathscr{C}_{\mathsf{pdeu}}$, $\mathscr{C}_{\mathsf{pdue}}$, $\mathscr{C}_{\mathsf{pdeue}}$, ... are also RRSs. Note that the classes $\mathscr{C}_{\mathsf{pde}}$, $\mathscr{C}_{\mathsf{pdue}}$, $\mathscr{C}_{\mathsf{pdeue}}$, ... are closed under existential quantification and disjunction.

Example 4.5.6. Let $\mathscr{P}$ be RRS with internal interpreter (W, S), where the function S is strict. Consider the 3-ary predicate P such that

$$P(x, y, z) \equiv x \in \mathsf{W}_y.$$

If we apply the RT (i) we get a total recursive function g such that

$$x \in \mathsf{W}_y \equiv x \in \mathsf{W}_{g(y)},$$

which means that $\mathsf{W}_y = \mathsf{W}_{g(y)}$ holds for all y. Since g is strict we have $y < g(y) < g(g(y))$, and so on, which means that the function g generates infinitely many (W, S)-indexes for the set W_y.

Let C be a set and f a unary function, not necessarily total. Then $f(C)$ is the set such that

$$x \in f(C) \equiv \exists y\colon y \in C \wedge f(y) \simeq x.$$

If B and A are two sets, and C is a set such that $B \subseteq C \subseteq \bar{A}$, we say that C is an *extension of* B *bounded by* A. The class of all extensions of B which are bounded by A is denoted by $[B, A]$. We say that $[B, A]$ is an *interval*. Note that $[A, B]$ is also an interval, and C is an element of $[A, B]$ if and only if $\bar{C}$ is an element of $[B, A]$.

Let $\mathscr{P}$ be a RRS with internal interpreter (W, S), and let [B, A] be some interval. If p is a unary total function that satisfies for all y the condition

$$p(y) \in (B \cap \overline{\mathsf{W}}_y) \cup (A \cap \mathsf{W}_y),$$

we say that p is a (W, S)-*productive function for the interval* $[B, A]$. It is clear that from this condition it follows that for no y is W_y an element of [B, A]; that is, $\mathscr{P} \cap [\mathsf{B}, \mathsf{A}] = \emptyset$.

Let $\mathscr{P}$ be a RRS and $\mathscr{C}$ a class of functions. We say that the interval [B, A] is $\mathscr{C}$-*productive* in $\mathscr{P}$, if there is an internal interpreter (W, S) for $\mathscr{P}$, and a $\mathscr{C}$-computable function p that is (W, S)-productive for [B, A]. Finally, [B, A] is *productive* in $\mathscr{P}$, if it is RC-productive in $\mathscr{P}$.

It is convenient to have a weak notion of productivity, where the function p is not necessarily total. If $\mathscr{P}$ is a RRS with internal interpreter (W, S), and [B, A] is some interval, we say that the unary function p is a *weak* (W, S)-*productive* function for [B, A] if whenever $\mathsf{W}_y \subseteq \bar{A}$ there is a v such that $p(y) \simeq v$ and $v \in B \cap \overline{\mathsf{W}}_y$. If $\mathscr{C}$ is a class of functions and there is an internal interpreter (W, S) for $\mathscr{P}$, and a $\mathscr{C}$-computable function p that is a weak (W, S)-productive function for [B, A], we say that [B, A] is *weakly* $\mathscr{C}$-*productive* in $\mathscr{P}$. If [B, A] is weakly RC-productive in $\mathscr{P}$, we say that it is *weakly productive* in $\mathscr{P}$.

Let $\mathscr{P}$ be a RRS with internal interpreter (W, S). If r is a total unary function such that $r(\mathsf{W}) \subseteq A$ and $r(\overline{\mathsf{W}}) \subseteq B$, we say that r is a (W, S)-*reduction* function for the interval [B, A]. If r is a function not necessarily total, but $r(y)$ is defined whenever $y \in \overline{\mathsf{W}}$, $r(\mathsf{W}) \subseteq A$, and $r(\overline{\mathsf{W}}) \subseteq B$, we say that r is a *weak* (W, S)-*reduction* function for [B, A].

Example 4.5.7. Let $\mathscr{P}$ be a RRS with internal interpreter (W, S), and A, B sets such that there is a number $\mathsf{m} \in A \cap B$. Then the function C^1_m is a (W, S)-productive function for [B, A] and also for [A, B]. It follows that both intervals are productive in any RRS $\mathscr{P}$.

EXAMPLE 4.5.8. Let $\mathscr{P}$ be a RRS with internal interpreter (W, S), and let K_W, $K_{\overline{W}}$ be sets such that

$$y \in K_W \equiv y \in W_y$$

$$y \in K_{\overline{W}} \equiv y \notin W_y;$$

hence K_W and $K_{\overline{W}}$ are one complement of the other. The function $p(y) = y$ is a (W, S)-productive function for the interval $[K_{\overline{W}}, K_W]$. For if $p(y) = y \notin K_W \cap W_y$ it follows that $y \in K_{\overline{W}} \cap \overline{W}_y$. This means that $[K_{\overline{W}}, K_W]$ is productive in $\mathscr{P}$.

EXAMPLE 4.5.9. Let $\mathscr{P}$ be a RRS with internal interpreter (W, S). Let K_W be the set of the preceding example, and let K'_W be the set such that

$$y \in K'_W \equiv K_W \cap W_y = \varnothing.$$

Note that $K'_W \subseteq K_{\overline{W}}$. The function $p(y) = y$ is a weak (W, S)-productive function for the interval $[K'_W, K_W]$, for if $W_y \subseteq K_{\overline{W}}$ then $y \in K'_W \cap \overline{W}_y$. Hence $[K'_W, K_W]$ is weakly productive in $\mathscr{P}$.

EXAMPLE 4.5.10. Let $\mathscr{P}$ be a RRS with internal interpreter (W, S). The total function $r(y) = y$ is a reduction function for $[\overline{W}, W]$. On the other hand, the nontotal function r', such that $r'(y) \simeq v$ if and only if $y \in \overline{W}$ and $v = y$, is a weak (W, S)-reduction function for $[\overline{W}, \varnothing]$.

Theorem 4.5.4. *Let $\mathscr{P}$ be a RRS with internal interpreter (W, S), $\mathscr{C}$ a class of functions such that $\mathscr{P}$ is closed under general substitution with $\mathscr{C}$-computable functions, and $[B, A]$ some interval. Then*

(i) *If p is a $\mathscr{C}$-computable weak (W, S)-productive function for $[B, A]$, and C is a set in $\mathscr{P}$, then there is a unary total recursive function g_1 such that $r = p \circ g_1$ is a total function, $r(C) \subseteq A$, and $r(\bar{C}) \subseteq B$.*
(ii) *If r is a $\mathscr{C}$-computable weak (W, S)-reduction function for $[B, A]$, then there is a unary total recursive function g_2 such that $p = r \circ g_2$ is a (W, S)-productive function for $[B, A]$.*

PROOF. To prove (i) consider the 3-ary predicate P_1 such that

$$P_1(x, y, z) \equiv y \in C \wedge \exists v\colon p(z) \simeq v \wedge v = x.$$

Clearly, P_1 is in $\mathscr{P}$, and using the RT (i) we get a unary total recursive function g_1 such that

$$\begin{aligned} x \in W_{g_1(y)} &\equiv P_1(x, y, g_1(y)) \\ &\equiv y \in C \wedge \exists v\colon p(g_1(y)) \simeq v \wedge v = x. \end{aligned}$$

Note that $W_{g_1(y)} = \varnothing$ in case $y \notin C$ or $p(g_1(y))$ is undefined. It follows that $r = p \circ g_1$ is a total function, for if $r(y)$ is undefined, then $W_{g_1(y)} \subseteq \bar{A}$ and $p(g_1(y)) \simeq r(y)$ is defined. We must show that $r(C) \subseteq A$ and $r(\bar{C}) \subseteq B$. If $y \in C$ then $W_{g_1(y)} = \{r(y)\}$; hence $r(y) \in A$, for otherwise $W_{g_1(y)} \subseteq \bar{A}$ and $r(y) \notin W_{g_1(y)}$, which is a contradiction. This proves that $r(C) \subseteq A$. If $y \notin C$ then $W_{g_1(y)} = \varnothing$; hence $r(y) \in B$. This proves $r(\bar{C}) \subseteq B$.

To prove (ii) we consider the binary predicate P_2 such that

$$P_2(y, z) \equiv \exists v\colon r(z) \simeq v \wedge v \in \mathsf{W}_y.$$

Clearly, P_2 is in $\mathscr{P}$, and using the RT (ii) we get a unary total recursive function g_2 such that

$$\begin{aligned} g_2(y) \in \mathsf{W} &\equiv P_2(y, g_2(y)) \\ &\equiv \exists v\colon r(g_2(y)) \simeq v \wedge v \in \mathsf{W}_y. \end{aligned}$$

It follows that $p = r \circ g_2$ is total, for if $p(y)$ is undefined then $g_2(y) \in \overline{\mathsf{W}}$; hence $r(g_2(y)) \simeq p(y)$ is defined. To prove that p is a (W, S)-productive function for $[B, A]$, assume that $p(y) \notin A \cap \mathsf{W}_y$. If $p(y) \in \mathsf{W}_y$ then $g_2(y) \in \mathsf{W}$; hence $p(y) = r(g_2(y)) \in A$, contradicting the assumption. It follows that $p(y) \notin \mathsf{W}_y$; hence $g_2(y) \in \overline{\mathsf{W}}$ and $p(y) \in B \cap \overline{\mathsf{W}}_y$. □

Theorem 4.5.5. *Let $\mathscr{P}$ be a* RRS *closed under general substitution with $\mathscr{C}$-computable functions, where $\mathscr{C}$ is closed under recursive operations. If $[B, A]$ is* an interval the following conditions are equivalent:

(i) *$[B, A]$ is $\mathscr{C}$-productive in $\mathscr{P}$.*
(ii) *$[B, A]$ is weakly $\mathscr{C}$-productive in $\mathscr{P}$.*
(iii) *If C is a set in $\mathscr{P}$, then there is a total $\mathscr{C}$-computable function r such that $r(C) \subseteq A$ and $r(\overline{C}) \subseteq B$.*
(iv) *If* (W, S) *is an internal interpreter for $\mathscr{P}$, then there is a $\mathscr{C}$-computable* (W, S)*-reduction function for $[B, A]$.*
(v) *If* (W, S) *is an internal interpreter for $\mathscr{P}$, then there is a $\mathscr{C}$-computable weak* (W, S)*-reduction function for $[B, A]$.*
(vi) *If* (W, S) *is an internal interpreter for $\mathscr{P}$, then there is a $\mathscr{C}$-computable* (W, S)*-productive function for $[B, A]$.*

PROOF. The implication from (i) to (ii) is trivial, and the implication from (ii) to (iii) follows from Theorem 4.5.4 (i). The implication from (iii) to (iv) is immediate if we take $C = \mathsf{W}$, the implication from (iv) to (v) is trivial, and the implication from (v) to (vi) follows from Theorem 4.5.4 (ii). Finally, the implication from (vi) to (i) is trivial. □

The preceding theorem applies, in particular, when $\mathscr{C} = \mathsf{RC}$. This version is obtained by replacing everywhere $\mathscr{C}$-computable by recursive and $\mathscr{C}$-productive by productive.

If an interval $[B, A]$ is $\mathscr{C}$-productive in a RRS $\mathscr{P}$, a number of total functions are induced, some of them productive functions, others reduction functions, but all depending on a particular internal interpreter (W, S) for the reflexive structure. We shall now show that these functions can be taken 1-1, provided the input function S in the interpreter is 1-1.

Theorem 4.5.6. *Let $[B, A]$ be a nonempty $\mathscr{C}$-productive interval in a* RRS *$\mathscr{P}$, where $\mathscr{C}$ is closed under recursive operations and $\mathscr{P}$ is closed under general* *substitution with $\mathscr{C}$-computable functions. If* (W, S) *is an internal interpreter for $\mathscr{P}$, then there is a* 1-1 *$\mathscr{C}$-computable weak* (W, S)*-productive function for* $[B, A]$.

Proof. Assume that p is a given total $\mathscr{C}$-computable (W, S)-productive function for $[B, A]$. We use recursive operations to specify a total $\mathscr{C}$-computable function p' that is a weak (W, S)-productive function for $[B, A]$ and satisfies the condition $p'(y) < p'(y + 1)$ for all y. We need only the given productive function p, recursive functions, and primitive recursive operations.

We set $p'(0) = p(0)$. To specify $p'(y + 1)$ we note that p' is intended to be only a weak productive function, so we must pay attention only to the case $\mathsf{W}_{y+1} \subseteq \bar{A}$. Using the abstraction property we introduce a total unary recursive function f such that

$$x \in \mathsf{W}_{f(y)} \equiv x \in \mathsf{W}_y \vee x = p(y).$$

Note that we use disjunction with a $\mathscr{C}$-decidable predicate, so Corollary 4.5.3.1 applies. If we put $g(y, n) = p(f^n(y + 1))$, and assume $\mathsf{W}_{y+1} \subseteq \bar{A}$, it follows that $g(y, n) \neq g(y, n')$ whenever $n \neq n'$, and furthermore since $A \cap B = \varnothing$ we have $g(y, n) \in B \cap \overline{\mathsf{W}}_y$ for all n. We put

$$h(y) = \max\{g(y, n): n \leq p'(y) + 1\}$$

$$p'(y + 1) = \max\{p'(y) + 1, g(y, h(y))\}.$$

The idea is that if all values $g(y, 0), g(y, 1), \ldots, g(y, p'(y) + 1)$ are different then there is n such that $p'(y) < g(y, n)$, and it is safe to take $p'(y + 1) = g(y, n)$. If there is a repetition this means that $\mathsf{W}_{y+1} \nsubseteq \bar{A}$ and it is sufficient to take $p'(y + 1) = p'(y) + 1$. □

Corollary 4.5.6.1. *Let $[B, A]$ be a nonempty $\mathscr{C}$-productive interval in a* RRS *$\mathscr{P}$, where $\mathscr{C}$ is closed under recursive operations and $\mathscr{P}$ is closed under general substitution with $\mathscr{C}$-computable functions. Let* (W, S) *be an internal interpreter for $\mathscr{P}$, where* S *is* 1-1. *Then*

(i) *If C is a set in $\mathscr{P}$, then there is a unary total* 1-1 *$\mathscr{C}$-computable function r such that $r(C) \subseteq A$ and $r(\bar{C}) \subseteq B$.*
(ii) *There is a* 1-1 *$\mathscr{C}$-computable* (W, S)*-productive function for* $[B, A]$.

Proof. From Theorem 4.5.6 we know that there is a 1-1 $\mathscr{C}$-computable weak (W, S)-productive function p for $[B, A]$. From Theorem 4.5.4 (i) we get a total recursive function g_1 such that $r = p \circ g_1$ satisfies the conditions $r(C) \subseteq A$ and $r(\bar{C}) \subseteq B$. Since g_1 is given by the RT and S is 1-1, it follows that r is 1-1. To prove part (ii) we use Theorem 4.5.4 (ii) with the function r given by part (i) for $C = \mathsf{W}$. □

Let $\mathscr{P}$ be a RRS and $\mathscr{C}$ a class of functions. If A is a set in $\mathscr{P}$ such that the interval $[\bar{A}, A]$ is $\mathscr{C}$-productive we say that A is *$\mathscr{C}$-creative* in $\mathscr{P}$. If A is RC-creative in $\mathscr{P}$ we say that A is *creative* in $\mathscr{P}$.

Theorem 4.5.7. *Let $\mathscr{P}$ be a* RRS *and $\mathscr{C}$ a class of functions closed under recursive operations. Assume that $\mathscr{P}$ is closed under general substitution with $\mathscr{C}$-computable functions, and A is a set in $\mathscr{P}$. The following conditions are equivalent:*

(i) *A is $\mathscr{C}$-creative in $\mathscr{P}$.*
(ii) *For each set C in $\mathscr{P}$ there is a total* 1-1 *$\mathscr{C}$-computable function f such that*

$$x \in C \equiv f(x) \in A.$$

(iii) *For each set C in $\mathscr{P}$ there is a total $\mathscr{C}$-computable function f such that $f(C) \subseteq A$ and $f(\bar{C}) \subseteq \bar{A}$.*

PROOF. The implication from (i) to (ii) follows from Theorem 4.5.4 (i) and Theorem 4.5.7, noting that from Theorem 4.5.2 it follows that $\mathscr{P}$ has an internal interpreter (W, S), where S is 1-1. The implication from (ii) to (iii) is trivial, and the implication from (iii) to (i) follows from Theorem 4.5.5 (iii). □

Theorem 4.5.8. *Let $\mathscr{P}$ be a* RRS *and $\mathscr{C}$ a class of functions closed under recursive operations. Assume $\mathscr{P}$ is closed under general substitution with $\mathscr{C}$-computable functions, and the sets A_1 and A_2 are both $\mathscr{C}$-creative. Then there is a $\mathscr{C}$-permutation h such that $h(A_1) = A_2$.*

PROOF. Immediate from Theorem 4.5.7 (ii) and Theorem 2.3.2. □

Examples of productive intervals in a RRS $\mathscr{P}$ are usually derived from a given internal interpreter (W, S). If $\mathscr{S}$ is a class of sets such that $\mathscr{S} \subseteq \mathscr{P}$, then $[\mathscr{S}]_{(\mathsf{W},\mathsf{S})}$ denotes the set that contains all (W, S)-indexes of sets in $\mathscr{S}$. In particular, $[\varnothing]_{(\mathsf{W},\mathsf{S})}$ is the set of all (W, S)-indexes of the empty set, and if C is an arbitrary set in $\mathscr{P}$ then $[C]_{(\mathsf{W},\mathsf{S})}$ is the set of all (W, S)-indexes for C.

Theorem 4.5.9. *Let $\mathscr{P}$ be a* RRS *with internal interpreter* (W, S). *We put $B = [\varnothing]_{(\mathsf{W},\mathsf{S})}$ and $A = [C]_{(\mathsf{W},\mathsf{S})}$, where C is a nonempty set in $\mathscr{P}$. Then the interval $[B, A]$* is productive in $\mathscr{P}$.

PROOF. Let g be the total unary recursive function given by the abstraction property such that

$$x \in \mathsf{W}_{g(y)} \equiv y \in \mathsf{W} \wedge x \in C.$$

If $y \in \mathsf{W}$ then $\mathsf{W}_{g(y)} = C$, hence $g(y) \in A$; if $y \notin \mathsf{W}$ then $\mathsf{W}_{g(y)} = \varnothing$, hence $g(y) \in B$. It follows that g is a (W, S)-reduction function for the interval $[B, A]$, which is productive in $\mathscr{P}$. □

Corollary 4.5.9.1. *Let $\mathscr{P}$ be a* RRS *with internal interpreter* (W, S). *If B is a set in $\mathscr{P}$ such that $[\varnothing]_{(\mathsf{W},\mathsf{S})} \subseteq B$, then B contains an infinite number of* (W, S)-*indexes for every set in $\mathscr{P}$.*

Proof. From Theorem 4.5.9 it follows that $[\varnothing]_{(\mathsf{W},\mathsf{S})}$ is infinite. If C is a nonempty set in $\mathscr{P}$, and B contains only a finite number of (W, S)-indexes for C, we can delete these indexes from B and obtain a set B' that is also in $\mathscr{P}$, contradicting Theorem 4.5.9 which implies that $[B', \bar{B}']$ is productive in $\mathscr{P}$. □

Let $\mathscr{P}$ be a RRS and $\mathscr{C}$ a class of functions. We say that a set B is *$\mathscr{C}$-productive* in $\mathscr{P}$ if the interval $[B, \bar{B}]$ is $\mathscr{C}$-productive in $\mathscr{P}$. Or equivalently, if there is a set A such that $[B, A]$ is $\mathscr{C}$-productive in $\mathscr{P}$, and $B \cap A = \varnothing$. Note that this condition is trivial unless some restriction is imposed on the class $\mathscr{C}$ (usually that $\mathscr{P}$ is closed under general substitution with $\mathscr{C}$-computable functions), for if B is not in $\mathscr{P}$ and (W, S) is an internal interpreter for $\mathscr{P}$, we can always define a (W, S)-productive function for $[B, \bar{B}]$.

Exercises

4.5.1. Let $\mathscr{P}$ be a class of predicates closed under general substitution with recursive functions, under conjunction, and under universal bounded quantification. Prove that the following conditions are equivalent:
(a) $\mathscr{P}$ is a RRS.
(b) There is a binary predicate Q in $\mathscr{P}$ such that whenever P is a unary predicate in $\mathscr{P}$ then there is a number z such that

$$P(x) \equiv Q(\mathrm{z}, x).$$

(c) There is a set W in $\mathscr{P}$ such that whenever P is a k-ary predicate in $\mathscr{P}$ then there is an elementary function f such that

$$P(\mathbf{x}) \equiv f(\mathbf{x}) \in \mathsf{W}.$$

(d) There is a set W in $\mathscr{P}$ such that whenever P is a k-ary predicate in $\mathscr{P}$ then there is a recursive function f such that

$$P(\mathbf{x}) \equiv f(\mathbf{x}) \in \mathsf{W}.$$

4.5.2. Let $\mathscr{P}$ be a RRS, and let A and B be sets, where B is finite. Prove that $A \cup B$ is in $\mathscr{P}$ if and only if A is in $\mathscr{P}$.

4.5.3. Prove that if (W, S) is an internal interpreter for a RRS $\mathscr{P}$, then W is infinite.

4.5.4. Let $\mathscr{P}$ be a RRS with internal interpreter (W, S). Prove the following:
(a) There is a total binary recursive function g_1 such that

$$\mathsf{W}_{g_1(y,z)} = \mathsf{W}_y \cap \mathsf{W}_z.$$

(b) There is a total binary recursive function g_2 such that

$$\mathsf{W}_{g_2}(y, z) = \mathsf{W}_y \cup \{z\}.$$

(c) There is a total unary recursive function g_3 such that

$$\mathsf{W}_{g_3(y)} = \mathsf{W}_y - \{y\}.$$

(d) There is a total unary recursive function g_4 such that

$$x \in \mathsf{W}_{g_4(y)} \equiv y \in \mathsf{W}_x.$$

(e) There is a total unary recursive function g_5 such that whenever $y \in \mathsf{W}_y$ then $\mathsf{W}_{g_5(y)} = \{y\}$, and whenever $y \notin \mathsf{W}_y$ then $\mathsf{W}_{g_5(y)} = \varnothing$.

4.5.5. Let $\mathscr{P}$ be a RRS with internal interpreter (W, S), and let $\mathscr{C}$ be a class of functions closed under recursive operations. Assume that $\mathscr{P}$ is closed under general substitution with $\mathscr{C}$-computable functions. Prove the following:
(a) If B is a $\mathscr{C}$-decidable set there is a total unary recursive function g_1 such that

$$\mathsf{W}_{g_1(y)} = \mathsf{W}_y \cup B.$$

(b) If f is a total $\mathscr{C}$-computable function there is a total unary recursive function g_2 such that

$$\mathsf{W}_{g_2(y)} = \mathsf{W}_y \cup \{f(y)\}.$$

4.5.6. Let $\mathscr{P}$ be a RRS with internal interpreter (W, S). Prove the following:
(a) There is a total binary recursive function g_1 such that if $v \in \mathsf{W}_y$ then $\mathsf{W}_{g_1(y,v)} = \{g_1(y,v)\}$, and if $v \notin \mathsf{W}_y$ then $\mathsf{W}_{g_1(y,v)} = \varnothing$.
(b) There is a total binary recursive function g_2 such that

$$x \in \mathsf{W}_{g_2(y,v)} \equiv \langle x, g_2(y,v)\rangle \in \mathsf{W}_y.$$

(c) There is a total binary recursive function g_3 such that

$$\mathsf{W}_{g_3(y,v)} = \mathsf{W}_y - \{g_3(y,v)\}.$$

4.5.7. Let $\mathscr{P}$ be a RRS with internal interpreter (W, S). Prove the following
(a) There is a number z such that $\mathsf{W}_z = \{z\}$.
(b) There is a total unary recursive function g such that

$$\mathsf{W}_{g(y)} = \mathsf{W}_y \cup \{g(y)\}.$$

(c) There is a total unary recursive function f such that

$$\mathsf{W}_{f(y)} = \{f(0), \ldots, f(y)\}.$$

4.5.8. Let $\mathscr{P}$ be a RRS and $\mathscr{C}$ a class of functions closed under recursive operations. Assume the interval $[B, A]$ is nonempty (i.e., $B \cap A = \varnothing$) and $\mathscr{C}$-productive in $\mathscr{P}$. Prove the following:
(a) B is not in $\mathscr{P}$, $\bar{A}$ is not in $\mathscr{P}$, and A is not $\mathscr{C}$-decidable.
(b) B has an infinite $\mathscr{C}$-decidable subset.
(c) A has an infinite $\mathscr{C}$-decidable subset.

4.5.9. Let $\mathscr{P}$ be a RRS closed under general substitution with $\mathscr{C}$-computable functions, where $\mathscr{C}$ is closed under recursive operations. Prove that $\mathscr{C} \subseteq \mathscr{P}_{\#}$.

4.5.10. Let $\mathscr{P}$ be a RRS with internal interpreter (W, S), closed under general substitution with $\mathscr{C}$-computable functions. Assume f is a unary $\mathscr{C}$-computable function, and let g be the unary total recursive function given by the abstraction property, such that

$$x \in \mathsf{W}_{g(y)} \equiv \exists v \colon f(x) \simeq v \wedge v \in \mathsf{W}_y.$$

Assume $[B, A]$ is an interval, and prove the following:
(a) If p is a weak (W, S)-productive function for $[B, A]$ and $B \subseteq \mathsf{D}_f$, then $f \circ p \circ g$ is a weak (W, S)-productive function for $[f(B), f(A)]$.
(b) If p is a (W, S)-productive function for $[B, A]$ and $B \subseteq \mathsf{D}_f$, then $f \circ p \circ g$ is a (W, S)-productive function for $[f(B), f(A)]$.

4.5.11. Let $\mathscr{P}$ be a RRS closed under general substitution with $\mathscr{C}$-computable functions, where $\mathscr{C}$ is closed under recursive operations. Assume f is a unary 1-1 $\mathscr{C}$-computable function. Prove the following:
(a) If B is $\mathscr{C}$-productive in $\mathscr{P}$ and $B \subseteq \mathsf{D}_f$ then $f(B)$ is $\mathscr{C}$-productive in $\mathscr{P}$.
(b) If A is $\mathscr{C}$-creative in $\mathscr{P}$ and f is total then $f(A)$ is $\mathscr{C}$-creative in $\mathscr{P}$.

4.5.12. Let $\mathscr{P}$ be a RRS with internal interpreter (W, S), and assume the set B' is in $\mathscr{P}$. Prove that there is a total unary recursive function g such that:
(a) If p is a weak (W, S)-productive function for an interval $[B, A]$ and $B \subseteq B'$, then $p \circ g$ is a weak (W, S)-productive function for the interval $[B, A \cap B']$.
(b) If p is a (W, S)-productive function for an interval $[B, A]$ and $B \subseteq B'$, then $p \circ g$ is a (W, S)-productive function for the interval $[B, A \cap B']$.

4.5.13. Let $\mathscr{P}$ be a RRS and $\mathscr{C}$ a class of functions closed under recursive operations. Assume B' is a set in $\mathscr{P}$. Prove the following:
(a) If B is $\mathscr{C}$-productive in $\mathscr{P}$ and $B \subseteq B'$, then $B \cup \bar{B}'$ is $\mathscr{C}$-productive in $\mathscr{P}$.
(b) If A is $\mathscr{C}$-creative in $\mathscr{P}$ and $A \cup B' = \mathbb{N}$, then $A \cap B'$ is $\mathscr{C}$-creative in $\mathscr{P}$.

4.5.14. Let $\mathscr{P}$ be a RRS with internal interpreter (W, S) closed under general substitution with $\mathscr{C}$-computable functions, where $\mathscr{C}$ is closed under recursive operations, and let A' be a set that is $\mathscr{C}$-decidable. Prove that there is a total unary recursive function g such that:
(a) If p is a weak (W, S)-productive function for an interval $[B, A]$ and $A \subseteq A'$, then $p \circ g$ is a weak (W, S)-productive function for the interval $[B \cap A', A]$.
(b) If p is a (W, S)-productive function for an interval $[B, A]$ and $A \subseteq A'$, then $p \circ g$ is a (W, S)-productive function for the interval $[B \cap A', A]$.

4.5.15. Let $\mathscr{P}$ be a RRS closed under general substitution with $\mathscr{C}$-computable functions, where $\mathscr{C}$ is closed under recursive operations. Assume A' is a $\mathscr{C}$-decidable set. Prove the following:
(a) If B is $\mathscr{C}$-productive in $\mathscr{P}$ and $\bar{B} \subseteq A'$, then $B \cap A'$ is $\mathscr{C}$-productive in $\mathscr{P}$.
(b) If A is $\mathscr{C}$-creative in $\mathscr{P}$ and $A \subseteq A'$, then $A \cup \bar{A}'$ is $\mathscr{C}$-creative in $\mathscr{P}$.

4.5.16. Let $\mathscr{P}$ be a RRS with internal interpreter (W, S). Assume $\mathscr{P}$ is closed under disjunction and C and C' are sets in $\mathscr{P}$ such that $C \subseteq C'$. Prove that if $B = [C]_{(\mathsf{W},\mathsf{S})}$ and $A = [C']_{(\mathsf{W},\mathsf{S})}$then the interval $[B, A]$ is productive in $\mathscr{P}$.

4.5.17. Let $\mathscr{P}$ be a RRS with internal interpreter (W, S), and assume C is a set in $\mathscr{P}$ such that $[C]_{(\mathsf{W},\mathsf{S})} \subseteq A$, where A is a set such that

$$z \in A \equiv \exists y Q(z, y),$$

and the predicate $\bar{Q}$ is in $\mathscr{P}$. Prove that there is a number $\mathbf{z} \in A$ that is a (W, S)-index for a finite subset of C.

4.5.18. Let $\mathscr{P}$ be a RRS with internal interpreter (W, S), closed under general substitution with $\mathscr{C}$-computable functions, where $\mathscr{C}$ is closed under elementary operations. Let P be a k-ary predicate in $\mathscr{P}$, and let A be the set of (W, S)-indexes for P. Prove that A is not $\mathscr{C}$-decidable.

4.5.19. Let $\mathscr{P}$ be a RRS with internal interpreter (W, S), and let A be the set such that

$$x \in A \equiv \exists y\colon (x = 2 \times y \wedge x \in \mathsf{W}_y) \vee (x = 2 \times y + 1 \wedge x \notin \mathsf{W}_y).$$

Prove that both $[A, \bar{A}]$ and $[\bar{A}, A]$ are productive in $\mathscr{P}$.

4.5.20. Let $\mathscr{P}$ be a RRS closed under existential quantification. Prove that $\mathscr{P} = \mathscr{C}_{\mathsf{pd}}$, where $\mathscr{C}$ is a class of functions closed under recursive operations.

4.5.21. Let $\mathscr{P}$ be a RRS closed under general substitution with $\mathscr{C}$-computable functions, where $\mathscr{C}$ is closed under basic operations and contains the functions σ, pd, and cd. Assume $\mathscr{C}$ has the selector property relative to $\mathscr{P}$. Prove that $\mathscr{C}$ is a FRS and $\mathscr{C} = \mathscr{P}_{\#}$.

4.5.22. Let $\mathscr{P}$ be a RRS with internal interpreter (W, S), and let h be a unary function such that

$$h(x) \simeq \sigma y\colon y \in \mathsf{W}_x \wedge 2 \times x < y.$$

Set $A = \mathsf{R}_h$, and prove the following:
(a) $\bar{A}$ is infinite.
(b) If B is a set in $\mathscr{P}$ and B is infinite then $B \nsubseteq \bar{A}$.

4.5.23. Let $\mathscr{P}$ be a RRS and assume A is a set such that $\bar{A}$ is infinite and whenever B is an infinite set in $\mathscr{P}$ then $B \nsubseteq \bar{A}$. Prove that if $\mathscr{C}$ is a class of functions such that $\mathscr{C}_{\mathsf{d}} \subseteq \mathscr{P}$ then $[\bar{A}, A]$ is not $\mathscr{C}$-productive.

4.5.24. Prove that there is a recursively enumerable set A such that:
(a) $\bar{A}$ is infinite.
(b) If B is an infinite recursively enumerable set then $B \nsubseteq \bar{A}$.
(c) $\bar{A}$ is not recursively enumerable and A is not creative.

Notes

A relational structure represents a weak form of reflexivity, which still supports a number of interesting constructions. This type of structure is preserved under enumeration, and we shall see later that it is also preserved under hyperenumeration.

Creative sets were introduced in Post [17]. Most of the results in this section are derived from Myhill [16] and Smullyan [26]. Note that we deal with intervals rather than sets, but the results are essentially the same.

§6. Uniform Structures

In this section we discuss briefly general functions, which are essentially functional transformations in the sense of Chapter 2. A k-ary *general function* is a function h_c that depends on a unary function c. So in order to evaluate $h_c(\mathbf{x})$ we need the value of the input $(\mathbf{x})$ and also the function c. By changing the function c the evaluation may be undefined or produce different values. There are many ways in which such a dependency on c can be defined. For example, there is a unary general function I_c such that $\mathsf{I}_c(x) \simeq c(x)$, or equivalently, $\mathsf{I}_c = c$ for any unary function c. Another example, very important from the point of view of the discussion in this section, is the function δ_c, which is specified by recursion in Section 2.

General functions include ordinary functions, for given a k-ary function h we can introduce the k-ary general function h_c such that $h_c(\mathbf{x}) \simeq h(\mathbf{x})$, and we say h_c is *induced* by h. Conversely, if h_c is a k-ary general function and c is a fixed unary function, we can introduce the ordinary k-ary function h such that $h(\mathbf{x}) \simeq h_c(\mathbf{x})$, and we say that h is *induced* by h_c with c fixed.

We have already mentioned that general functions are functionals and belong properly to second-order computability, which is beyond the scope of this book. Our purpose here is to identify some uniform relations, where a number z or a function g (usually recursive) has a property that applies to an arbitrary function c. The use of general functions is a convenient tool to identify such relations.

We use the letter $\mathscr{F}$ to denote classes of general functions. The elements of $\mathscr{F}$ are said to be $\mathscr{F}$-computable. If h_c is $\mathscr{F}$-computable, and we fix the unary function c, then h_c induces an ordinary function h. The class of all ordinary functions that are induced in this way is denoted by $\mathscr{F}_c$, where c here denotes the particular fixed function.

Example 4.6.1. Let $\mathscr{F} = \{\mathsf{I}_c\}$. Then $\mathscr{F}_c = \{c\}$ for any fixed function c. In particular, $\mathscr{F}_\sigma = \{\sigma\}$.

Example 4.6.2. Consider the unary general function h_c such that $h_c(x) \simeq c(x) + 1$, and set $\mathscr{F} = \{h_c, \sigma\}$. If we fix $c(x) = x$, that is, $c = \mathsf{I}_1^1$, we get $\mathscr{F}_c = \{\sigma\}$.

We denote by $\mathscr{F}_0$ the class of all ordinary functions that are elements of $\mathscr{F}$, that is, the ordinary functions that induce general functions in $\mathscr{F}$.

A k-ary *general predicate* is a k-ary predicate P_c which depends on the unary function c. Note that P_c must be total for every function c. For example, if we specify $P_c(x) \equiv c(x) = 0$, this is undefined whenever $c(x)$ is undefined and P_c is not a general predicate. On the other hand, the specification $P_c'(x) \equiv c(x) \simeq 0$ is well defined, and P_c' is a general predicate.

The finitary rules of Chapter 1 can be applied with given general functions and predicates, and the result is also a general function or predicate. This extension is straightforward and there is no need to write again the details. Similarly, recursive specifications can be applied with general functions, the result being another general function.

If $\mathscr{F}$ is a class of general functions we can define closure under any of the finitary rules, or under recursive specifications, exactly as was done for ordinary functions and predicates. Note that in the definition of closure general predicates enter via the general characteristic function, which is denoted by χ_{P_c} whenever P_c is a general predicate.

Note that if $\mathscr{F}$ is a class closed under one of the rules, and c is a fixed unary function, then $\mathscr{F}_c$ is closed under the same rule applied to ordinary functions. The class $\mathscr{F}_0$ is also closed under the rule.

From the definition of closure under a rule we move to the definition of closure under basic (elementary) (primitive recursive) (recursive) operations. The definition is the same given for ordinary functions, with one minor change, namely, we require that the class $\mathscr{F}$ contain the identity general function I_c. For example, we say that $\mathscr{F}$ is closed under *basic operations* if $\mathscr{F}$ contains I_c, the identity functions, the constant functions, and is closed under full substitution. By the identity functions and the constant functions we mean the general functions induced by such ordinary functions.

Note that if a class $\mathscr{F}$ is closed under some operations, and c is a fixed unary function, then $\mathscr{F}_c$ and $\mathscr{F}_0$ are closed under the same operations. Furthermore, the class $\mathscr{F}_c$ contains the function c.

Minimal closure is defined again in the same manner, that is, via $\mathscr{F}$-derivations which make explicit the particular operations involved in the definition. Hence if $\mathscr{F}$ is an arbitrary class of general functions, there is a class $\mathsf{GRC}(\mathscr{F})$ that is the minimal class of general functions that contains $\mathscr{F}$ and it is closed under recursive operations. Similarly, there is a class $\mathsf{GEL}(\mathscr{F})$ for elementary operations and a class $\mathsf{GPR}(\mathscr{F})$ for primitive recursive operations. The only case we are interested in is when $\mathscr{F} = \varnothing$, hence $\mathsf{GRC} = \mathsf{GRC}(\varnothing)$, $\mathsf{GEL} = \mathsf{GEL}(\varnothing)$, and $\mathsf{GPR} = \mathsf{GPR}(\varnothing)$.

Theorem 4.6.1. *Let $\mathscr{F}$ be a class of general functions. Then*

(i) $\mathsf{RC}(\mathscr{F}_0) \subseteq \mathsf{GRC}_0(\mathscr{F})$.
(ii) *If c is a unary fixed function then* $\mathsf{RC}(\mathscr{F}_c \cup \{c\}) = \mathsf{GRC}_c(\mathscr{F})$.

PROOF. To prove (i) note that $\mathscr{F} \subseteq \mathsf{GRC}(\mathscr{F})$, hence $\mathscr{F}_0 \subseteq \mathsf{GRC}_0(\mathscr{F})$, and since $\mathsf{GRC}_0(\mathscr{F})$ is closed under recursive operations we have $\mathsf{RC}(\mathscr{F}_0) \subseteq \mathsf{GRC}_0(\mathscr{F})$. To prove (ii) we note first that $\mathscr{F} \cup \{\mathsf{I}_c\} \subseteq \mathsf{GRC}(\mathscr{F})$, hence $\mathscr{F}_c \cup \{c\} \subseteq \mathsf{GRC}_c(\mathscr{F})$, and $\mathsf{RC}(\mathscr{F}_c \cup \{c\}) \subseteq \mathsf{GRC}_c(\mathscr{F})$. To prove the converse relation we consider a recursive $\mathscr{F}$-derivation for a general function h_c. Such a derivation may contain elements of $\mathscr{F}$ and also the general function I_c. If we fix the unary function c we get a recursive derivation relative to $\mathscr{F}_c \cup \{c\}$, hence $\mathsf{GRC}_c(\mathscr{F}) \subseteq \mathsf{RC}(\mathscr{F}_c \cup \{c\})$. □

Corollary 4.6.1.1. *Let c be a fixed unary function. Then*

(i) $\mathsf{RC} = \mathsf{GRC}_0$.
(ii) $\mathsf{RC}(c) = \mathsf{GRC}_c$.

PROOF. Part (ii) is immediate from Theorem 4.6.1 (ii). To prove (i) we need only show that the inclusion $\mathsf{GRC}_0 \subseteq \mathsf{RC}$ is valid. A recursive derivation in GRC for an ordinary function h may involve occurrences of the general function I_c. We eliminate such occurrences by setting $c = \sigma$. In this way we get an ordinary recursive derivation for h, so h is a recursive function. □

Let $\mathscr{F}$ be a class of general functions. An interpreter for $\mathscr{F}$ is a pair $(\Phi_c. \mathsf{S})$, where Φ_c is a unary general function and S is an ordinary total binary function, and for every k-ary general function h_c in $\mathscr{F}$ there is a number $\mathbf{z}$ that satisfies the relation

$$h_c(\mathbf{x}) \simeq \Phi_c(\mathsf{S}^k(\mathbf{z}, \mathbf{x})).$$

We say that a number z satisfying such a condition is a (Φ_c, S)-index for the general function h_c.

Note that this means that whenever c is a fixed unary function then (Φ_c, S) is an interpreter for $\mathscr{F}_c$, and the number z above is a (Φ_c, S)-index for the ordinary function h_c. On the other hand, it is possible to have a pair (Φ_c, S), where Φ_c is a unary general function and S is a total binary recursive function and whenever c is a fixed unary function then (Φ_c, S) is an interpreter for $\mathscr{F}_c$, without (Φ_c, S) being an interpreter for $\mathscr{F}$. The crucial part of the condition is that the number z above is independent of c. To make this situation explicit we say sometimes that (Φ_c, S) is a *uniform* interpreter for $\mathscr{F}$, and the number z is a *uniform* (Φ_c, S)-index for the general function h_c.

An interpreter (Φ_c, S) for the class $\mathscr{F}$ is *internal* if both Φ_c and S are $\mathscr{F}$-computable.

A class $\mathscr{F}$ of general functions is a *general functional reflexive structure* (GFRS) if the following conditions are satisfied:

(i) $\mathscr{F}$ is closed under basic operations.
(ii) The functions (i.e., the general functions induced by) σ, pd, and cd are $\mathscr{F}$-computable.
(iii) $\mathscr{F}$ has an internal interpreter.

Note that closure under basic operations implies that the class $\mathscr{F}$ contains the general function I_c.

It follows from this definition that whenever c is a fixed unary function then $\mathscr{F}_c$ is a FRS. This does not mean that $\mathscr{F}_c = \mathsf{RC}(c)$. In fact, an example is given in Chapter 5 where $\mathscr{F}_c$ is not e-total for every function c. On the other hand, we prove later that GRC is a GFRS and we know that $\mathsf{GRC}_c = \mathsf{RC}(c)$.

In dealing with a general reflexive structure with internal interpreter (Φ_c, S) we use the standard notation Φ_c^k, $k \geq 0$, where

$$\Phi_c^k(z, \mathbf{x}) \simeq \Phi_c(\mathsf{S}^k(z, \mathbf{x})).$$

Theorem 4.6.2. *Let $\mathscr{F}$ be a* GFRS *with internal interpreter (Φ_c, S). Then*

(i) *If h_c is a $(k + m)$-ary general function that is $\mathscr{F}$-computable, there is a k-ary total $\mathscr{F}_0$-computable function g such that*

$$\Phi_c^m(g(\mathbf{x}), \mathbf{y}) \simeq h_c(\mathbf{x}, \mathbf{y}).$$

(ii) *If h_c is a $(k + 1)$-ary general function that is $\mathscr{F}$-computable, there is a number* z *such that*

$$\Phi_c^k(\mathbf{z}, \mathbf{x}) \simeq h_c(\mathbf{z}, \mathbf{x}).$$

(iii) *The class $\mathscr{F}$ is closed under recursive operations, and $\mathscr{F} = \mathsf{GRC}(\{\Phi_c, \mathsf{S}\})$.*

Proof. Part (i) is a reformulation of the abstraction property and it is derived from the factorization properties of the function S^{k+m}. Part (ii) is a general version of the recursion theorem, and it is proved as in Theorem 4.1.4. To prove part (iii) we follow the procedure of Section 1 and use the recursion theorem to prove closure under primitive recursion and unbounded minimalization with functions. □

Note that in part (i) of Theorem 4.6.2 the total function g is independent of the function c, so we say that g is *uniformly* given by the abstraction property. Similarly, in part (ii) we say that the number **z** is *uniformly* given by the recursion theorem.

We recall now that in Section 2 we introduced a general function δ_c and derived from it the general function ϕ_c, proving that whenever c is a fixed unary function then $(\phi_c, \mathsf{S}_\delta)$ is an internal interpreter for $\mathsf{RC}(c)$.

Theorem 4.6.3. *The class* GRC *is a* GFRS *with internal interpreter* $(\phi_c, \mathsf{S}_\delta)$.

PROOF. The same proof given for Theorem 4.2.1. □

The introduction of general functions throws new light on the theory of reflexive domains, and it is in fact a useful tool that we apply in Chapter 5, particularly via the uniform abstraction property. On the other hand, general functions are essentially functionals and belong to second-order computability theory, which is beyond the scope of this volume.

EXAMPLE 4.6.3. Let c_0 be a fixed unary function and $\mathscr{F} = \mathsf{GRC}(c_0)$. Then $\mathscr{F}_0 = \mathsf{RC}(c_0)$ and $\mathscr{F}_c = \mathsf{RC}(\{c_0, c\})$ for any unary function c. Furthermore, $\mathscr{F}$ is a GFRS. To obtain an interpreter for $\mathscr{F}$ we extend the general function δ_c of Section 2 to a general function $\delta_{c_0,c}$. The recursive specification of this function is derived from the specification of δ_c by adding a new equation:

$$\delta_{c_0,c}(z, x) \simeq c_0([x]_1) \quad \text{if } (z)_0 = 12.$$

If we take $\Phi_{c_0,c}(z) \simeq \delta_{c_0,c}(z, 0)$, it follows that $(\Phi_{c_0,c}, \mathsf{S}_\delta)$ is an internal interpreter for $\mathscr{F}$ (same proof as Theorem 4.2.1).

Theorem 4.6.4. *Let c_0 be a fixed total unary function. The class $\mathscr{F} = \mathsf{GRC}(c_0)$ is a* GFRS *with internal interpreter $(\Phi_{c_0,c}, \mathsf{S}_\delta)$ and there is a 4-ary elementary predicate Q such that*

$$\Phi_{c_0,c}(z) \simeq v \equiv \exists w Q(z, v, \bar{c}_0(w), \bar{c}(w))$$

whenever c is a total unary function.

PROOF. The interpreter is obtained in the manner described in Example 4.6.3. The predicate Q is obtained by the same procedure used in the proof of Theorem 4.2.5 and Corollary 4.2.5.1. □

Some important points are missing in our definition, which is too weak for the requirements of second-order computability. In fact, we have failed to include some form of functional substitution, which is essential in the discussion of computation on higher types. Such a rule can be defined along the following lines: given a k-ary general function f_c, and a $(k+1)$-ary general function g_c, we introduce a new k-ary general function h_c such that

$$h_c(\mathbf{x}) \simeq f_{c'}(\mathbf{x}),$$

where $c'(y) \simeq g_c(\mathbf{x}, y)$.

In applications it is convenient to impose some restrictions on the general functions under consideration. For example, we say that a k-ary general function h_c is *monotonic* if whenever $c \subseteq_1 c'$ then $h_c \subseteq_k h_{c'}$. We can define continuity in a similar way, following the example in functional transformations. All finitary rules, including functional substitution, preserve the property of being monotonic and also of being continuous.

If $\mathscr{F}$ is a class of general functions we can define $\mathscr{F}_{\mathsf{pd}\#}$ in a manner that generalizes the definition of $\mathscr{C}_{\mathsf{pd}\#}$ for classes of ordinary functions. The class $\mathscr{F}_{\mathsf{pd}\#}$ contains a k-ary general function h_c if and only if there is a $(k+2)$-ary general function f_c that is $\mathscr{F}$-computable and furthermore

$$h_c(\mathbf{x}) \simeq v \equiv \exists y f_c(\mathbf{x}, v, y) \simeq 0.$$

This definition imposes a single-valued condition on a functional relation. The crucial point here is that the single-valuedness must hold for an arbitrary unary function c.

If $\mathscr{F}$ is closed under recursive operations we have $\mathscr{F} \subseteq \mathscr{F}_{\mathsf{pd}\#}$; hence if c is a fixed unary function we have $\mathscr{F}_c \subseteq \mathscr{F}_{\mathsf{pd}\#c}$. We show in the next theorem that the relation $\mathsf{RC}_{\mathsf{pd}\#}(c) \subseteq \mathscr{F}_{\mathsf{pd}\#c}$ may fail in some situations.

Theorem 4.6.5. *Let $\mathscr{F}$ be a* GFRS *where all general functions are continuous. There are unary functions f and c, where f is total and partially recursively enumerable in c but does not belong to the class $\mathscr{F}_{\mathsf{pd}\#c}$.*

Proof. Let (Φ_c, S) be an internal interpreter for $\mathscr{F}$. We define two infinite sequences of finite functions: $f_0, f_1, \ldots, f_n, \ldots$ and $c_0, c_1, \ldots, c_n, \ldots$, where $f_n \subseteq_1 f_{n+1}$, $c_n \subseteq_1 c_{n+1}$, and $f_n(x)$ is defined if and only if $x < n$. We set $f_0 = c_0 = \mathsf{UD}_1$. To define f_{n+1} and c_{n+1} we denote by x_0 the least number, such that whenever $c_n(x)$ is defined then $x < x_0$, and consider two cases:

Case 1. There is a finite function c' and there are numbers v and y such that $c_n \subseteq_1 c'$ and $\Phi_{c'}^3(n, n, v, y) \simeq 0$. We put $f_{n+1}(n) \simeq v + 1$, and extend c_n to c_{n+1} by putting $c_{n+1}(x) \simeq \langle f_{n+1}(0), \ldots, f_{n+1}(n) \rangle$, where x is the least number such that whenever $c'(x')$ is defined then $x' < x$.

Case 2. Otherwise. We put $f_{n+1}(n) = 0$ and $c_{n+1}(x_0) = \langle f_{n+1}(0), \ldots, f_{n+1}(n) \rangle$.

The functions f and c are obtained as follows:

$$f = f_0 \cup f_1 \cup f_2 \cup \cdots \cup f_n \cup \cdots$$

$$c = c_0 \cup c_1 \cup c_2 \cup \cdots \cup c_n \cup \cdots.$$

Note that f is total and the function c' in case 1 is consistent with c; hence the union $c' \cup c$ is well defined. It is clear that the function f is partially recursively enumerable in c, for we have

$$f(x) = v \equiv \exists y \exists w\colon c(y) \simeq w \wedge \ell w = x + 1 \wedge v = \nabla w.$$

To prove that f is not in the class $\mathscr{F}_{\mathsf{pd}\#c}$ we derive a contradiction from the assumption that there is a general function f_c in $\mathscr{F}_{\mathsf{pd}\#}$ such that $f = f_c$ when c is fixed as above. For if there is such a general function, then there is a number n such that

$$f_c(x) \simeq v \equiv \exists y \Phi_c^3(n, x, v, y) \simeq 0$$

for all functions c. Since Φ_c^3 is continuous it follows that in the definition case 1 applies, so there are finite function c' and number y' such that $\Phi_{c'}^3(n, n, v', y') \simeq 0$ and $f(n) = v' + 1$. If we put $c'' = c \cup c'$ we get by monotonicity that $f_{c''}(n) \simeq v'$ and $f_{c''}(n) \simeq v' + 1$, which is a contradiction. □

We present now the construction of a GFRS $\mathscr{F}$ which is closely related to GRC, but it is more inclusive, and on occasions it has been considered the proper formalization of general computability. In fact, the structure $\mathscr{F}$ is similar to GRC in the sense that $\mathscr{F}_c = \mathsf{GRC}_c = \mathsf{RC}(c)$ whenever c is a total function. On the other hand, $\mathscr{F}_{c\mathsf{pd}}$ is closed under existential quantification and disjunction so the relation above does not hold in some cases where c is a nontotal function.

We say that a 4-ary predicate Q is *consistent* if whenever $Q(x, v, u, y)$ and $Q(x, v', u', y')$ both hold with $\mathsf{cn}(u, u')$, then $v = v'$. If Q is consistent and $k \geq 0$, then Q *induces* a k-ary general function h_c such that

$$h_c(\mathbf{x}) \simeq v \equiv \exists u \exists y\colon \mathsf{pr}_c(u) \wedge Q(\langle \mathbf{x} \rangle, v, u, y).$$

We denote by GCN the class of all general functions that are induced by consistent elementary predicates. We shall prove that GCN is a GFRS such that $\mathsf{GCN}_c = \mathsf{RC}(c)$ whenever c is a total function.

Theorem 4.6.6. *The class* GCN *is closed under basic operations and* $\mathsf{GCN}_0 = \mathsf{RC}$.

Proof. From Theorem 3.2.4 we know that $\mathsf{RC} = \mathsf{EL}_{\mathsf{d}\#}$ and this implies that $\mathsf{RC} \subseteq \mathsf{GCN}_0$. To prove the converse note that if c is a unary recursive function then $\mathsf{GCN}_0 \subseteq \mathsf{GCN}_c \subseteq \mathsf{RC}$. It is easy to show that the general function I_c is in GCN. To complete the proof we assume that a k-ary general function h_c is given by full substitution in the form

$$h_c(\mathbf{x}) \simeq f_c(g_c'(\mathbf{x}), g_c''(\mathbf{x})),$$

where f_c, g_c', and g_c'' are in GCN. Let Q, Q_1, and Q_2 be consistent elementary predicates that induce the general functions f_c, g_c', and g_c'', respectively. It follows that

$$\begin{aligned} h_c(\mathbf{x}) \simeq v \equiv \exists u \exists y\colon & \mathsf{pr}_c(u) \wedge \exists u' \leq u\colon \mathsf{ex}(u', u) \wedge Q_1(\langle \mathbf{x} \rangle, (y)_0, u', (y)_2) \\ & \wedge \exists u' \leq u\colon \mathsf{ex}(u', u) \wedge Q_2(\langle \mathbf{x} \rangle, (y)_1, u', (y)_3) \\ & \wedge \exists u' \leq u\colon \mathsf{ex}(u', u) \wedge Q(\langle (y)_0, (y)_1 \rangle, v, u', (y)_4), \end{aligned}$$

and this implies that h_c is in GCN. □

We want to prove that GCN is a GFRS. To construct an internal interpreter for GCN we introduce several functions. We start with the binary function δ'_{e}, which is given by the following recursive specification:

$$
\begin{array}{lll}
(0) & \delta'_{\mathsf{e}}(z,x) \simeq [x]_1 + 1 & \text{if } (z)_0 = 0 \\
(1) & \quad \simeq \mathsf{cd}([x]_1, [x]_2, [x]_3) & \text{if } (z)_0 = 1 \\
(2) & \quad \simeq \varepsilon([x]_1, [x]_2) & \text{if } (z)_0 = 2 \\
(3) & \quad \simeq \delta'_{\mathsf{e}}((z)_1, x\S) & \text{if } (z)_0 = 3 \\
(4) & \quad \simeq \delta'_{\mathsf{e}}((z)_1, \S x) & \text{if } (z)_0 = 4 \\
(5) & \quad \simeq \delta'_{\mathsf{e}}((z)_1, x \,\square\, \langle \delta'_{\mathsf{e}}((z)_2, x) \rangle) & \text{if } (z)_0 = 5 \\
(6a) & \quad \simeq 0 & \text{if } (z)_0 = 6,\ \nabla x = 0 \\
(6b) & \quad \simeq \delta'_{\mathsf{e}}(z, x\S \,\square\, \langle \nabla x \dot{-} 1 \rangle) + \delta'_{\mathsf{e}}((z)_1, x\S \,\square\, \langle \nabla x \dot{-} 1 \rangle) & \text{if } (z)_0 = 6,\ \nabla x \neq 0 \\
(7a) & \quad \simeq 1 & \text{if } (z)_0 \geq 7,\ \nabla x = 0 \\
(7b) & \quad \simeq \delta'_{\mathsf{e}}(z, x\S \,\square\, \langle \nabla x \dot{-} 1 \rangle) \times \delta'_{\mathsf{e}}((z)_1, x\S \,\square\, \langle \nabla x \dot{-} 1 \rangle) & \text{if } (z)_0 \geq 7,\ \nabla x \neq 0.
\end{array}
$$

Theorem 4.6.7. *The function δ'_{e} is recursive and total. A k-ary function h is elementary if and only if there is a number* $\mathbf{z}$ *such that* $h(\mathbf{x}) = \delta'_{\mathsf{e}}(\mathbf{z}, \langle \mathbf{x} \rangle)$.

PROOF. It is clear that δ'_{e} is recursive. To prove that it is total we show by induction on z that $\delta'_{\mathsf{e}}(z, x)$ is defined for every x. All cases are trivial, but in case $(z)_0 = 6$ or $(z)_0 \geq 7$ we use a secondary induction on x, noting that whenever $\nabla x \neq 0$ then $x\S \,\square\, \langle \nabla x \dot{-} 1 \rangle < x$.

To prove the second part we show first, by induction on z, that whenever h is a k-ary function such that $h(\mathbf{x}) = \delta'_{\mathsf{e}}(z, \langle \mathbf{x} \rangle)$ then h is elementary. When $k = 0$ the function h is always a constant function. If $k > 0$ we get familiar elementary functions. When $(z)_0 = 6$ we use bounded sum and the induction hypothesis. If $(z)_0 \geq 7$ we use bounded product and the induction hypothesis.

To complete the proof we show that whenever h is a k-ary elementary function then there is a number $\mathbf{z}$ such that $h(\mathbf{x}) = \delta'_{\mathsf{e}}(\mathbf{z}, \langle \mathbf{x} \rangle)$. Here again all cases are trivial. Note that C_0^0 is obtained by equation (6a). Closure under bounded sum is derived from equations (6a) and (6b), and closure under bounded product is derived from equations (7a) and (7b). □

The function δ'_{e} enumerates all the elementary functions. We use δ'_{e} to introduce by course-of-values recursion a binary function q that enumerates all the consistent elementary predicates.

$$
\begin{array}{ll}
q(z, v) \simeq 1 & \text{if } \exists w < v \colon q(z, w) \simeq 0 \wedge \mathsf{cn}([w]_3, [v]_3) \wedge [w]_2 \neq [v]_2 \\
\quad \simeq \mathsf{sg}(\delta'_{\mathsf{e}}(z, v)) & \text{otherwise.}
\end{array}
$$

It is clear that q is a total recursive function. Furthermore, a 4-ary elementary predicate Q is consistent if and only if there is a number $\mathbf{z}$ such that

$$\chi_Q(x, v, u, y) = q(\mathsf{z}, \langle x, v, u, y\rangle).$$

Since q is recursive there is a 5-ary elementary predicate Q' such that

$$q([\nabla x]_1, \langle \S\nabla x, v, u, y\rangle) = 0 \equiv \exists y' Q'(x, v, u, y, y').$$

We set $Q''(x, v, u, y) \equiv Q'(x, v, u, (y)_0, (y)_1)$. Since Q'' is consistent we can introduce a unary general function Φ_c such that

$$\Phi_c(x) \simeq v \equiv \exists u \exists y \colon \mathsf{pr}_c(u) \wedge Q''(\langle x\rangle, v, u, y).$$

Next we take an internal interpreter (ϕ, S) for RC and prove that $(\Phi_c \circ \phi, \mathsf{S})$ is an internal interpreter for GCN. In fact, let h_c be a k-ary general function induced by a 4-ary consistent elementary predicate Q; hence there is a number z' such that

$$\chi_Q(x, v, u, y) = q(\mathsf{z}', \langle x, v, u, y\rangle).$$

Let z be a (ϕ, S)-index for the function f such that $f(\mathbf{x}) = (\langle \mathsf{z}', \langle \mathbf{x}\rangle)$. To prove that z is a $(\Phi_c \circ \phi, \mathsf{S})$-index for the general function h_c we compute

$$\begin{aligned}
\Phi_c(\phi(\mathsf{S}^k(\mathsf{z}, \mathbf{x}))) \simeq v &\equiv \exists u \exists y \colon \mathsf{pr}_c(u) \wedge Q''(\langle \phi(\mathsf{S}^k(\mathsf{z}, \mathbf{x}))\rangle, v, u, y)\\
&\equiv \exists u \exists y \colon \mathsf{pr}_c(u) \wedge Q''(\langle\langle z', \mathbf{x}\rangle\rangle, v, u, y)\\
&\equiv \exists u \exists y \colon \mathsf{pr}_c(u) \wedge q(\mathsf{z}', \langle\langle \mathbf{x}\rangle, v, u, y\rangle) = 0\\
&\equiv \exists u \exists y \colon \mathsf{pr}_c(u) \wedge Q(\langle \mathbf{x}\rangle, v, u, y)\\
&\equiv h_c(\mathbf{x}) \simeq v,
\end{aligned}$$

and this implies that $h_c(\mathbf{x}) \simeq \Phi_c(\phi(\mathsf{S}^k(\mathsf{z}, \mathbf{x})))$.

Theorem 4.6.8. *The class* GCN *is a* GFRS *where all general functions are continuous. If c is a fixed unary function then* $\mathsf{RC}(c) \subseteq \mathsf{GCN}_c \subseteq \mathsf{RC}_{\mathsf{pd}\#}(c)$. *If* $\mathsf{RC}(c)$ *is closed under enumeration operations then* $\mathsf{RC}(c) = \mathsf{GCN}_c$.

PROOF. From Theorem 4.6.6 and the above construction of an internal interpreter it follows that GCN is a GFRS. The elements of GCN are clearly continuous, and the inclusion $\mathsf{RC}(c) \subseteq \mathsf{GCN}_c$ holds in general. The inclusion $\mathsf{GCN}_c \subseteq \mathsf{RC}_{\mathsf{pd}\#}(c)$ follows from the form of the definition of GCN. If $\mathsf{RC}(c) = \mathsf{RC}_{\mathsf{pd}\#}(c)$ then $\mathsf{RC}(c) = \mathsf{GCN}_c$. □

EXERCISES

4.6.1. Let $\mathscr{F}$ be a GFRS with internal interpreter (Φ_c, S) and $k \geq 0$. Prove the following:

(a) There is a total unary $\mathscr{F}_0$-computable function g_1 such that if z is a (Φ_c, S)-index for an $\mathscr{F}$-computable $(k + 1)$-ary general function f_c, then $g_1(\mathsf{z})$ is a (Φ_c, S)-index for the k-ary general function h_c such that

$$h_c(\mathbf{x}) \simeq \mu y(f_c(\mathbf{x}, y) \simeq 0).$$

(b) There is a total binary $\mathscr{F}_0$-computable function g_2 such that if z_1 is a

(Φ_c, S)-index for an $\mathscr{F}$-computable k-ary general function f_c, and z_2 is a (Φ_c, S)-index for an $\mathscr{F}$-computable $(k+2)$-ary general function g_c, then $g_2(\mathsf{z}_1, \mathsf{z}_2)$ is a (Φ_c, S)-index for the $(k+1)$-ary general function h_c specified by primitive recursion from f and g.

4.6.2. Let $\mathscr{F}$ be a GFRS and assume (Φ_c, S) and (Φ'_c, S') are both interpreters for $\mathscr{F}$. Prove that if $k \geq 0$ then there is a total unary $\mathscr{F}_0$-computable function g such that whenever z is (Φ_c, S)-index for the k-ary general function h_c then $g(\mathsf{z})$ is (Φ'_c, S')-index for h_c.

4.6.3. Let $\mathscr{F}$ be a GFRS with internal interpreter (Φ_c, S). Consider the general set W_c such that

$$z \in \mathsf{W}_c \equiv \Phi_c(z)\downarrow.$$

Prove that if h_c is a given k-ary $\mathscr{F}$-computable general function, $k \geq 0$, then there is a k-ary total $\mathscr{F}_0$-computable function g such that

$$h_c(\mathbf{x})\downarrow \equiv g(\mathbf{x}) \in \mathsf{W}_c.$$

4.6.4. Let $\mathscr{C}$ be a class of ordinary functions. Prove $\mathsf{GRC}_0(\mathscr{C}) = \mathsf{RC}(\mathscr{C})$.

4.6.5. Let $\mathscr{F}$ be a class of general functions closed under the functional substitution rule. Prove that if c' is a unary $\mathscr{F}_0$-computable function then $\mathscr{F}_0 = \mathscr{F}_{c'}$.

4.6.6. Let $\mathscr{F}$ be a class of general functions. Prove that if all elements of $\mathscr{F}$ are monotonic (continuous) then all elements of $\mathsf{GRC}(\mathscr{F})$ are also monotonic (continuous).

4.6.7. Let GPR be the minimal class of general functions which is closed under primitive recursive operations. Define a general interpreter (Φ_c, S) for GPR where Φ_c is a recursive general function, S is an elementary function, and Φ_c is total whenever c is a total function.

4.6.8. Let c be a fixed unary function. Prove that $\mathsf{GCN}_{\mathsf{cpd}}$ is closed under existential quantification and disjunction.

4.6.9. Prove that if h_c is a k-ary general function which is GCN-computable there is a $(k+2)$-ary elementary predicate Q such that

$$h_c(\mathbf{x}) \simeq v \equiv \exists w Q(\mathbf{x}, v, \bar{c}(w))$$

holds whenever c is a total unary function.

4.6.10. Find a consistent 4-ary elementary predicate Q such that

$$\mathsf{I}_c(x) \simeq v \equiv \exists u \exists y\colon \mathsf{pr}_c(u) \wedge Q(\langle x \rangle, v, u, y).$$

4.6.11. Let $\mathscr{F}$ be a GFRS with internal interpreter (Φ_c, S). Set $\mathscr{C} = \mathscr{F}_0$, and assume there is a 3-ary $\mathscr{C}$-decidable predicate Q such that

$$\Phi_\beta(z) \simeq v \equiv \exists w Q(z, v, \bar{\beta}(w))$$

holds whenever β is a unary total function. Prove the following:
(a) If β is a total unary function then $\mathscr{F}_\beta$ is e-total and $\mathscr{F}_\beta = \mathsf{RC}(\mathscr{C} \cup \{\beta\})$
(b) If c is a total unary recursive function then $\mathscr{C} = \mathscr{F}_c$.
(c) There is a total unary function c_0 such that $\mathsf{C} = \mathsf{RC}(c_0)$ and $\mathscr{F}_\beta = \mathsf{RC}(\{c_0, \beta\})$ whenever β is a total unary function.

Notes

The discussion of uniform structures in this section is fragmentary, and it is introduced to provide the necessary results for the treatment of hyper-hyperenumeration in Chapter 5. In particular, the rule of functional substitution is only mentioned in the text.

For a systematic study of these structures the reader is referred to Hinman [8] and Tourlakis [29], but note that some of the results asserted in the latter are valid only when restricted to total functions. In fact, Hinman makes this restriction from the start and avoids altogether nontotal functions. This restriction is traditional, and it is enforced in many places. There are good reasons for this approach, but we do not think that in the long run it is good policy to ignore the underlying partial structure, which determines to a great extent the properties of the total structure.

CHAPTER 5

Hyperenumeration

In this chapter we introduce new closure conditions, much stronger than those considered before. These conditions apply primarily to classes of predicates but can be extended to classes of functions via graph predicates. They are nonfinitary, for they involve different forms of universal quantification. Our main interest is to find reflexive structures that are closed under this type of condition. While they are not total in the sense defined in Chapter 2, we shall see that they may satisfy important selector properties.

§1. Function Quantification

Closure under universal quantification is possible for some relational structures, but we have not seen any example of a functional structure $\mathscr{C}$ where $\mathscr{C}_{\mathsf{pd}}$ is closed under such a form of quantification. We shall show that such functional structures exist, in fact that there are structures $\mathscr{C}$ where $\mathscr{C}_{\mathsf{pd}}$ is closed under a form of universal function quantification.

To express function quantification we need function variables. For this purpose we shall use greek letters $\alpha, \beta, \gamma, \ldots$, with the understanding that such variables range over *unary total functions*. If we have a boolean context $\cdots\alpha\cdots$ containing the function variable α, then $\forall\alpha{:}\cdots\alpha\cdots$ means that the context takes value T for every possible value of the variable α. And $\exists\alpha{:}\cdots\alpha\cdots$ means that the context takes value T for at least one value of the variable α.

We postpone a formal definition of what we consider to be a "context," and for the time being we only mention that we mean terms where the variable α occurs in the form $\alpha(U)$, where U is a basic numerical term, as defined in Chapter 1, which may contain α or some other function variable. Although this restriction applies in most cases, on special occasions we may allow expressions of the form $\alpha(U)$, where U is not a basic term.

EXAMPLE 5.1.1. Consider the following expression involving universal quantification over a function variable α:

$$\forall\alpha\colon \alpha(x) = 0 \vee (\exists y\colon \alpha(y) \neq 0 \wedge (y = 0 \vee \exists z\colon y = z + 2 \wedge \alpha(z) = 0)).$$

This expression satisfies the above restriction, because the variable α appears only in basic terms. The value of the expression depends on the value of the numerical variable x, which is free. It can be proved that the expression takes value T if and only if x is an even number.

This notion of function quantification is clear enough, at least in the context of standard set theory. Unfortunately, it cannot be applied to predicates with numerical arguments. It is possible, of course, to extend the definition and to allow for predicates with functional arguments, but we prefer to avoid this direction. We overcome the difficulty by introducing a special form of function quantification which can be applied to numerical predicates. This special quantification we call *barred quantification*, and it may take two forms: *existential barred quantification* and *universal barred quantification.*

The formal definition involves some elementary operations that were introduced in Chapter 3. We recall that if α is a unary total function then

$$\bar{\alpha}(x) = \left(\prod_{v<x} \mathrm{exp}(\mathrm{p}_v, \alpha(v) + 1)\right) \dot{-} 1.$$

Hence $\bar{\alpha}(x)$ encodes values of α up to $\alpha(x)$. In particular, $\bar{\alpha}(0) = 0$, and $\bar{\alpha}(x + 1) = \langle\alpha(0), \ldots, \alpha(x)\rangle = \bar{\alpha}(x) \,\square\, \langle\alpha(x)\rangle$.

Existential Barred Quantification. Let Q be a $(k + 1)$-ary predicate, $k \geq 0$. The k-ary predicate P such that

$$P(\mathbf{x}) \equiv \forall\alpha\exists v Q(\mathbf{x}, \bar{\alpha}(v))$$

is obtained from Q by *existential barred quantification.*

Universal Barred Quantification. Let Q be a $(k + 1)$-ary predicate, $k \geq 0$. The k-ary predicate P such that

$$P(\mathbf{x}) \equiv \exists\alpha\forall v Q(\mathbf{x}, \bar{\alpha}(v))$$

is obtained from Q by *universal barred quantification.*

Although barred quantification is defined as a combination of function quantification with numerical quantification, it is convenient to consider it as a unitary quantification, stronger than numerical quantification but weaker than function quantification. The two forms in which barred quantification appears, existential and universal, are related by the same connections that apply to numerical or function quantification. For example,

$$\sim\forall\alpha\exists v Q(\mathbf{x}, \bar{\alpha}(v)) \equiv \exists\alpha\forall v \sim Q(\mathbf{x}, \bar{\alpha}(v))$$

$$\sim\exists\alpha\forall v Q(\mathbf{x}, \bar{\alpha}(v)) \equiv \forall\alpha\exists v \sim Q(\mathbf{x}, \bar{\alpha}(v)).$$

Furthermore, the rules for change of bound variables apply as usual, and a barred quantifier can be exported over disjunction and conjunction. On the other hand, two barred quantifiers of the same kind cannot be permuted, as is the case with numerical and function quantifiers.

Theorem 5.1.1. *There are unary elementary functions d_1 and d_2 such that whenever P is a $(k+2)$-ary predicate then*

$$\forall\alpha\exists v\forall\beta\exists y P(\mathbf{x}, \bar{\alpha}(v), \bar{\beta}(y)) \equiv \forall\gamma\exists z P(\mathbf{x}, d_1(\bar{\gamma}(z)), d_2(\bar{\gamma}(z))).$$

PROOF. We set

$$d_1'(w, x) = \mathsf{exp}(\mathsf{p}_x, ((w)_{\mathsf{exp}(2,x)} \dot{-} 1)_0 + 1)$$

$$d_2'(w, y) = \mathsf{exp}(\mathsf{p}_y, ((w)_{\mathsf{exp}(2,(\ell w \dot{-} 1)_0) \times \mathsf{exp}(3,y)} \dot{-} 1)_1 + 1)$$

$$d_1(w) = \left(\prod_{x < (\ell w \dot{-} 1)_0} d_1'(w, x)\right) \dot{-} 1$$

$$d_2(w) = \left(\prod_{y < (\ell w \dot{-} 1)_1} d_2'(w, y)\right) \dot{-} 1.$$

To prove that the equivalence holds we assume first the left side, that is,

$$\forall\alpha\exists v\forall\beta\exists y P(\mathbf{x}, \bar{\alpha}(v), \bar{\beta}(y)),$$

and in order to prove the right side we take a fixed total unary function γ and define unary functions α and β_v, where v is a number, as follows:

$$\alpha(x) = (\gamma(\mathsf{exp}(2, x)))_0$$

$$\beta_v(y) = (\gamma(\mathsf{exp}(2, v) \times \mathsf{exp}(3, y)))_1.$$

From the assumption it follows that there is a number v_0 such that

$$\forall\beta\exists y P(\mathbf{x}, \bar{\alpha}(v_0), \bar{\beta}(y)).$$

In particular, if we take $\beta = \beta_{v_0}$ there is a number y_0 such that

$$P(\mathbf{x}, \bar{\alpha}(v_0), \bar{\beta}_{v_0}(y_0)).$$

We set $z_0 = \mathsf{exp}(2, v_0) \times \mathsf{exp}(3, y_0)$ and it follows that $d_1(\bar{\gamma}(z_0)) = \bar{\alpha}(v_0)$, and $d_2(\bar{\gamma}(z_0)) = \bar{\beta}_{v_0}(y_0)$; hence

$$\exists z P(\mathbf{x}, d_1(\bar{\gamma}(z)), d_2(\bar{\gamma}(z))).$$

To prove the converse we assume the left side is false, so there is a total unary function α such that

$$\forall v\exists\beta\forall y \sim P(\mathbf{x}, \bar{\alpha}(v), \bar{\beta}(y)).$$

Hence with every number v we can associate a function β_v such that

$$\forall v\forall y \sim P(\mathbf{x}, \bar{\alpha}(v), \bar{\beta}_v(y)).$$

To prove the right side is false we introduce a function γ such that

$$\gamma(x) = (\mathsf{exp}(2, \alpha((x \dot{-} 1)_0)) \times \mathsf{exp}(3, \beta_{(x \dot{-} 1)_0}((x \dot{-} 1)_1))) \dot{-} 1.$$

It follows that for any z

$$d_1(\overline{\gamma}(z)) = \overline{\alpha}((z \dot{-} 1)_0)$$
$$d_2(\overline{\gamma}(z)) = \overline{\beta}_{(z \dot{-} 1)_0}((z \dot{-} 1)_1)$$
$$\forall z \sim P(\mathbf{x}, d_1(\overline{\gamma}(z)), d_2(\overline{\gamma}(z))). \qquad \square$$

Theorem 5.1.2. *There is a binary elementary function d such that whenever P is a $(k + 1 + m)$-ary predicate, $m \geq 1$, then*

$$\forall y < z \exists w P(\mathbf{x}, y, \overline{\alpha}_1(w), \ldots, \overline{\alpha}_m(w))$$
$$\equiv \exists w \forall y < z P(\mathbf{x}, y, d(y, \overline{\alpha}_1(w)), \ldots, d(y, \overline{\alpha}_m(w))).$$

Proof. We set

$$d'(y, w) = (\ell w \dot{-} 1)_y$$
$$d(y, w) = \left(\prod_{v < d'(y, w)} \mathsf{exp}(\mathsf{p}_v, (w)_v) \right) \dot{-} 1.$$

During the proof $\mathbf{x}$, z, $\alpha_1, \ldots, \alpha_m$ are fixed. Assume the left side is true, hence

$$\forall y < z \exists w P(\mathbf{x}, y, \overline{\alpha}_1(w), \ldots, \overline{\alpha}_m(w)).$$

Hence for every $y < z$ there is a number w_y such that $P(\mathbf{x}, y, \overline{\alpha}_1(w_y), \ldots, \overline{\alpha}_m(w_y))$. We take $w = \prod_{y<x} \mathsf{exp}(\mathsf{p}_y, w_y)$. Note that if $y < z$ then $d'(y, \overline{\alpha}_i(w)) = (w \dot{-} 1)_y = w_y$. Hence

$$d(y, \overline{\alpha}_i(w)) = \left(\prod_{v < w_y} \mathsf{exp}(\mathsf{p}_v, (\overline{\alpha}_i(w))_v) \right) \dot{-} 1 = \overline{\alpha}_i(w_y),$$

for if $y < z$ then $w_y < \mathsf{exp}(\mathsf{p}_y, w_y) \leq w$. It follows that

$$\exists w \forall y < z P(\mathbf{x}, y, d(y, \overline{\alpha}_1(w)), \ldots, d(y, \overline{\alpha}_m(w))).$$

Conversely, if the right side holds there is w such that

$$\forall y < z P(\mathbf{x}, y, d(y, \overline{\alpha}_1(w)), \ldots, d(y, \overline{\alpha}_m(w))).$$

So it is sufficient to prove that $d(y, \overline{\alpha}_i(w)) = \overline{\alpha}_i(w')$ for some w', $i = 1, \ldots, m$. If $w = 0$ then $d(y, \overline{\alpha}_i(w)) = 0 = \overline{\alpha}_i(0)$. If $w \neq 0$ then $d'(y, \overline{\alpha}_i(w)) = (w \dot{-} 1)_y = w'$. Since $w' < w$ it follows that $d(y, \overline{\alpha}_i(w)) = \overline{\alpha}_i(w')$. $\square$

Let $\mathscr{P}$ be a class of predicates. We say that a k-ary predicate P is $\mathscr{P}$-*hyperenumerable* if there is a $(k + 1)$-ary predicate Q in $\mathscr{P}$ such that

$$P(\mathbf{x}) \equiv \forall \alpha \exists v Q(\mathbf{x}, \overline{\alpha}(v)).$$

The class of all predicates that are $\mathscr{P}$-hyperenumerable is denoted by $\mathscr{P}_{\mathsf{he}}$. We put $\mathscr{P}_{\mathsf{hu}} = \mathscr{P}_{\mathsf{nhen}}$, so $\mathscr{P}_{\mathsf{hu}}$ contains all predicates obtained by universal barred quantification on predicates in $\mathscr{P}$.

These subscripts can be combined in the same manner as those introduced in Chapter 3. For example, $\mathscr{C}_{\mathsf{pdhe}}$ denotes the class of all predicates P such that $P(\mathbf{x}) \equiv \forall\alpha\exists v Q(\mathbf{x}, \bar{\alpha}(v))$ and Q is partially $\mathscr{C}$-decidable.

The predicates in $\mathsf{RC}_{\mathsf{dhe}} = \mathsf{RC}_{\mathsf{dehe}} = \mathsf{RC}_{\mathsf{pdhe}}$ are said to be *recursively hyperenumerable*. The predicates in $\mathsf{RC}_{\mathsf{dhe}}(\mathscr{C})$ are *recursively hyperenumerable in* $\mathscr{C}$, and the predicates in $\mathsf{RC}_{\mathsf{pdhe}}(\mathscr{C})$ are *partially recursively hyperenumerable in* $\mathscr{C}$.

Theorem 5.1.3. *Let $\mathscr{P}$ be a class of predicates closed under substitution with total $\mathscr{C}$-computable functions, where $\mathscr{C}$ is closed under elementary operations. Assume $\mathsf{EL_d} \subseteq \mathscr{P}$. Then*

(i) *$\mathscr{P} \subseteq \mathscr{P}_{\mathsf{he}}$ and $\mathscr{P}_{\mathsf{he}}$ is closed under substitution with total $\mathscr{C}$-computable functions.*
(ii) *$\mathscr{P}_{\mathsf{he}}$ is closed under existential barred quantification and existential unbounded quantification.*
(iii) *If $\mathscr{P}$ is closed under conjunction then $\mathscr{P}_{\mathsf{he}}$ is closed under conjunction, universal unbounded quantification, and existential bounded quantification.*
(iv) *If $\mathscr{P}$ is closed under disjunction then $\mathscr{P}_{\mathsf{he}}$ is closed under disjunction.*
(v) *If $\mathscr{P}$ is closed under universal bounded quantification then $\mathscr{P}_{\mathsf{he}}$ is closed under universal bounded quantification.*

Proof. Part (i) is trivial. To prove (ii) we note that closure under existential barred quantification follows from Theorem 5.1.1, and closure under existential unbounded quantification follows from the relation

$$\exists y \forall\alpha\exists v Q(\mathbf{x}, y, \bar{\alpha}(v)) \equiv \forall\beta\exists y\forall\alpha\exists v Q(\mathbf{x}, \ell\bar{\beta}(y), \bar{\alpha}(v)).$$

In part (iii) we note that closure under conjunction follows by exportation of quantifiers and closure under existential barred quantification. Regarding closure under universal unbounded quantification and existential bounded quantification, we note the following relations:

$$\forall y\forall\alpha\exists v Q(\mathbf{x}, y, \bar{\alpha}(v)) \equiv \forall\beta\exists y\forall\alpha\exists v\colon Q(\mathbf{x}, [\bar{\beta}(y)]_1, \bar{\alpha}(v)) \wedge \ell\bar{\beta}(y) \neq 0$$

$$\exists y < z\forall\alpha\exists v Q(\mathbf{x}, y, \bar{\alpha}(v)) \equiv \exists y\forall\alpha\exists v\colon y < z \wedge Q(\mathbf{x}, y, \bar{\alpha}(v)).$$

The proof of part (iv) is straightforward using exportation of quantifiers, and part (v) follows from Theorem 5.1.2. □

From the preceding result and using duality we can derive dual closure properties for the class $\mathscr{P}_{\mathsf{hu}}$.

Corollary 5.1.3.1. *Let $\mathscr{P}$ be a class of predicates closed under substitution with elementary functions, conjunction, disjunction, and existential barred quantification. Assume $\mathsf{EL_d} \subseteq \mathscr{P}$. Then $\mathscr{P}$ is closed under bounded and unbounded numerical quantification.*

Proof. We can apply Theorem 5.1.3 because we have $\mathscr{P} = \mathscr{P}_{\mathsf{he}}$. The proof of closure under universal bounded quantification uses closure under disjunction. □

EXAMPLE 5.1.2. Let $\mathscr{C}$ be a class of functions closed under elementary operations. Then $\mathscr{C}_{\mathsf{d}}$ satisfies all assumptions in Theorem 5.1.3, and it follows that $\mathscr{C}_{\mathsf{dhe}}$ is closed under substitution with total $\mathscr{C}$-computable functions, conjunction, disjunction, bounded and unbounded numerical quantification, and existential barred quantification. Note that any boolean combination involving predicates in $\mathscr{C}_{\mathsf{d}}$, negation, conjunction, disjunction, and numerical quantification is in $\mathscr{C}_{\mathsf{dhe}}$.

EXAMPLE 5.1.3. Let $\mathscr{C}$ be a class closed under recursive operations. It follows that $\mathscr{C}_{\mathsf{pd}}$ is closed under substitution with elementary functions and also under conjunction and universal bounded quantification. Furthermore, $\mathsf{EL}_{\mathsf{d}} \subseteq \mathscr{C}_{\mathsf{pd}}$. Note that $\mathscr{C}_{\mathsf{pd}} \subseteq \mathscr{C}_{\mathsf{pde}} \subseteq \mathscr{C}_{\mathsf{pdhe}}$; hence $\mathscr{C}_{\mathsf{pdhe}} \subseteq \mathscr{C}_{\mathsf{pdehe}} \subseteq \mathscr{C}_{\mathsf{pdhehe}} = \mathscr{C}_{\mathsf{pdhe}}$, and $\mathscr{C}_{\mathsf{pdehe}}$ is closed under conjunction, disjunction, existential barred quantification, unbounded numerical quantification, and bounded numerical quantification.

Let $\mathscr{P}$ be a class of predicates. If h is a k-ary function and G_h is $\mathscr{P}$-hyperenumerable, we say that h is *$\mathscr{P}$-hyperenumerable.* The class of all functions that are $\mathscr{P}$-hyperenumerable is denoted by $\mathscr{P}_{\mathsf{h}\#} = \mathscr{P}_{\mathsf{heg}}$.

The functions in $\mathsf{RC}_{\mathsf{dh}\#} = \mathsf{RC}_{\mathsf{pdh}\#}$ are said to be *recursively hyperenumerabe.* A function in $\mathsf{RC}_{\mathsf{dh}\#}(\mathscr{C})$ is said to be *recursively hyperenumerable in $\mathscr{C}$*, and a function in $\mathsf{RC}_{\mathsf{pdh}\#}(\mathscr{C})$ is said to be *partially recursively hyperenumerable in $\mathscr{C}$.*

Theorem 5.1.4. *Let $\mathscr{P}$ be a class of predicates closed under substitution with elementary functions and conjunction. Assume $\mathsf{EL}_{\mathsf{d}} \subseteq \mathscr{P}$. Then*

(i) $\mathscr{P}_{\mathsf{e}} \subseteq \mathscr{P}_{\mathsf{he}} = \mathscr{P}_{\mathsf{ehe}} = \mathscr{P}_{\mathsf{hee}} = \mathscr{P}_{\mathsf{hehe}}$.
(ii) $\mathscr{P}_{\mathsf{h}\#\mathsf{d}} \subseteq \mathscr{P}_{\mathsf{h}\#\mathsf{de}} \subseteq \mathscr{P}_{\mathsf{h}\#\mathsf{dhe}} \subseteq \mathscr{P}_{\mathsf{h}\#\mathsf{pd}} = \mathscr{P}_{\mathsf{h}\#\mathsf{pde}} = \mathscr{P}_{\mathsf{h}\#\mathsf{pdhe}} = \mathscr{P}_{\mathsf{he}}$.
(iii) $\mathscr{P}_{\mathsf{h}\#\mathsf{d}\#} \subseteq \mathscr{P}_{\mathsf{h}\#\mathsf{dh}\#} \subseteq \mathscr{P}_{\mathsf{h}\#\mathsf{pd}\#} = \mathscr{P}_{\mathsf{h}\#\mathsf{pdh}\#} = \mathscr{P}_{\mathsf{h}\#}$.
(iv) *If $\mathscr{P}_{\mathsf{he}}$ is closed under disjunction then $\mathscr{P}_{\mathsf{h}\#}$ is closed under recursive operations and $\mathsf{RC}_{\mathsf{pdh}\#}(\mathscr{P}_{\mathsf{pd}}) = \mathscr{P}_{\mathsf{h}\#}$.*

PROOF. The proof of (i) uses only closure under substitution with an elementary function, and it is easily derived from the known inclusions $\mathscr{P} \subseteq \mathscr{P}_{\mathsf{e}}$, $\mathscr{P} \subseteq \mathscr{P}_{\mathsf{he}}$, $\mathscr{P}_{\mathsf{hee}} \subseteq \mathscr{P}_{\mathsf{he}}$, and $\mathscr{P}_{\mathsf{hehe}} \subseteq \mathscr{P}_{\mathsf{he}}$, using monotonicity.

To prove (ii) we proceed as in Theorem 3.1.5 and prove first the relation $\mathscr{P}_{\mathsf{h}\#\mathsf{d}} \subseteq \mathscr{P}_{\mathsf{h}\#\mathsf{pd}} = \mathscr{P}_{\mathsf{he}}$ and derive the rest using monotonicity. Part (iii) is immediate from (ii).

To prove (iv) we note that $\mathscr{P}_{\mathsf{he}} = \mathscr{P}_{\mathsf{hee}}$ by (i), and $\mathscr{P}_{\mathsf{he}}$ is closed under conjunction and disjunction. From Theorem 5.1.3 (iii) it follows that $\mathscr{P}_{\mathsf{he}}$ is closed under universal unbounded quantification, and using disjunction we get universal bounded quantification. Hence, from Theorem 3.1.5 (iv), it follows that $\mathscr{P}_{\mathsf{h}\#} = \mathscr{P}_{\mathsf{he}\#}$ is closed under recursive operations. The proof of the equality $\mathsf{RC}_{\mathsf{pdh}\#}(\mathscr{P}_{\mathsf{pd}}) = \mathscr{P}_{\mathsf{h}\#}$ is exactly as in Theorem 3.1.5 (iv). □

Theorem 5.1.5. *Let $\mathscr{P}$ be a class of predicates closed under substitution with elementary functions and under boolean operations. Assume that* $\mathsf{EL_d} \subseteq \mathscr{P}$. *Then*

(i) $\mathscr{P} \subseteq \mathscr{P}_{\mathsf{he}} \cap \mathscr{P}_{\mathsf{hu}} = \mathscr{P}_{\mathsf{h\#d}}$.
(ii) $\mathscr{P}_{\mathsf{h\#dhe}} = \mathscr{P}_{\mathsf{h\#pd}} = \mathscr{P}_{\mathsf{h\#pde}} = \mathscr{P}_{\mathsf{h\#pdhe}} = \mathscr{P}_{\mathsf{he}}$.
(iii) $\mathscr{P}_{\mathsf{h\#dh\#}} = \mathscr{P}_{\mathsf{h\#pd\#}} = \mathscr{P}_{\mathsf{h\#pdh\#}} = \mathscr{P}_{\mathsf{h\#}}$.
(iv) $\mathscr{P}_{\mathsf{h\#}}$ *is closed under recursive operations and* $\mathsf{RC_{dh\#}}(\mathscr{P}_{\mathsf{d}}) = \mathscr{P}_{\mathsf{h\#}}$.

Proof. Similar to the proof of Theorem 3.1.6 noting that universal bounded quantification follows from disjunction and universal unbounded quantification. □

Theorem 5.1.6. *Let $\mathscr{C}$ be a class of functions closed under elementary operations such that* $\mathscr{C}_{\mathsf{dh\#}} \subseteq \mathscr{C}$. *Then*

(i) $\mathscr{C}_{\mathsf{dhe}} \subseteq \mathscr{C}_{\mathsf{pd}}$.
(ii) $\mathscr{C}_{\mathsf{dh\#}} \subseteq \mathscr{C}_{\mathsf{pd\#}}$.
(iii) $\mathscr{C}_{\mathsf{d}} = \mathscr{C}_{\mathsf{dhe}} \cap \mathscr{C}_{\mathsf{dhu}} = \mathscr{C}_{\mathsf{dh\#d}}$.
(iv) $\mathscr{C}_{\mathsf{d}}$ *is closed under unbounded quantification.*

Proof. To prove (i) note that if P is a predicate in $\mathscr{C}_{\mathsf{dhe}}$ then ψ_P is in $\mathscr{C}_{\mathsf{dh\#}} \subseteq \mathscr{C}$; hence P is partially $\mathscr{C}$-decidable. Part (ii) follows from Theorem 3.1.4 (ii). In part (iii) we note that $\mathscr{C}_{\mathsf{d}} \subseteq \mathscr{C}_{\mathsf{dhe}} \cap \mathscr{C}_{\mathsf{dhu}} = \mathscr{C}_{\mathsf{dh\#d}}$ is a consequence of Theorem 5.1.5 (i), and $\mathscr{C}_{\mathsf{dh\#d}} \subseteq \mathscr{C}_{\mathsf{d}}$ follows from the assumption. □

Example 5.1.4. The class $\mathsf{RC_d}$ satisfies the conditions in Theorem 5.1.4 and Theorem 5.1.5, but not the condition in Theorem 5.1.6. It follows that:

(i) $\mathsf{RC_{dh\#}}$ is closed under recursive operations.
(ii) $\mathsf{RC_{dhe}} = \mathsf{RC_{dh\#pd}} = \mathsf{RC_{dh\#pdhe}} = \mathsf{RC_{dh\#dhe}}$.
(iii) $\mathsf{RC_{dh\#}} = \mathsf{RC_{dh\#pdh\#}} = \mathsf{RC_{dh\#dh\#}}$.
(iv) $\mathsf{RC_{dh\#d}} = \mathsf{RC_{dhe}} \cap \mathsf{RC_{dhu}}$.

Let $\mathscr{C}$ be a class of functions. If $\mathscr{C}$ is closed under recursive operations and $\mathscr{C} = \mathscr{C}_{\mathsf{pdh\#}}$, we say that $\mathscr{C}$ is closed under *hyperenumeration operations.* For example, if $\mathscr{C}$ is closed under recursive operations, then $\mathscr{C}_{\mathsf{pdh\#}}$ is closed under hyperenumeration operations (Theorem 5.1.4 (iii) and (iv)).

Theorem 5.1.7. *Assume $\mathscr{C}$ is a class of functions closed under hyperenumeration operations. Then*

(i) $\mathscr{C}_{\mathsf{dh\#}} \subseteq \mathscr{C}$.
(ii) $\mathscr{C}_{\mathsf{dhe}} \subseteq \mathscr{C}_{\mathsf{pd}} = \mathscr{C}_{\mathsf{pde}} = \mathscr{C}_{\mathsf{pdhe}}$.
(iii) $\mathscr{C}_{\mathsf{d}} = \mathscr{C}_{\mathsf{dhe}} \cap \mathscr{C}_{\mathsf{dhu}}$.
(iv) $\mathscr{C}_{\mathsf{d}}$ *is closed under existential and universal unbounded quantification.*
(v) *A k-ary predicate P is $\mathscr{C}$-decidable if and only if both P and $\bar{P}$ are $\mathscr{C}_{\mathsf{pd}}$-hyperenumerable.*

Proof. To prove (i) note that $\mathscr{C}_{\mathsf{d}} \subseteq \mathscr{C}_{\mathsf{pd}}$, hence $\mathscr{C}_{\mathsf{dh}\#} \subseteq \mathscr{C}_{\mathsf{pdh}\#} = \mathscr{C}$. The inclusion in (ii) follows from Theorem 5.1.6 (i), and the equality from Theorem 5.1.4 (ii) with $\mathscr{P} = \mathscr{C}_{\mathsf{pd}}$, noting the equality $\mathscr{C}_{\mathsf{pdh}\#} = \mathscr{C}$. Parts (iii) and (iv) are simply reformulations of Theorem 5.1.6 (iii) and (iv). Finally, to prove (v) we note that it is clear that if P is $\mathscr{C}$-decidable then both P and $\bar{P}$ are $\mathscr{C}_{\mathsf{pd}}$-hyperenumerable. Conversely, if both P are $\bar{P}$ are $\mathscr{C}_{\mathsf{pd}}$- hyperenumerable it follows that χ_P is in $\mathscr{C}_{\mathsf{pdh}\#} = \mathscr{C}$, and hence P is $\mathscr{C}$-decidable. □

Let $\mathscr{C}$ be a class of functions. If $\mathscr{C}$ is closed under recursive operations and $\mathscr{C} = \mathscr{C}_{\mathsf{dh}\#}$, we say that $\mathscr{C}$ is he-*total.* For example, if $\mathscr{C}$ is closed under elementary operations, then $\mathscr{C}_{\mathsf{dh}\#}$ is he-total (Theorem 5.14 (iv) and Theorem 5.1.5 (iii)).

Theorem 5.1.8. *Let $\mathscr{C}$ be a class of functions. Assume $\mathscr{C}$ is* he-*total. Then*

(i) $\mathscr{C}_{\mathsf{dhe}} = \mathscr{C}_{\mathsf{pdhe}}$.
(ii) $\mathscr{C}_{\mathsf{dh}\#} = \mathscr{C}_{\mathsf{pdh}\#}$.
(iii) *$\mathscr{C}$ is closed under hyperenumeration operations.*

Proof. To prove (i) we compute

$$\mathscr{C} = \mathscr{C}_{\mathsf{dh}\#}$$

$$\begin{aligned}\mathscr{C}_{\mathsf{pdhe}} &= \mathscr{C}_{\mathsf{dh}\#\mathsf{pdhe}}\\ &= \mathscr{C}_{\mathsf{dhe}} \quad \text{(Theorem 5.1.4 (ii)).}\end{aligned}$$

Part (ii) is immediate from (i), and part (iii) is immediate from (ii). □

Theorem 5.1.9. *Let $\mathscr{C}$ be a class containing only total functions. Then*

(i) $\mathsf{EL}_{\mathsf{dhe}}(\mathscr{C}) = \mathsf{RC}_{\mathsf{dhe}}(\mathscr{C}) = \mathsf{RC}_{\mathsf{pdhe}}(\mathscr{C})$.
(ii) $\mathsf{EL}_{\mathsf{dh}\#}(\mathscr{C}) = \mathsf{RC}_{\mathsf{dh}\#}(\mathscr{C}) = \mathsf{RC}_{\mathsf{pdh}\#}(\mathscr{C})$.

Proof. To prove (i) we need only show that $\mathsf{RC}_{\mathsf{pdhe}}(\mathscr{C}) \subseteq \mathsf{EL}_{\mathsf{dhe}}(\mathscr{C})$. We know that $\mathsf{RC}(\mathscr{C}) = \mathsf{EL}_{\mathsf{d}\#}(\mathscr{C}) \subseteq \mathsf{EL}_{\mathsf{dh}\#}(\mathscr{C})$ (Theorem 3.2.4); hence

$$\begin{aligned}\mathsf{RC}_{\mathsf{pdhe}}(\mathscr{C}) &\subseteq \mathsf{EL}_{\mathsf{dh}\#\mathsf{pdhe}}(\mathscr{C})\\ &= \mathsf{EL}_{\mathsf{dhe}}(\mathscr{C}) \quad \text{(Theorem 5.1.4 (ii)).}\end{aligned}$$

Part (ii) is immediate from (i). □

Theorem 5.1.10. *Let $\mathscr{C}$ be a class closed under recursive operations. The following conditions are equivalent*:

(i) *$\mathscr{C}$ is* he-*total.*
(ii) $\mathscr{C} = \mathsf{RC}_{\mathsf{dh}\#}(\mathscr{C}_{\mathsf{dd}})$.
(iii) *There is a class $\mathscr{C}'$ containing only total functions such that* $\mathscr{C} = \mathsf{RC}_{\mathsf{dh}\#}(\mathscr{C}')$.

(iv) *There is a class* $\mathscr{C}'$ *containing only total functions such that* $\mathscr{C} = \mathsf{EL}_{\mathsf{dh}\#}(\mathscr{C}')$

(v) *There is a class* $\mathscr{C}'$ *of functions closed under elementary operations such that* $\mathscr{C} = \mathscr{C}'_{\mathsf{dh}\#}$.

Proof. The implication from (i) to (ii) follows from Theorem 5.1.5 (iv) with $\mathscr{P} = \mathscr{C}_{\mathsf{d}}$. The implication from (ii) to (iii) is trivial, and (iv) follows from (iii) by Theorem 5.1.9 (ii). The implication from (iv) to (v) is trivial. To get (i) from (v) we note that

$$\begin{aligned}\mathscr{C}_{\mathsf{dh}\#} &= \mathscr{C}'_{\mathsf{dh}\#\mathsf{dh}\#} \\ &= \mathscr{C}'_{\mathsf{dh}\#} \quad \text{(Theorem 5.1.5 (iii))} \\ &= \mathscr{C}.\end{aligned}$$

□

Let $\mathscr{C}$ be a class of functions. We say that a k-ary predicate P is $\mathscr{C}$-*hyperarithmetical* if both P and $\bar{P}$ and $\mathscr{C}_{\mathsf{d}}$-hyperenumerable. We denote by $\mathscr{C}_{\mathsf{ha}}$ the class of all $\mathscr{C}$-hyperarithmetical predicates. Clearly, we have $\mathscr{C}_{\mathsf{ha}} = \mathscr{C}_{\mathsf{dhe}} \cap \mathscr{C}_{\mathsf{dhen}} = \mathscr{C}_{\mathsf{dhe}} \cap \mathscr{C}_{\mathsf{dnhu}}$. If $\mathscr{C}_{\mathsf{d}}$ is closed under negation we have $\mathscr{C}_{\mathsf{ha}} = \mathscr{C}_{\mathsf{dhe}} \cap \mathscr{C}_{\mathsf{dhu}}$.

If $\mathscr{C}$ is an arbitrary class of functions, the predicates in $\mathsf{RC}_{\mathsf{ha}}(\mathscr{C})$ are said to be *hyperarithmetical* in $\mathscr{C}$. The predicates in $\mathsf{RC}_{\mathsf{ha}}$ are the *hyperarithmetical* predicates.

Theorem 5.1.11. *Let* $\mathscr{C}$ *be a class of functions closed under elementary operations. Then*

(i) $\mathscr{C}_{\mathsf{d}} \subseteq \mathscr{C}_{\mathsf{ha}}$.

(ii) $\mathscr{C}_{\mathsf{dh}\#\mathsf{dhe}} = \mathscr{C}_{\mathsf{dhe}}$, $\mathscr{C}_{\mathsf{dh}\#\mathsf{dh}\#} = \mathscr{C}_{\mathsf{dh}\#}$, *and* $\mathscr{C}_{\mathsf{dh}\#\mathsf{dhu}} = \mathscr{C}_{\mathsf{dhu}}$.

(iii) $\mathscr{C}_{\mathsf{ha}} = \mathscr{C}_{\mathsf{dh}\#\mathsf{d}}$.

(iv) $\mathscr{C}_{\mathsf{ha}}$ *is closed under negation, disjunction, conjunction, and bounded quantification.*

(v) $\mathscr{C}_{\mathsf{ha}}$ *is closed under unbounded quantification.*

(vi) $\mathscr{C}_{\mathsf{hahe}} = \mathscr{C}_{\mathsf{dhe}}$ *and* $\mathscr{C}_{\mathsf{hahu}} = \mathscr{C}_{\mathsf{dhu}}$.

(vii) *An infinite set* A *is* $\mathscr{C}$*-hyperarithmetical if and only if there is a total* $\mathscr{C}_{\mathsf{d}}$*-hyperenumerable function* f *such that* $A = \mathsf{R}_f$ *and* $f(y) < f(y+1)$ *holds for all* $y \in \mathbb{N}$.

Proof. To prove (i) we note that $\mathscr{C}_{\mathsf{d}} \subseteq \mathscr{C}_{\mathsf{dhe}} \cap \mathscr{C}_{\mathsf{dhu}} = \mathscr{C}_{\mathsf{ha}}$. Part (ii) follows from Theorem 5.1.5 (ii) and (iii). To prove (iii) we note that Theorem 5.1.6 applies to $\mathscr{C}_{\mathsf{dh}\#}$; hence $\mathscr{C}_{\mathsf{dh}\#\mathsf{d}} = \mathscr{C}_{\mathsf{dh}\#\mathsf{dhe}} \cap \mathscr{C}_{\mathsf{dh}\#\mathsf{dhu}} = \mathscr{C}_{\mathsf{dhe}} \cap \mathscr{C}_{\mathsf{dhu}} = \mathscr{C}_{\mathsf{ha}}$. Part (iv) follows from (iii), noting that $\mathscr{C}_{\mathsf{dh}\#}$ is closed under recursive operations, and part (v) also follows from (iii) with Theorem 5.1.6 (iv) applied to $\mathscr{C}_{\mathsf{dh}\#}$. To prove (vi) we note that $\mathscr{C}_{\mathsf{dhe}} \subseteq \mathscr{C}_{\mathsf{hahe}} \subseteq \mathscr{C}_{\mathsf{dhehe}} = \mathscr{C}_{\mathsf{dhe}}$; the second equation follows using negation and noting that $\mathscr{C}_{\mathsf{ha}}$ and $\mathscr{C}_{\mathsf{d}}$ are closed under negation. Finally, (vii) is a reformulation of Theorem 3.2.1 (vii) applied to $\mathscr{C}_{\mathsf{dh}\#}$. □

Corollary 5.1.11.1. *Let $\mathscr{C}$ be a class of functions closed under elementary operations. If $\mathscr{C}_{\mathsf{dh\#}} \subseteq \mathscr{C}$ then $\mathscr{C}_{\mathsf{d}} = \mathscr{C}_{\mathsf{ha}}$.*

PROOF. From Theorem 5.1.11 (i) we have $\mathscr{C}_{\mathsf{d}} \subseteq \mathscr{C}_{\mathsf{ha}}$. Conversely, $\mathscr{C}_{\mathsf{ha}} = \mathscr{C}_{\mathsf{dh\#d}} \subseteq \mathscr{C}_{\mathsf{d}}$. □

It is useful to extend this notion to functions. We say that a k-ary function h is *$\mathscr{C}$-hyperarithmetical* if G_h is $\mathscr{C}$-hyperarithmetical. Note that in case $\mathscr{C}$ is closed under elementary operations then $\mathscr{C}_{\mathsf{ha}} = \mathscr{C}_{\mathsf{hae}}$; hence $\mathscr{C}_{\mathsf{hag}} = \mathscr{C}_{\mathsf{haeg}} = \mathscr{C}_{\mathsf{ha\#}}$ denotes the class of all $\mathscr{C}$-hyperarithmetical functions.

Theorem 5.1.12. *Let $\mathscr{C}$ be a class of functions closed under elementary operations. Then*

(i) *$\mathscr{C}_{\mathsf{ha\#}}$ is* e*-total.*
(ii) *$\mathscr{C}_{\mathsf{ha\#d}} = \mathscr{C}_{\mathsf{ha\#de}} = \mathscr{C}_{\mathsf{ha}}$.*
(iii) *$\mathscr{C}_{\mathsf{ha}}$ is not* RRS, *and $\mathscr{C}_{\mathsf{ha\#}}$ is not a* FRS.
(iv) *If c is a unary $\mathscr{C}$-hyperarithmetical function, then δ_c and c_{j} are $\mathscr{C}$-hyperarithmetical functions.*
(v) *If $\mathscr{C}' \subseteq \mathscr{C}_{\mathsf{ha\#}}$ and $\mathscr{C}'$ is a* FRS, *then $\mathscr{C}'_{\mathsf{j}} \subseteq \mathscr{C}_{\mathsf{ha\#}}$.*

PROOF. Part (i) follows from Theorem 3.1.5 (iv) and Theorem 3.1.6 (iii). To prove (ii) we note that the inclusion $\mathscr{C}_{\mathsf{ha\#d}} \subseteq \mathscr{C}_{\mathsf{ha\#de}}$ is trivial, and $\mathscr{C}_{\mathsf{ha\#de}} = \mathscr{C}_{\mathsf{hae}} = \mathscr{C}_{\mathsf{ha}}$ follows from Theorem 3.1.6 (ii) and Theorem 5.1.11 (v). Finally, the inclusion $\mathscr{C}_{\mathsf{ha}} \subseteq \mathscr{C}_{\mathsf{ha\#d}}$ follows because if P is a $\mathscr{C}$-hyperarithmetical predicate then χ_P is a $\mathscr{C}$-hyperarithmetical function.

In proving (iii) we note that $\mathscr{C}_{\mathsf{ha}}$ is closed under negation and hence is not a RRS (Corollary 4.5.5.1 (iii)). It follows that $\mathscr{C}_{\mathsf{ha\#}}$ is not FRS, since if this is the case then $\mathscr{C}_{\mathsf{ha\#pd}} = \mathscr{C}_{\mathsf{ha}}$ is a RRS (Example 4.5.4).

Part (iv) follows from Theorem 4.2.4 (i), noting that the function δ is recursive, hence it is $\mathscr{C}$-hyperarithmetical, and the specification of c_{j} uses total definition with the $\mathscr{C}$-hyperarithmetical cases. Part (v) is immediate from (iv), since $\mathscr{C}' = \mathsf{RC}(c)$ for some unary $\mathscr{C}$-hyperarithmetical function c. □

Theorem 5.1.13. *Let $\mathscr{P}$ be a* RRS. *Then*

(i) *$\mathscr{P}_{\mathsf{he}}$ is a* RRS.
(ii) *If $\mathscr{P}$ is closed under disjunction then $\mathscr{P}_{\mathsf{hu}}$ is a* RRS.

PROOF. To prove (i) note that closure under conjunction and universal bounded quantification follows from Theorem 5.1.2 (iii) and (v), and closure under general substitution with recursive functions follows by exportation of quantifiers. If (W, S) is an internal interpreter for $\mathscr{P}$, then $(\mathsf{W}', \mathsf{S})$ is an internal interpreter for $\mathscr{P}_{\mathsf{he}}$, where

$$z \in \mathsf{W}' \equiv \forall\alpha\exists v \mathsf{S}(z, \bar{\alpha}(v)) \in \mathsf{W}.$$

The proof of (ii) is similar, but here closure under universal bounded quantification follows using closure under disjunction and universal unbounded quantification. □

Corollary 5.1.13.1. *Let $\mathscr{C}$ be a* FRS *such that $\mathscr{C} \subseteq \mathscr{C}_{\mathsf{ha\#}}$. Then*

(i) $\mathscr{C}_{\mathsf{pd}} \subseteq \mathscr{C}_{\mathsf{ha}}$.
(ii) $\mathscr{C}_{\mathsf{pdhe}} = \mathscr{C}_{\mathsf{dhe}}$.
(iii) $\mathscr{C}_{\mathsf{dhe}}$ *is a* RRS *and* $\mathscr{C}_{\mathsf{ha}} \subset \mathscr{C}_{\mathsf{dhe}}$.

PROOF. The inclusion in part (i) is clear from the assumption. To prove (ii) note that

$$\mathscr{C}_{\mathsf{pdhe}} \subseteq \mathscr{C}_{\mathsf{hahe}} = \mathscr{C}_{\mathsf{dhe}} \subseteq \mathscr{C}_{\mathsf{pdhe}}.$$

The inclusion in part (iii) is proper because $\mathscr{C}_{\mathsf{dhe}}$ is not closed under negation. □

EXAMPLE 5.1.5. If $\mathscr{C}$ is e-total then $\mathscr{C} = \mathscr{C}_{\mathsf{d\#}} \subset \mathscr{C}_{\mathsf{ha\#}}$. In particular, $\mathsf{RC} \subset \mathsf{RC}_{\mathsf{ha\#}}$.

EXERCISES

5.1.1. Let $\mathscr{P}$ be a class of predicates closed under substitution with total $\mathscr{C}$-computable functions, where $\mathscr{C}$ is closed under elementary operations. Assume $\mathsf{EL_d} \subseteq \mathscr{P}$. Prove the following:
(a) $\mathscr{P} \subseteq \mathscr{P}_{\mathsf{hu}}$ and $\mathscr{P}_{\mathsf{hu}}$ is closed under substitution with total $\mathscr{C}$-computable functions.
(b) $\mathscr{P}_{\mathsf{hu}}$ is closed under universal barred quantification and universal unbounded quantification.
(c) If $\mathscr{P}$ is closed under disjunction, then $\mathscr{P}_{\mathsf{hu}}$ is closed under disjunction, existential unbounded quantification, and universal bounded quantification.
(d) If $\mathscr{P}$ is closed under conjunction, then $\mathscr{P}_{\mathsf{hu}}$ is closed under conjunction.
(e) If $\mathscr{P}$ is closed under existential bounded quantification, then $\mathscr{P}_{\mathsf{hu}}$ is closed under existential bounded quantification.

5.1.2. Let $\mathscr{P}$ be a class of predicates closed under substitution with elementary functions, conjunction, disjunction, and universal barred quantification. Assume $\mathsf{EL_d} \subseteq \mathscr{P}$. Prove that $\mathscr{P}$ is closed under bounded and unbounded numerical quantification.

5.1.3. Let $\mathscr{P}$ be a class of predicates closed under substitution with elementary functions. Prove the following:
(a) $\mathscr{P}_{\mathsf{hugpd}} \subseteq \mathscr{P}_{\mathsf{hu}}$.
(b) $\mathscr{P}_{\mathsf{hu}} = \mathscr{P}_{\mathsf{uhu}} = \mathscr{P}_{\mathsf{huu}} = \mathscr{P}_{\mathsf{huhu}}$.

5.1.4. Let $\mathscr{P}$ be a class of predicates closed under substitution with elementary functions and conjunction. Assume $\mathsf{EL_d} \subseteq \mathscr{P}$. Prove that $\mathscr{P}_{\mathsf{he}} = \mathscr{P}_{\mathsf{heu}} = \mathscr{P}_{\mathsf{uhe}}$.

5.1.5. Let $\mathscr{P}$ be a class of predicate closed under substitution with elementary functions and disjunction. Assume $\mathsf{EL_d} \subseteq \mathscr{P}$. Prove that $\mathscr{P}_{\mathsf{hu}} = \mathscr{P}_{\mathsf{hue}} = \mathscr{P}_{\mathsf{ehu}}$.

5.1.6. Let $\mathscr{C}$ be a class of functions closed under elementary operations. Assume P_1 and P_2 are k-ary predicates, P_1 is $\mathscr{C}_{\mathsf{d}}$-hyperenumerable, and P_2 is $\mathscr{C}_{\mathsf{pd}}$-hyperenumerable. Prove that $P_1 \vee P_2$ is $\mathscr{C}_{\mathsf{pd}}$-hyperenumerable.

5.1.7. Let $\mathscr{C}$ be a class of functions closed under elementary operations. Prove that a total function h is $\mathscr{C}_{\mathsf{d}}$-hyperenumerable if and only if it is $\mathscr{C}$-hyperarithmetical.

5.1.8. Let $\mathscr{C}$ be a class of functions closed under hyperenumeration operations. Prove there is a class $\mathscr{C}'$ of functions such that:
(a) $\mathscr{C}'$ is closed under recursive operations.
(b) $\mathscr{C}'$ is bounded by RC.
(c) $\mathscr{C} = \mathscr{C}'_{\mathsf{pdh}\#}$.

5.1.9. Let $\mathscr{C}$ be a FRS. Assume $\mathscr{C}$ is e-total. Prove the following:
(a) $\mathscr{C}_{\mathsf{dh}\#} \nsubseteq \mathscr{C}$.
(b) $\mathscr{C}$ is not closed under hyperenumeration operations.
(c) $\mathscr{C}$ is not he-total.

5.1.10. Let U be the expression in Example 5.1.1. Prove that U holds if and only if x is an even number.

5.1.11. Let $\mathscr{C}$ be a class closed under elementary operations. Prove that $\mathscr{C}_{\mathsf{ha}\#\mathsf{j}} = \mathscr{C}_{\mathsf{ha}\#}$.

5.1.12. Let $\mathscr{C}$ be a class closed under elementary operations. Prove that $\mathscr{C}_{\mathsf{dh}\#\mathsf{ha}} = \mathscr{C}_{\mathsf{ha}}$.

5.1.13. Define the hyperenumeration hierarchy induced by a class $\mathscr{P}$ of predicates in a manner analogous to the enumeration hierarchy in Chapter 3, Section 1. Prove the proper version of Theorem 3.1.6.

Notes

Hyperenumeration (= existential barred quantification) is a natural extension of enumeration (= existential unbounded quantification). This is more than a mere verbal relation, and we show in this section that many of the results in Chapter 3 extend to barred quantification. In fact, a case can be made that barred quantification is a weaker predicative form of function quantification, but we do not elaborate this direction. On the other hand, we note the important fact that the proof of Theorem 5.1.1 involves a strong form of the axiom of choice.

The study of hyperenumeration was initiated by Kleene [11], and we follow to a great extent that approach in this chapter. Different approaches can be found in Rogers [20] and Hinman [8].

§2. Nonfinitary Induction

In the preceding section we discussed the closure properties induced by hyperenumeration. Now we show that closure under some forms of function quantification are also valid, and we apply these results to the theory of nonfinitary induction, that is, induction involving universal quantification, numerical or functional. This requires introducing boolean terms in which such operations are admissible.

We recall that basic terms, as defined in Chapter 1, involve only numerical variables, constants, and function symbols. Now we allow also function variables, that is, letters α, β, γ, ..., which range over total unary functions. We treat such symbols as function symbols in the definition of basic terms. Hence, if U is a basic term and α is a function variable, then $\alpha(U)$ is a basic term. Note that U may contain α or other function variables.

We proceed to define A-terms, which are terms involving basic terms as defined above, and predicate symbols of given arity. The construction rules are as follows:

A 1: If Q is a predicate symbol of arity n, $n \geq 0$, and $U_1, \ldots, U_n$ are basic terms, then $Q(U_1, \ldots, U_n)$ is an A-term.

A 2: If U_1 and U_2 are A-terms, then $(U_1) \vee (U_2)$ and $(U_1) \wedge (U_2)$ are also A-terms.

A 3: If U is an A-term, V is a basic term, and y is a numerical variable not occurring in V, then $\forall y < V$: U and $\exists y < V$: U are also A-terms.

A 4: If U is an A-term and y is a numerical variables, then $\forall y$: U and $\exists y$: U are also A-terms.

A 5: If U is an A-term and α is a functional variable then $\forall \alpha$: U is an A-term.

This definition extends E-terms, introduced in Chapter 3, by allowing universal quantification, numerical and functional. Note that existential functional quantification is not allowed. The discussion in Chapter 3 can be extended in an obvious way and there is no need to repeat it here. Most of the time we shall consider A-terms in which the only free variables are numerical variables; that is, the function variables, if any, are bound. If U is such a term, and $\mathbf{x}$ is a list of k variables containing all free variables in U, we can introduce a k-ary predicate P in the form

$$P(\mathbf{x}) \equiv U.$$

This is, of course, another form of *explicit specification.*

The study of closure under this type of specification, in the context of this work, presents some obvious difficulties, since we do not allow for predicates with functional arguments. To overcome this problem we shall prove that function quantification in A-terms can be eliminated in terms of existential barred quantification.

Let U be an A-term, which may contain free function variables. The notation $U[\bar{\alpha}(v)]$ denotes the substitution of the expression $\bar{\alpha}(v)$ for all free occurrences of the variable v in the term U. The expression $U[\bar{\alpha}(v)]$ is not an A-term, but it is semantically clear and can be used without problems. On the other hand, note that

$$\exists v U[\bar{\alpha}(v)] \equiv \exists v\colon U \wedge \forall y < \ell v(v)_y = \alpha(y) + 1,$$

and the expression on the right side is an A-term.

We are interested in the substitution $U[\overline{\alpha}(v)]$ when the variable α does not occur free in U. In particular, if U contains no free function variable we can introduce a predicate Q such that

$$Q(\mathbf{x}, v) \equiv U,$$

and it follows that

$$\forall\alpha\exists v Q(\mathbf{x}, \overline{\alpha}(v)) \equiv \forall\alpha\exists v U[\overline{\alpha}(v)].$$

Theorem 5.2.1. *Let U be an* A*-term, y a numerical variable that is not bound in U, and α a function variable that is not bound in U. Let U' be the* A*-term obtained by replacing in U every subterm of the form $\alpha(W)$ by $\alpha(\langle y, W\rangle)$. Then*

$$\exists y\forall\alpha U \equiv \forall\alpha\exists y U'.$$

Proof. The implication from left to right is trivial. Assume the left side is false; that is, $\exists y\forall\alpha U \equiv$ F. This means that for every y there is a function α_y such that $U \equiv$ F when α is interpreted as α_y. If we put $\alpha(x) = \alpha_{[x]_1}([x]_2)$, it follows that $U' \equiv$ F, and hence $\exists\alpha\forall y \sim U'$; that is the right side is false. □

We say that an A-term U is *simple* if the only operations in U are conjunction and disjunction. This means that U contains no quantifier, bounded or unbounded. A term U' is an *elementary extension* of a term U if any function or predicate that occurs in U' but not in U is elementary.

Theorem 5.2.2. *Let U be a simple* A*-term, y a numerical variable, and α a function variable. There is a simple* A*-term U' and a numerical variable v, such that*

$$\exists y U \equiv \exists v U'[\overline{\alpha}(v)],$$

and the following conditions are satisfied:

(i) *The variable α does not occur in U' and the variable v does not occur in U.*
(ii) *The term U' is an elementary extension of U.*

Proof. The A-term U' is obtained from U by a sequence of replacements that introduce elementary functions and predicates. We take a new variable v that does not occur in U and replace all free occurrences of y in U by the term $(\ell v)_0$. The term so obtained we call U_0. Next we consider a subterm in U_0 of the form $\alpha(V_1)$, where V_1 does not contain the variable α, and replace this subterm by $(v)_{V_1} \dot{-} 1$. The term so obtained we call U_1. Next we consider a subterm in U_1 of the form $\alpha(V_2)$, where V_2 does not contain the variable α, and replace this subterm by $(v)_{V_2} \dot{-} 1$. The term so obtained we call U_2. In this way we generate terms U_1, U_2, ... until we get to a term U_m, $m \geq 0$, which does not contain α. We take U' as the term

$$U_m \wedge V_1 < \ell v \wedge V_2 < \ell v \wedge \cdots \wedge V_m < \ell v.$$

In order to prove the equivalence $\exists y U \equiv \exists v U'[\overline{\alpha}(v)]$ we assume first that the left side takes value T; hence there is some y such that $U \equiv \mathsf{T}$. Since U is finitary the evaluation of U requires only a finite number of values of α. Let v be a number such that $\bar{\alpha}(v)$ encodes all those values. By taking v sufficiently higher we may assume that $(v)_0 = y$. From the construction of U' it follows that $U'[\bar{\alpha}(v)] \equiv \mathsf{T}$. Conversely, if $\exists v U'[\bar{\alpha}(v)] \equiv \mathsf{T}$ there is a number v such that $U'[\bar{\alpha}(v)] \equiv \mathsf{T}$. If we take $y = (\ell\bar{\alpha}(v))_0 = (v)_0$ it follows that $U \equiv \mathsf{T}$. □

Theorem 5.2.3. *Let U be an* A*-term. There is a simple* A*-term U' such that*

$$U \equiv \forall\alpha\exists v U'[\bar{\alpha}(v)],$$

and the following conditions are satisfied:

(i) *The variable α does not occur in U' and the variable v does not occur in U.*
(ii) *The term U' is an elementary extension of U.*

Proof. We execute a number of transformations in U until we get an equivalent term $\forall\alpha\exists y U_1$, where U_1 is simple. Then we apply Theorem 5.2.2.

First, we eliminate all bounded quantifiers in terms of unbounded quantifiers. This requires only the elementary predicates $<$ and $>$ combined with conjunction and disjunction. Second, we eliminate the universal numerical quantifiers in terms of universal function quantifiers. For example, a context of the form

$$\forall x : \cdots x \cdots$$

is replaced by

$$\forall\alpha : \cdots \alpha(0) \cdots .$$

In this way we obtain an equivalent A-term in which the only operations are conjunction, disjunction, existential numerical quantification, and universal function quantification.

The next step is to export all quantifiers over conjunction and disjunction. This is a simple logical transformation, which may require in some cases a change of bound variables. By using Theorem 5.2.1 we obtain an A-term in prenex form where all universal function quantifiers precede the existential numerical quantifiers which precede a simple A-term.

We proceed now to contract all quantifiers of the same type. For example,

$$\forall\alpha_1\forall\alpha_2\cdots\forall\alpha_m : \cdots \alpha_1(V_1) \cdots \alpha_2(V_2) \cdots \alpha_m(V_m) \cdots$$

can be replaced by

$$\forall\alpha : \cdots (\alpha(V_1))_1 \cdots (\alpha(V_2))_2 \cdots (\alpha(V_m))_m \cdots$$

and similarly for existential quantifiers. The result is a simple term U_1 such that

$$U \equiv \forall\alpha\exists y U_1 .$$

The term U' is obtained using Theorem 5.2.2. □

Corollary 5.2.3.1. *Let $\mathscr{P}$ be a class of predicates closed under substitution with total $\mathscr{C}$-computable functions, where $\mathscr{C}$ is closed under elementary operations. Assume $\mathsf{EL_d} \subseteq \mathscr{P}$, and $\mathscr{P}$ is closed under conjunction and disjunction. Let U be an* A*-term in which all functions are $\mathscr{C}$-computable and all predicates are in $\mathscr{P}$, and assume all free numerical variables in U occur in the list* $\mathbf{x}$ *of length $k \geq 0$, and all free function variables in U occur in the list $\beta_1, \ldots, \beta_m$, $m \geq 0$. There is a $(k + m - 1)$-ary predicate Q in $\mathscr{P}$ such that*

$$U \equiv \forall\alpha\exists v Q(\mathbf{x}, \bar{\beta}_1(v), \ldots, \bar{\beta}_m(v), \bar{\alpha}(v)).$$

Proof. Using Theorem 5.2.3 we get a simple term U_0 such that

$$U \equiv \forall\alpha\exists v U_0[\bar{\alpha}(v)],$$

where α does not occur in U_0. Using Theorem 5.2.2 we get simple terms U_1, $\ldots$, U_m such that

$$U_0 \equiv \exists y_1 U_1[\bar{\beta}_1(y_1)]$$
$$U_1 \equiv \exists y_2 U_2[\bar{\beta}_2(y_2)]$$
$$\vdots$$
$$U_{m-1} \equiv \exists y_m U_m[\bar{\beta}_m(y_m)].$$

We set Q' the $(k + m + 1)$-ary predicate such that

$$U_m \equiv Q'(\mathbf{x}, y_1, \ldots, y_m, v),$$

so Q' is in $\mathscr{P}$ and

$$U \equiv \forall\alpha\exists v\exists y_1 \cdots \exists y_m Q'(\mathbf{x}, \bar{\beta}_1(y_1), \ldots, \bar{\beta}_m(y_m), \bar{\alpha}(v)).$$

In order to contract the existential quantifiers we introduce unary elementary functions $g_1, \ldots, g_m, g_{m+1}$ such that

$$g_i(u) = \left(\prod_{v < [\ell u]_i} \mathsf{exp}(\mathsf{p}_v, (u)_v)\right) \dot{-} 1 \qquad i = 1, \ldots, m + 1,$$

and we set Q the $(k + m + 1)$-ary predicate such that

$$Q(\mathbf{x}, y_1, \ldots, y_m, v) \equiv Q'(\mathbf{x}, g_1(y_1), \ldots, g_m(y_m), g_{m+1}(v)),$$

and it follows that

$$U \equiv \forall\alpha\exists v Q(\mathbf{x}, \bar{\beta}_1(v), \ldots, \bar{\beta}_m(v), \bar{\alpha}(v)). \qquad \square$$

Example 5.2.1. Typical applications of Corollary 5.2.3.1 are classes of the form $\mathscr{C}_{\mathsf{d}}$, where $\mathscr{C}$ is closed under elementary operations. Note that closure under disjunction may fail in $\mathscr{C}_{\mathsf{pd}}$ even when $\mathscr{C}$ is closed under recursive operations.

The theory of induction in Chapter 2 can easily be extended to include A-terms *without free function variables.* If $U_1, \ldots, U_m$ are such terms, inductive specification consists of m statements of the form

(l) If U_1 then $P(\mathbf{x})$.

$\vdots$

(m) If U_m then $P(\mathbf{x})$.

Here the A-terms $U_1, \ldots, U_m$ may contain the uninterpreted predicate symbol P. A *solution* is a predicate P that satisfies all the statments.

Following the approach in Chapter 2 we associate with an inductive specification a predicate transformation F such that

$$F(P)(\mathbf{x}) \equiv U_1 \vee U_2 \vee \cdots \vee U_m.$$

In general, this transformation is not continuous but it is monotonic. The minimal fixed point of F is the *minimal solution* of the inductive specification.

This type of specification is nonfinitary in the sense that the evaluation of the minimal solution may require an infinite number of steps.

EXAMPLE 5.2.2. The following example has no intrinsic interest, but it provides an indication of the type of induction we shall use later. A set A is introduced as the minimal solution of an inductive specification involving the following two statements:

(a) If $x = 0$ then $x \in A$.
(b) If $\forall y(\mathsf{exp}(x, y) \dot{-} 1)_0 \in A$ then $x \in A$.

From statement (1) it follows that $0 \in A$, and from statement (2) it follows that all odd numbers are elements of A. Below we prove that $A = \{0, 1, 3, \ldots\}$.

Every inductive specification induces an associated method of proof by induction. This matter was discussed in Chapter 3 and there is no need for further discussion here.

EXAMPLE 5.2.3. In relation with Example 5.2.2 we introduce the set $B = \{0, 1, 3, \ldots\}$, that is, $x \in B$ if and only if $x = 0$ or x is an odd number. We want to prove that $A = B$. We have already shown that $B \subseteq A$. The converse is equivalent to the assertion "if $x \in A$ then $x \in B$" and can be proved using induction on the specification of the set A. This requires replacing the symbol A by the symbol $A \cap B$ in the two inductive statements, so we get:

(1′) If $x = 0$ then $x \in B$.
(2′) If $\forall y(\mathsf{exp}(x, y) \dot{-} 1)_0 \in A \cap B$, then $x \in B$.

It is obvious that statement (1′) is valid. The validity of (2′) requires some argument. It is clear that whenever x is an odd number the assertion is true. If x is an even number then $(\mathsf{exp}(x, 2) \dot{-} 1)_0$ is an even number different from 0. This means that $\forall y(\mathsf{exp}(x, y) \dot{-} 1)_0 \in B$ is false, hence (2′) is true.

Theorem 5.2.4. *Let $\mathscr{P}$ be a class of predicates closed under substitution with total $\mathscr{C}$-computable functions, conjunction, disjunction, and existential barred* *quantification, where $\mathscr{C}$ is a class of functions closed under elementary operations. Assume $\mathsf{EL_d} \subseteq \mathscr{P}$. If the k-ary predicate P is the minimal solution of an inductive specification involving* A*-terms in which the functions are $\mathscr{C}$-computable and the predicates are in $\mathscr{P}$, then P is in $\mathscr{P}$.*

PROOF. First we assume that $k = 1$ so P is a unary predicate, and the specification has the form

$$\text{If } U \text{ then } P(x).$$

During the proof the symbol P denotes the minimal solution of the specification, which means that P is the minimal fixed point of the associated predicate transformation F.

We take a function variable α that does not occur in U and replace every part of U of the form $P(V)$ by $\alpha(V) = 0$. We call U' the term so obtained. We use U' to introduce a unary predicate P_0 such that

$$P_0(x) \equiv \forall\alpha\colon \alpha(x) = 0 \vee \exists y\colon U' \wedge \alpha(y) \neq 0.$$

From Corollary 5.2.3.1 it follows that P_0 is in $\mathscr{P}$. We proceed to prove that $P = P_0$.

To prove that $P_0 \subseteq_k P$ we assume that $P_0(x) \equiv \mathsf{T}$. If we fix $\alpha = \chi_P$ it follows that

$$\exists y(U' \wedge \alpha(y) \neq 0) \vee \alpha(x) = 0.$$

Since P is a solution it is impossible that there is y such that $U' \wedge \alpha(y) \neq 0$; hence $\alpha(x) = 0$, that is, $P(x) \equiv \mathsf{T}$.

To prove that $P \subseteq_k P_0$ we assume that $P(x) \equiv \mathsf{T}$, and to prove that $P_0(x) \equiv \mathsf{T}$ we take an arbitrary unary function α. Assume that

$$\sim\exists y(U' \wedge \alpha(y) \neq 0).$$

If we put $P'(y) \equiv \alpha(y) = 0$, this means that P' is a solution, hence $P \subseteq_k P'$, so $P'(x) \equiv \mathsf{T}$, that is, $\alpha(x) = 0$. This means that $P_0(x) \equiv \mathsf{T}$.

We consider now the case in which $k > 1$, so P is a k-ary predicate, which is the minimal solution of the inductive specification

$$\text{If } U \text{ then } P(\mathbf{x}).$$

We introduce a new unary predicate symbol P_1 and perform in U the following replacements: replace each free occurrence of the variable x_i by $[x]_i$, $i = 1, \ldots, k$, where x is a new variable (we assume here that $\mathbf{x} = x_1, \ldots, x_k$); replace each occurrence of a subterm $P(V_1, \ldots, V_k)$ by $P_1(\langle V_1, \ldots, V_k\rangle)$. The term so obtained we call U'. Now P_1 denotes the minimal solution of the specification

$$\text{If } U' \text{ then } P_1(x).$$

We introduce a k-ary predicate P' such that $P'(\mathbf{x}) \equiv P_1(\langle\mathbf{x}\rangle)$. Clearly, since P_1 is in $\mathscr{P}$ by the argument in the first part of the proof, it follows that P' is also in $\mathscr{P}$. We shall prove that $P = P'$ by proving that P' is the minimal solution of the original specification.

To show that P' is a solution assume $U \equiv \mathsf{T}$ when $P = P'$. It follows that $U' \equiv \mathsf{T}$ with $x = \langle \mathbf{x} \rangle$; hence $P_1(\langle \mathbf{x} \rangle) \equiv \mathsf{T}$, that is, $P'(x) \equiv \mathsf{T}$. To prove P' is the minimal solution assume P'' is another solution and let P_1'' be the unary predicate such that $P_1''(x) \equiv P''([x]_1, \ldots, [x]_k)$. To prove that P_1'' is a solution of the second specification we assume that $U' \equiv \mathsf{T}$ with $P_1 = P_1''$. It follows that $U \equiv \mathsf{T}$ with $P = P''$ and $x_1 = [x]_1, \ldots, x_k = [x]_k$; hence $P''([x]_1, \ldots, [x]_k) \equiv \mathsf{T}$, that is, $P_1''(x) \equiv \mathsf{T}$. This means that $P_1 \subseteq_k P_1''$, so $P' \subseteq_1 P''$, and P' is the minimal solution. □

Corollary 5.2.4.1. *Let $\mathscr{C}$ be a class of functions closed under elementary operations. Then $\mathscr{C}_{\mathsf{dhe}}$ is closed under inductive specifications with* A*-terms in which the functions are $\mathscr{C}$-computable and the predicates are $\mathscr{C}_{\mathsf{d}}$-hyperenumerable.*

Proof. See discussion in Example 5.1.2. □

Corollary 5.2.4.2. *Let $\mathscr{C}$ be a class of functions closed under recursive operations. Then $\mathscr{C}_{\mathsf{pdhe}}$ is closed under inductive specifications with* A*-terms in which the functions are $\mathscr{C}$-computable and the predicates are $\mathscr{C}_{\mathsf{pd}}$-hyperenumerable.*

Proof. See discussion in Example 5.1.3. □

Example 5.2.4. Let Q be a $(k + 1)$-ary predicate, $k \geq 0$, and consider the $(k + 1)$-ary predicate P introduced by the following induction:

(1) If $Q(\mathbf{x}, y)$ then $P(\mathbf{x}, y)$.
(2) If $\forall v P(\mathbf{x}, y \,\square\, \langle v \rangle)$ then $P(\mathbf{x}, y)$.

It follows that

$$\forall \alpha \exists v Q(\mathbf{x}, \bar{\alpha}(v)) \equiv P(\mathbf{x}, 0).$$

Example 5.2.5. Let $\mathscr{C}$ be a FRS with internal interpreter (ϕ, S), and assume the function S is strict. We introduce a set A_ϕ by the following induction:

(1) If $w = 0$ then $w \in \mathsf{A}_\phi$.
(2) If $\forall v \exists v' \colon \phi^1(w, v) \simeq v' \wedge v' \in \mathsf{A}_\phi$ then $w \in \mathsf{A}_\phi$.

It follows that A is $\mathscr{C}_{\mathsf{pd}}$-hyperenumerable. Now consider a $(k + 1)$-ary predicate Q, $k \geq 0$, assume Q is $\mathscr{C}$-decidable and introduce using total definition by cases the total $\mathscr{C}$-computable $(k + 2)$-ary function f such that

$$\begin{aligned} f(z, \mathbf{x}, y) &= 0 && \text{if } Q(\mathbf{x}, y) \\ &= g(z, \mathbf{x}, y) && \text{otherwise,} \end{aligned}$$

where g is the total $\mathscr{C}$-computable function given by the abstraction property such that

$$\phi^1(g(z, \mathbf{x}, y), v) \simeq \phi^{k+1}(z, \mathbf{x}, y \,\square\, \langle v \rangle).$$

Now fix a number z such that $f(\mathrm{z}, \mathbf{x}, y) = \phi^{k+1}(\mathrm{z}, \mathbf{x}, y)$, and let h be the total

$\mathscr{C}$-computable k-ary function such that

$$h(\mathbf{x}) = f(\mathbf{z}, \mathbf{x}, 0).$$

It follows that

$$\forall\alpha\exists v Q(\mathbf{x}, \bar{\alpha}(v)) \equiv h(\mathbf{x}) \in \mathsf{A}_\phi.$$

EXERCISES

5.2.1. In relation to Example 5.2.4 prove the following assertions:
(a) If $\sim P(\mathbf{x}, y)$ then $\sim Q(\mathbf{x}, y)$ and there is some v such that $\sim P(\mathbf{x}, y \,\square\, \langle v \rangle)$.
(b) If $\sim P(\mathbf{x}, 0)$ there is a total unary function α such that $\sim Q(\mathbf{x}, \bar{\alpha}(v))$ and $\sim P(\mathbf{x}, \bar{\alpha}(v))$ hold for all v.
(c) Let α be a total unary function. If $P(\mathbf{x}, y)$ and $\exists v \bar{\alpha}(v) = y$ then $\exists v \colon \ell y \leq v \wedge Q(\mathbf{x}, \bar{\alpha}(v))$.
(d) If $P(\mathbf{x}, 0)$ then $\forall\alpha\exists v Q(\mathbf{x}, \bar{\alpha}(v))$.

5.2.2. In relation to Example 5.2.5 note that the function S is strict and the number $\mathbf{z}$ is fixed by the RT. Prove the following assertions:
(a) If $f(\mathbf{z}, \mathbf{x}, y) \notin A_\phi$ then $\sim Q(\mathbf{x}, y)$ and there is some v such that $f(\mathbf{z}, \mathbf{x}, y \,\square\, \langle v \rangle) \notin A_\phi$.
(b) If $h(\mathbf{x}) \notin A_\phi$ then there is a total unary function α such that $\sim Q(\mathbf{x}, \bar{\alpha}(v))$ and $f(\mathbf{z}, \mathbf{x}, \bar{\alpha}(v)) \notin A_\phi$ hold for all v.
(c) Assume α is a total unary function and $(\mathbf{x}) \in \mathbb{N}^k$. If $w \in A_\phi$ and $\exists v f(\mathbf{z}, \mathbf{x}, \bar{\alpha}(v)) \simeq w$ then $\exists v Q(\mathbf{x}, \bar{\alpha}(v))$.
(d) If $h(\mathbf{x}) \in A_\phi$ then $\forall\alpha\exists v Q(\mathbf{x}, \bar{\alpha}(v))$.

5.2.3. Let U be a simple term and β a function variable. Prove there is a simple term U' such that

$$U \equiv \exists y U'[\bar{\beta}(y)],$$

the variable β does not occur in U', and U' is an elementary extension of U.

5.2.4. Let U be an A-term and β a function variable. Prove there is a simple term U' such that

$$U \equiv \forall\alpha\exists v\exists y U'[\bar{\beta}(y)][\bar{\alpha}(v)],$$

the variables α and β do not occur free in U', and U' is an elementary extension of U.

5.2.5. Consider an inductive specification of the form: If U then $P(\mathbf{x})$, where U is an A-term, the free numerical variables of U are in the list $\mathbf{x}$, and furthermore U contains free occurrences of a function variable β. Denote by P_β the minimal solution of the specification for a given fixed function β. Prove there is a simple term U' such that

$$P_\beta(\mathbf{x}) \equiv \forall\alpha\exists y\exists v U'[\bar{\beta}(y)][\bar{\alpha}(v)].$$

5.2.6. Let P be a k-ary predicate that is the minimal solution of an inductive specification of the form: If U then $P(\mathbf{x})$, where U is an A-term. Assume P' is a k-ary predicate that satisfies the condition: If $P(\mathbf{x})$ and U' then $P'(\mathbf{x})$, where U' is the result of replacing all occurrences of P in U with $P \wedge P'$. Prove that $P \subseteq_k P'$.

Notes

Induction was initially considered in Chapter 3, and the basic theory presented there is adequate for the study of nonfinitary induction. Our main concern here is the formalization of induction via A-terms, which include function quantifiers. The main result is that the minimal solution of such an induction can be expressed using hyperenumeration. This situation was clarified, after some initial confusion, by Kleene [11].

The formalization of inductive specifications induces the formalization of proofs by induction. Again this technique was introduced in Chapter 3, and the same theory applies here. The formalization provides an exact procedure to carry out inductive proofs, particularly to identify properly the induction hypothesis in each case of the induction.

§3. Functional Induction

Induction is a nondeterministic procedure that is used to specify predicates. It can also be used to specify functions, via the graph predicate, but in this case the single-valuedness property must be checked. In this section we give an example in which induction is used to specify the internal interpreter for a FRS, which depends on a given unary function c and satisfies special closure properties.

To discuss this construction we need two new specification rules. These are nonfinitary rules, essentially generalizations of the standard unbounded numerical quantification, and are denoted in the same way.

Universal Quantification on Functions. Let f be a $(k + 1)$-ary function. We introduce a k-ary function h such that

$$h(\mathbf{x}) \simeq v \equiv \forall y \exists w f(\mathbf{x}, y) \simeq w$$
$$\wedge ((v = 0 \wedge \forall y f(\mathbf{x}, y) \simeq 0) \vee (v = 1 \wedge \exists y \exists w f(\mathbf{x}, y) \simeq w + 1)).$$

We say that h is obtained from f by *universal quantification.* In applications the function h is introduced with the following notation:

$$h(\mathbf{x}) \simeq \forall y f(\mathbf{x}, y).$$

Existential Quantification on Functions. Let f be a $(k + 1)$-ary function. We introduce a k-ary function h such that

$$h(\mathbf{x}) \simeq v \equiv \forall y \exists w f(\mathbf{x}, y) \simeq w$$
$$\wedge ((v = 0 \wedge \exists y f(\mathbf{x}, y) \simeq 0) \vee (v = 1 \wedge \forall y \exists w f(\mathbf{x}, y) \simeq w + 1)).$$

We say that h is obtained from f by *existential quantification.* In applications the function h is introduced with the following notation:

$$h(\mathbf{x}) \simeq \exists y f(\mathbf{x}, y).$$

Both operations will be used with terms, in particular with basic terms. For example, if U is a term and the variables in U are contained in the list $\mathbf{x}$, y, we write $h_1(\mathbf{x}) \simeq \forall y U$, or $h_2(\mathbf{x}) \simeq \exists y U$, which is equivalent to $h_1(\mathbf{x}) \simeq \forall y f(\mathbf{x}, y)$, or to $h_2(\mathbf{x}) \simeq \exists y f(\mathbf{x}, y)$, where $f(\mathbf{x}, y) \simeq U$.

Note that if Q is $(k+1)$-ary predicate, and we set

$$P_1(\mathbf{x}) \equiv \forall y Q(\mathbf{x}, y)$$

$$P_2(\mathbf{x}) \equiv \exists y Q(\mathbf{x}, y).$$

It follows that

$$\chi_{P_1}(\mathbf{x}) = \forall y \chi_Q(\mathbf{x}, y)$$

$$\chi_{P_2}(\mathbf{x}) = \exists y \chi_Q(\mathbf{x}, y)$$

$$\psi_{P_1}(\mathbf{x}) \simeq \forall y \psi_Q(\mathbf{x}, y).$$

These two forms of quantification are related in the usual way, which means that one can be obtained from the other by substitution. For example,

$$\forall y f(\mathbf{x}, y) \simeq \mathsf{csg}(\exists y \mathsf{csg}(f(\mathbf{x}, y)))$$

$$\exists y f(\mathbf{x}, y) \simeq \mathsf{csg}(\forall y \mathsf{csg}(f(\mathbf{x}, y))).$$

Each one of the rules we are discussing induces the corresponding notion of closure, which applies to classes of functions. Under very general conditions it follows that closure under one implies closure under the other. This is the case, for example, if a class $\mathscr{C}$ is closed under elementary operations. Closure under both rules will be referred to simply as *closure under quantification*.

Theorem 5.3.1. *Let $\mathscr{C}$ be a class of functions. Then*

(i) *If $\mathscr{C}$ is closed under quantification, then $\mathscr{C}_{\mathsf{d}}$ is closed under unbounded numerical quantification and $\mathscr{C}_{\mathsf{pd}}$ is closed under universal unbounded quantification.*
(ii) *If $\mathscr{C}$ is closed under elementary operations, then $\mathscr{C}_{\mathsf{ha}\#}$ and $\mathscr{C}_{\mathsf{dh}\#}$ are closed under quantification.*
(iii) *If $\mathscr{C}$ is closed under recursive operations, then $\mathscr{C}_{\mathsf{pdh}\#}$ is closed under quantification.*

Proof. Part (i) follows from the equations above, which relate the characteristic functions and partial characteristic functions of predicates derived by universal and existential numerical quantification. Part (ii) follows from the discussion in Example 5.1.2 and Theorem 5.1.10 (iv) and (v). Part (iii) follows from the discussion in Example 5.1.3. □

Let c be a unary function. We use induction with A-terms to introduce a 3-ary predicate H_c. The statements of the induction are as follows:

H0 If $(z)_0 = 0 \wedge \mathsf{H}_c([x]_1, [x]_2, v)$ then $\mathsf{H}_c(z, x, v)$.
H1 If $(z)_0 = 1 \wedge \mathsf{H}_c((z)_1, \langle (z)_2 \rangle \square x, v)$ then $\mathsf{H}_c(z, x, v)$.
H2 If $(z)_0 = 2 \wedge v = \mathsf{S}_\delta([x]_1, [x]_2)$ then $\mathsf{H}_c(z, x, v)$.
H3 If $(z)_0 = 3 \wedge v = (z)_1$ then $\mathsf{H}_c(z, x, v)$.
H4 If $(z)_0 = 4 \wedge v = [x]_1$ then $\mathsf{H}_c(z, x, v)$.
H5 If $(z)_0 = 5 \wedge v = [x]_1 + 1$ then $\mathsf{H}_c(z, x, v)$.
H6 If $(z)_0 = 6 \wedge v = [x]_1 \dot{-} 1$ then $\mathsf{H}_c(z, x, v)$.
H7 If $(z)_0 = 7 \wedge v = \mathsf{cd}([x]_1, [x]_2, [x]_3)$ then $\mathsf{H}_c(z, x, v)$.
H8 If $(z)_0 = 8 \wedge \mathsf{H}_c((z)_1, x\S, v)$ then $\mathsf{H}_c(z, x, v)$.
H9 If $(z)_0 = 9 \wedge \mathsf{H}_c((z)_1, \S x, v)$ then $\mathsf{H}_c(z, x, v)$.
H10 If $(z)_0 = 10 \wedge \exists w\colon \mathsf{H}_c((z)_2, x, w) \wedge \mathsf{H}_c((z)_1, x \square \langle w \rangle, v)$ then $\mathsf{H}_c(z, x, v)$.
H11 If $(z)_0 = 11 \wedge \mathsf{G}_c([x]_1, v)$ then $\mathsf{H}_c(z, x, v)$.
H12a If $(z)_0 = 12 \wedge v = 0 \wedge \forall y \mathsf{H}_c((z)_1, x \square \langle y \rangle, 0)$ then $\mathsf{H}_c(z, x, v)$.
H12b If $(z)_0 = 12 \wedge v = 1 \wedge \forall y \exists w \mathsf{H}_c((z)_1, x \square \langle y \rangle, w) \wedge \exists y \exists w \mathsf{H}_c((z)_1, x \square \langle y \rangle, w + 1)$ then $\mathsf{H}_c(z, x, v)$.

This specification involves only elementary functions and the graph predicate G_c. It follows that if $\mathscr{C}$ is closed under elementary operations and c is $\mathscr{C}_{\mathsf{d}}$-hyperenumerable, then H_c is $\mathscr{C}_{\mathsf{d}}$-hyperenumerable. On the other hand, if $\mathscr{C}$ is closed under recursive operations and c is $\mathscr{C}_{\mathsf{pd}}$-hyperenumerable, then H_c is $\mathscr{C}_{\mathsf{pd}}$-hyperenumerable.

Theorem 5.3.2. *If* $\mathsf{H}_c(z, x, v)$ *and* $\mathsf{H}_c(z, x, v')$ *then* $v' = v$.

Proof. We introduce a predicate P such that

$$P(z, x, v) \equiv \forall v'\colon \sim \mathsf{H}_c(z, x, v') \vee v' = v,$$

and we prove by induction on the specification of H_c that $\mathsf{H}_c(z, x, v)$ implies $P(z, x, v)$. There are 14 cases in the induction with case H12 branching into two subcases. We consider only a few of these cases. In each case we assume $\mathsf{H}_c(z, x, v')$ and prove $v' = v$. We use systematically the inversion property on $\mathsf{H}_c(z, x, v')$.

Case H0: By the induction hypothesis we have $(z)_0 = 0$, $\mathsf{H}_c([x]_1, [x]_2, v)$, and $P([x]_1, [x]_2, v)$. By inversion on $\mathsf{H}_c(z, x, v')$ we have $\mathsf{H}_c([x]_1, [x]_2, v')$. From $P([x]_1, [x]_2, v)$ we get $v' = v$.

Case H10: By the induction hypothesis $(z)_0 = 10$ and there is w such that

$$\mathsf{H}_c((z)_2, x, w), \quad P((z)_2, x, w), \quad \mathsf{H}_c((z)_1, x \square \langle w \rangle, v), \quad P((z)_1, x \square \langle w \rangle, v).$$

By inversion we have that there is w' such that

$$\mathsf{H}_c((z)_2, x, w'), \quad \mathsf{H}_c((z)_1, x \square \langle w' \rangle, v').$$

From $P((z)_2, x, w)$ we get $w' = w$, and from $P((z)_1, x \square \langle w \rangle, v)$ we get $v' = v$.

Case H12a: By the induction hypothesis we have $(z)_0 = 12$, $v = 0$, and

$$\forall y\colon \mathsf{H}_c((z)_1, x \square \langle y \rangle, 0) \wedge P((z)_1, x \square \langle y \rangle, 0).$$

If $v' = 0$ we are done. Otherwise by inversion on $\mathsf{H}_c(z, x, v')$ it follows that there are y and w such that

$$\mathsf{H}_c((z)_1, x \,\Box\, \langle y \rangle, w + 1)$$

and this contradicts $P((z)_1, x \,\Box\, \langle y \rangle, 0)$. □

We have proved that the predicate H_c is single-valued. It follows that we can introduce a binary function Δ_c such that

$$\Delta_c(z, x) \simeq v \equiv \mathsf{H}_c(z, x, v).$$

The inductive specification of H_c induces a number of recursive relations on the function Δ_c, which it is convenient to make explicit for future applications. To avoid unnecessary repetitions we introduce two inverting functions κ and π_c.

$$\begin{aligned}
\kappa(z, x) &= [x]_1 && \text{if } (z)_0 = 0\\
&= (z)_1 && \text{otherwise}\\
\pi_c(z, x) &\simeq [x]_2 && \text{if } (z)_0 = 0\\
&\simeq \langle (z)_2 \rangle \,\Box\, x && \text{if } (z)_0 = 1\\
&\simeq x\S && \text{if } (z)_0 = 8\\
&\simeq \S x && \text{if } (z)_0 = 9\\
&\simeq x \,\Box\, \langle \Delta_c((z)_2, x) \rangle && \text{if } (z)_0 = 10.
\end{aligned}$$

Note that κ is an elementary function independent of c. On the other hand, the specification of π_c involves partial definition by cases and depends on Δ_c.

Theorem 5.3.3. *Let c be a unary function. Then*

(i) $\delta_c \subseteq_2 \Delta_c$.
(ii) $\delta_c(z, x) \simeq \Delta_c(z, x)$ *if* $(z)_0 = 2, 3, 4, 5, 6, 7,$ *or* 11.
(iii) $\Delta_c(z, x) \simeq \Delta_c(\kappa(z, x), \pi_c(z, x))$ *if* $(z)_0 = 0, 1, 8, 9,$ *or* 10.
(iv) $\Delta_c(z, x) \simeq \forall y \Delta_c((z)_1, x \,\Box\, \langle y \rangle)$ *if* $(z)_0 = 12$.

Proof. Part (i) follows by induction on the evaluation of $\delta_c(z, x)$, and part (ii) follows from (i) and inversion. To prove (iii) and (iv) we consider the corresponding statements in the specification of H_c. □

Let c be a unary function. We denote by $\mathsf{HC}(c)$ the class of all k-ary functions h such that there is a number $\mathbf{z}$ and

$$h(\mathbf{x}) \simeq \Delta_c(\mathbf{z}, \langle \mathbf{x} \rangle).$$

Let Φ_c be the unary function such that

$$\Phi_c(z) \simeq \Delta_c(z, 0).$$

Theorem 5.3.4. *Let c be a unary function. Then*

(i) $\Delta_c(\mathsf{S}_\delta^k(z, \mathbf{x}), 0) \simeq \Delta_c(z, \langle \mathbf{x} \rangle)$.
(ii) $\mathsf{HC}(c)$ *is a* FRS *with internal interpreter* $(\Phi_c, \mathsf{S}_\delta)$.
(iii) *The functions c and* Δ_c *are in* $\mathsf{HC}(c)$.
(iv) $\mathsf{RC}(c) \subseteq \mathsf{HC}(c)$, *and* $\mathsf{RC}(\Delta_c) = \mathsf{HC}(c)$.
(v) $\mathsf{HC}(c)$ *is closed under universal and existential quantification.*

Proof. Part (i) follows easily by induction on k using statement H1. The proof of (ii) is exactly the same as that given in Theorem 4.2.1. To prove (iii) note that

$$c(x) \simeq \Delta_c(\exp(2, 11) \dot{-} 1, \langle x \rangle)$$

$$\Delta_c(z, x) \simeq \Delta_c(0, \langle z, x \rangle).$$

Part (iv) is immediate from (iii) and the definition of $\mathsf{HC}(c)$. Part (v) follows from Theorem 5.3.3. (iv). □

From Theorem 5.3.4 (v) it follows that $\mathsf{HC}(c)$ is not e-total. The main result of this section is that in case c is a total function then $\mathsf{HC}(c)$ is closed under enumeration operations; that is, $\mathsf{HC}(c) = \mathsf{HC}_{\mathsf{pd}\#}(c)$. To get this result we prove that $\mathsf{HC}(c)$ has the selector property relative to $\mathsf{HC}_{\mathsf{pd}}(c)$, and this follows from Theorem 4.4.6 (i) and (iv). We must show that there is a Φ_c-preorder function that is in the FRS $\mathsf{HC}(c)$, where $(\Phi_c, \mathsf{S}_\delta)$ is the internal interpreter for $\mathsf{HC}(c)$ defined above.

The construction of the preorder function is rather complicated. It involves 15 auxiliary functions that are used in the specification of a function f. The preorder is derived from f by substitution. All these functions are in the class $\mathsf{HC}(c)$, where c is a unary function that is fixed in the construction. To prove the crucial properties of the function f we must assume c is a total function.

The auxiliary functions are introduced in two groups. The first group contains nine 5-ary functions $r_1, \ldots, r_9$, which are introduced by explicit specification in terms of some fixed internal interpreter (ϕ, S) for $\mathsf{HC}(c)$:

$$r_1(w, z, x, z', x') \simeq \phi^4(w, \kappa(z, x), \pi_c(z, x), \kappa(z', x'), \pi_c(z', x'))$$

$$r_2(w, z, x, z', x') \simeq \phi^4(w, (z)_2, x, \kappa(z', x'), \pi_c(z', x'))$$

$$r_3(w, z, x, z', x') \simeq \phi^4(w, \kappa(z, x), \pi_c(z, x), (z')_2, x')$$

$$r_4(w, z, x, z', x') \simeq \exists y \phi^4(w, \kappa(z, x), \pi_c(z, x), (z')_1, x' \mathbin{\square} \langle y \rangle)$$

$$r_5(w, z, x, z', x') \simeq \forall y \phi^4(w, (z)_1, x \mathbin{\square} \langle y \rangle, \kappa(z', x'), \pi_c(z', x'))$$

$$r_6(w, z, x, z', x') \simeq \phi^4(w, (z)_2, x, (z')_2, x')$$

$$r_7(w, z, x, z', x') \simeq \exists y \phi^4(w, (z)_2, x, (z')_1, x' \mathbin{\square} \langle y \rangle)$$

$$r_8(w, z, x, z', x') \simeq \forall y \phi^4(w, (z)_1, x \mathbin{\square} \langle y \rangle, (z')_2, x')$$

$$r_9(w, z, x, z', x') \simeq \forall y \exists y' \phi^4(w, (z)_1, x \mathbin{\square} \langle y \rangle, (z')_1, x' \mathbin{\square} \langle y' \rangle).$$

It is difficult to explain at this stage the meaning of these functions. It should be clear that they are related to the specification of the predicate H_c. Furthermore, they are used later to specify the 5-ary function f. At that point the variable w will be fixed by the RT in such a way that the following equation is valid:

$$f(w, z, x, z', x') \simeq \phi^4(w, z, x, z', x').$$

The remaining six auxiliary functions are 6-ary functions denoted $q_1, \ldots, q_6$. They are introduced using partial definition by cases, and the extra argument is used to control the cases. The specifications are as follows:

$$\begin{aligned}
q_1(u, w, z, x, z', x') &\simeq \mathsf{C}_0^1(\Delta_c(z, x)) && \text{if } u = 0\\
&\simeq \mathsf{C}_1^1(\Delta_c(z', x') && \text{if } u \neq 0\\
q_2(u, w, z, x, z', x') &\simeq \mathsf{C}_0^1(\Delta_c(z, x)) && \text{if } u = 0\\
&\simeq q_1(r_1(w, z, x, z', x'), w, z, x, z', x') && \text{if } u \neq 0\\
q_3(u, w, z, x, z', x') &\simeq q_1(r_1(w, z, x, z', x'), w, z, x, z', x') && \text{if } u = 0\\
&\simeq \mathsf{C}_1^1(\Delta_c(z', x')) && \text{if } u \neq 0\\
q_4(u, w, z, x, z', x') &\simeq q_2(r_3(w, z, x, z', x'), w, z, x, z', x') && \text{if } u = 0\\
&\simeq q_3(r_2(w, z, x, z', x'), w, z, x, z', x') && \text{if } u \neq 0\\
q_5(u, w, z, x, z', x') &\simeq q_1(r_4(w, z, x, z', x'), w, z, x, z', x') && \text{if } u = 0\\
&\simeq \mathsf{C}_1^1(\Delta_c(z', x')) && \text{if } u \neq 0\\
q_6(u, w, z, x, z', x') &\simeq \mathsf{C}_0^1(\Delta_c(z, x)) && \text{if } u = 0\\
&\simeq q_1(r_5(w, z, x, z', x'), w, z, x, z', x') && \text{if } u \neq 0.
\end{aligned}$$

Note that there is no recursion in these equations, since each function is specified in terms of the preceding functions. Note also that whenever $q_i(u, w, z, z', x') \simeq v$, $i = 1, \ldots, 6$, then either $v \simeq \mathsf{C}_0^1(\Delta_c(z, x))$ or $v \simeq \mathsf{C}_1^1(\Delta_c(z', x'))$, that is, either $v = 0$ and $\Delta_c(z, x)\downarrow$ or $v = 1$ and $\Delta_c(z', x')\downarrow$ (it is not excluded that both $\Delta_c(z, x)\downarrow$ and $\Delta_c(z', x')\downarrow$).

We proceed now to specify the crucial function f. This is a 5-ary function introduced using partial definition by cases as follows:

$$\begin{aligned}
f(w, z, x, z', x') &\simeq \mathsf{C}_0^1(\Delta_c(z, x)) && \text{if } (z)_0 = 2, 3, 4, 5, 6, 7, \text{ or } 11\\
&\simeq \mathsf{C}_1^1(\Delta_c(z', x')) && \text{if } (z')_0 = 2, 3, 4, 5, 6, 7, \text{ or } 11\\
&\simeq q_1(r_1(w, z, x, z', x'), w, z, x, z', x') && \text{if } (z)_0 = 0, 1, 8, \text{ or } 9 \text{ and } (z')_0 = 0, 1, 8, \text{ or } 9\\
&\simeq q_2(r_3(w, z, x, z', x'), w, z, x, z', x') && \text{if } (z)_0 = 0, 1, 8, \text{ or } 9 \text{ and } (z')_0 = 10.
\end{aligned}$$

$$
\begin{aligned}
f(w,z,x,z',x') &\simeq q_1(r_4(w,z,x,z',x'),w,z,x,z',x') && \text{if } (z)_0 = 0, 1, 8, \text{ or } 9 \text{ and } (z')_0 = 12\\
&\simeq q_3(r_2(w,z,x,z',x'),w,z,x,z',x') && \text{if } (z)_0 = 10 \text{ and } (z')_0 = 0, 1, 8, \text{ or } 9\\
&\simeq q_4(r_6(w,z,x,z',x'),w,z,x,z',x') && \text{if } (z)_0 = 10 \text{ and } (z')_0 = 10\\
&\simeq q_5(r_7(w,z,x,z',x'),w,z,x,z',x') && \text{if } (z)_0 = 10 \text{ and } (z')_0 = 12\\
&\simeq q_1(r_5(w,z,x,z',x'),w,z,x,z',x') && \text{if } (z)_0 = 12 \text{ and } (z')_0 = 0, 1, 8, \text{ or } 9\\
&\simeq q_6(r_8(w,z,x,z',x'),w,z,x,z',x') && \text{if } (z)_0 = 12 \text{ and } (z')_0 = 10\\
&\simeq q_1(r_9(w,z,x,z',x'),w,z,x,z',x') && \text{if } (z)_0 = 12 \text{ and } (z')_0 = 12\\
&\simeq \mathsf{C}^1_0(\Delta_c(z,x)) && \text{if } 0 \le (z)_0 \le 12 \text{ and } (z')_0 > 12\\
&\simeq \mathsf{C}^1_1(\Delta_c(z',x')) && (z)_0 > 12 \text{ and } 0 \le (z')_0 \le 12.
\end{aligned}
$$

This completes the specification of the function f. The multiplicity of cases is unavoidable, because the whole idea is to compare the evaluation of $\Delta_c(z,x)$ and the evaluation of $\Delta_c(z',x')$, which depends on the values of $(z)_0$ and $(z')_0$ as determined by the specification of the predicate H_c. There are in fact $13 \times 13 \dot{-} 1$ cases, which we have been able to contract to just 13 cases.

Theorem 5.3.5. *If $f(w,z,x,z',x') \simeq v$ then either $v = 0$ and $\Delta_c(z,x)\downarrow$ or $v = 1$ and $\Delta_c(z',x')\downarrow$.*

Proof. We have already remarked that this property holds for each of the functions q_i, $i = 1, \ldots, 6$. By simple inspection it is clear that the property extends to the function f. □

The preceding results show that whenever $f(w,z,x,z',x')$ is defined then either $\Delta_c(z,x)$ is defined or $\Delta_c(z',x')$ is defined. Next we show that whenever $f(\mathsf{w},z,x,z',x')$ is undefined and c is a total function then both $\Delta_c(z,x)$ and $\Delta_c(z',x')$ are undefined, where the value w is given by the RT in such a way that

$$f(\mathsf{w},z,x,z',x') \simeq \phi^4(\mathsf{w},z,x,z',x').$$

From now on the variable w denotes this fixed value.

Theorem 5.3.6. *Assume c is a total function. If $\Delta_c(z,x)\downarrow$ then $f(\mathsf{w},z,x,z',x')\downarrow$ for all z',x'.*

Proof. We prove by induction on the specification of the predicate H_c that whenever $\mathsf{H}_c(z,x,v)$ holds then $f(\mathsf{w},z,x,z',x')\downarrow$ for all z',x'. The proof induces several cases in the specification of H_c, which are determined by the value $(z)_0$.

In each such case there are several subcases determined by the value $(z')_0$. We discuss only a few of these combinations. We write r_i for $r_i(\mathsf{w}, z, x, z', x')$ and $q_j(r_i)$ for $q_j(r_i, \mathsf{w}, z, x, z', x')$, $i = 1, \ldots, 9, j = 1, \ldots, 6$.

Case $(z)_0 = 10$, $(z')_0 = 0, 1, 8$, or 9. It follows that $f(\mathsf{w}, z, x, z', x') \simeq q_3(r_2)$. By the induction hypothesis it follows that r_2 is defined; hence $r_2 \simeq 0$ or $r_2 \simeq 1$. Note also that by the induction hypothesis and the fact that $\Delta_c(z, x)$ is defined it follows that r_1 is defined; hence $q_3(r_2)$ is defined in case $r_2 \simeq 0$. If $r_2 \simeq 1$ it follows that $\Delta_c(z', x')$ is defined; hence $q_3(r_2)$ is defined.

Case $(z)_0 = 10$ and $(z')_0 = 10$. Here $f(\mathsf{w}, z, x, z', x') = q_4(r_6)$. From the induction hypothesis it follows that r_6 is defined. If $r_6 \simeq 0$ we have $q_4(r_6) \simeq q_2(r_3)$, where r_3 is also defined by the induction hypothesis. If $r_3 \simeq 0$ then $q_2(r_3)$ is clearly defined. If $r_3 \simeq 1$ we note that this means that $\pi_c(z', x')$ is defined; hence r_1 is defined by the induction hypothesis, and $q_2(r_3)$ is defined. We consider now the case $r_6 \simeq 1$, hence $q_4(r_6) \simeq q_3(r_2)$. Note that if $r_6 \simeq 1$ then $\pi_c(z', x')$ is defined; hence r_2 and r_1 are also defined by the induction hypothesis. If $r_2 \simeq 0$ it is clear that $q_3(r_2)$ is defined. If $r_2 \simeq 1$ we note that also $r_6 \simeq 1$; hence $\Delta_c(z', x')$ is defined and $q_3(r_2)$ is defined.

Case $(z)_0 = 12$ and $(z')_0 = 12$. Hence $f(\mathsf{w}, z, x, z', x') \simeq q_1(r_9)$. From the induction hypothesis it follows that r_9 is defined. If $r_9 \simeq 0$ it is clear that $q_1(r_9)$ is defined. If $r_9 \simeq 1$ it follows that there is y such that for every y' $f(\mathsf{w}, z, (z)_1, x \,\square\, \langle y \rangle, (z')_1, x' \,\square\, \langle y' \rangle) \simeq 1$; hence $\Delta_c(z', x')$ is defined and $q_1(r_9)$ is defined.

Case $(z')_0 = 1, 2, 3, 4, 5, 6$, or 7. Since c is a total function it follows that $\Delta_c(z', x')$ is defined; hence $f(\mathsf{w}, z, x, z', x')$ is defined.

Case $(z)_0 = 0, 1, 8$, or 9 and $(z')_0 = 10$. Hence $f(\mathsf{w}, z, x, z', x') \simeq q_2(r_3)$. By the induction hypothesis it follows that r_3 is defined. If $r_3 \simeq 0$ it is clear that $q_2(r_3)$ is defined. If $r_3 \simeq 1$ it follows that the function $\pi_c(z', x')$ is defined; hence r_1 is defined by the induction hypothesis. Furthermore, if $r_1 \simeq 1$ it follows that $\Delta_c(z', x')$ is defined; hence $q_2(r_3)$ is defined. □

Theorem 5.3.7. *Assume c is a total function. If $\Delta_c(z', x')\downarrow$ then $f(\mathsf{w}, z, x, z', x')\downarrow$ for all z, x.*

Proof. We prove by induction on the specification of H_c that $\mathsf{H}_c(z', x', v)$ implies $f(\mathsf{w}, z, x, z', x')\downarrow$ for all z, x. The cases are similar to the ones considered in the preceding theorem. □

Corollary 5.3.7.1. *Let c be a total function. Then*

(i) *The class $\mathsf{HC}(c)$ contains a Φ_c-preorder function.*
(ii) *$\mathsf{HC}(c)$ has the selector property relative to $\mathsf{HC}_{\mathsf{pd}}(c)$.*
(iii) *$\mathsf{HC}_{\mathsf{pd}\#}(c) = \mathsf{HC}(c)$.*

Proof. To prove (i) we set $\mathsf{po}(z, z') \simeq f(\mathsf{w}, z, 0, z', 0)$. Noting that $\Phi_c(z) \simeq \Delta_c(z, 0)$ it follows from Theorem 5.3.5, Theorem 5.3.6, and Theorem 5.3.7 that po is a Φ_c-preorder. Part (ii) follows by Theorem 4.4.6, and part (iii) is immediate from part (i). □

In the specification of the predicate H_c we have taken the position that c is a fixed unary function. If we assume c is rather a function variable, we obtain a general 3-ary predicate H_c. This predicate induces a general binary function Δ_c, and $(\Phi_c, \mathsf{S}_\delta)$ is an internal interpreter for a class of general functions, which we denote by GHC. A general k-ary function h_c is in GHC if and only if there is a number z such that

$$h_c(\mathbf{x}) \simeq \Phi_c(\mathsf{S}_\delta^k(\mathsf{z}, \mathbf{x})).$$

As usual GHC_0 denotes the class of all ordinary functions in the class GHC. Note that RC is a proper subset of GHC_0, for the latter is closed under quantification. Furthermore, if c is a fixed unary function then $\mathsf{GHC}_c = \mathsf{HC}(c)$.

Theorem 5.3.8. *There is a 5-ary elementary predicate Q_{H} such that the relation*

$$\mathsf{H}_\beta(z, x, v) \equiv \forall\alpha\exists w Q_{\mathsf{H}}(z, x, v, \bar{\beta}(w), \bar{\alpha}(w))$$

holds whenever β is a total unary function.

Proof. We rewrite the inductive specification of H_c using the symbol β rather than c. Using the technique of Theorem 5.2.4 we get an A-term U, which contains the free variable β, such that

$$\mathsf{H}_\beta(z, x, v) \equiv U,$$

and all functions and predicates in U are elementary. Using Corollary 5.2.3.1 we get an elementary predicate Q_{H} such that

$$U \equiv \forall\alpha\exists w Q_{\mathsf{H}}(z, x, v, \bar{\beta}(w), \bar{\alpha}(w)). \qquad \square$$

Corollary 5.3.8.1. *Let h_c be a k-ary general function that is GHC-computable. There is an elementary $(k + 3)$-ary predicate Q such that*

$$h_\beta(\mathbf{x}) \simeq v \equiv \forall\alpha\exists w Q(\mathbf{x}, v, \bar{\beta}(w), \bar{\alpha}(w))$$

holds whenever β is a total unary function.

Proof. If z is a $(\Phi_c, \mathsf{S}_\delta)$-index for the general function h_c we take Q the predicate such that

$$Q(\mathbf{x}, v, u, w) \equiv Q_{\mathsf{H}}(\mathsf{z}, \langle\mathbf{x}\rangle, v, u, w). \qquad \square$$

Exercises

5.3.1. Let $\mathscr{C}$ be a FRS with internal interpreter (ϕ, S), and assume $\mathscr{C}$ is closed under universal quantification. Prove that the following predicates are partially $\mathscr{C}$-decidable:
(a) $P_1(z) \equiv z$ is a (ϕ, S)-index for a total unary function.
(b) $P_2(z) \equiv z$ is a (ϕ, S)-index for the characteristic function of a unary predicate.

5.3.2. In relation with the function f introduced in the text assume that the value w has been fixed according to the RT. Prove that if $(z)_0 = 0$, 1, 8, 9, or 10 and

$(z')_0 = 0$, 1, 8, 9, or 10, and $r_1(\mathsf{w}, z, x, z', x') \simeq v$ then either $v = 0$ and $\Delta_c(z, x)\downarrow$ or $v = 1$ and $\Delta_c(z', x')\downarrow$.

5.3.3. With the same assumptions as in the preceding exercise prove the following:
(a) If $(z)_0 = 12$ and $r_9(\mathsf{w}, z, x, z', x') \simeq 0$ then $\Delta_c(z, x)\downarrow$.
(b) If $(z')_0 = 12$ and $r_9(\mathsf{w}, z, x, z', x') \simeq 1$ then $\Delta_c(z', x')\downarrow$.

5.3.4. In relation with the function f introduced in the text assume the value w has been fixed according to the RT, c is a total unary function, and $\Delta_c(z, x)\downarrow$. Prove the following:
(a) $f(\mathsf{w}, z, x, z', x') \simeq 0$ or $f(\mathsf{w}, z', x', z, x) \simeq 1$ for all z', x'.
(b) $f(\mathsf{w}, z, x, z, x) \simeq 0$.

5.3.5. Assume f, w, and c are as in the preceding exercise. Prove that if $f(\mathsf{w}, z', x', z, x) \simeq 1$ then $f(\mathsf{w}, z, x, z', x') \simeq 0$ for all z, x, z', x'.

5.3.6. Let c be a total unary function. Prove that $\mathsf{HC}_{\mathsf{ha}}(c) = \mathsf{RC}_{\mathsf{ha}}(c)$.

Notes

The material in this section is rather a digression from the main topic. It provides a nice example (derived from Gabrovsky [6]) of a reflexive structure satisfying nonfinitary closure properties. The definition is by nonfinitary induction, which can be reduced to hyperenumeration. In this structure we have an occasion to apply the selector theorem of Chapter 4. This requires the introduction of a preorder function, and this turns out to be a major effort. Here we follow the presentation of Hinman [8]. The proof of the relevant properties of the preorder function involves induction on the specification and gives a good illustration of the advantages of formalizing inductive proofs.

Since the construction is relative to a given unary function c, the result is rather a general reflexive structure in the sense of Chapter 4. While this is an interesting example of uniformity, as far as it goes, note that closure under functional substitution appears to be beyond the properties of the construction.

§4. Ordinal Notations

The example in the preceding section gives a good indication of the power of hyperenumeration, particularly in the form of nonfinitary induction. We proceed now to study the deep structure of this operation, and this requires specific set-theoretical concepts, namely, ordinals. We start with a short explanation, intended as a motivation for this approach.

An application of hyperenumeration takes the form

$$P(\mathbf{x}) \equiv \forall\alpha\exists v Q(\mathbf{x}, \overline{\alpha}(v)),$$

where Q is a given predicate. This expression suggests a procedure where given a total unary function α we generate code-numbers $\overline{\alpha}(0)$, $\overline{\alpha}(1)$, $\overline{\alpha}(2)$, ... and whenever we generate $\bar{\alpha}(n)$ we check whether $Q(\mathbf{x}, \bar{\alpha}(n))$ holds, and in case it does hold we stop the procedure. It should be clear now that $P(\mathbf{x})$ holds if and only if the procedure stops for every function α. The same idea can be reformulated by introducing an order relation between code-numbers, where $\bar{\alpha}(n+1)$ precedes $\bar{\alpha}(n)$. With this notation the sequence above becomes what is called a descending chain, and the condition that $P(\mathbf{x})$ holds is equivalent to asserting that every descending chain halts. A relation satisfying this condition is said to be well-founded, and a linear ordering that is well-founded is a well-ordering. We conclude that it appears to be possible to reduce hyperenumeration to a well-ordering relation.

This reduction can be carried out at different levels. For example, it is possible to introduce a notion of $\mathscr{C}$-well-ordering, where $\mathscr{C}$ is a class of functions, which refers to those well-orderings determined by $\mathscr{C}$-decidable or partially $\mathscr{C}$-decidable predicates. While this approach is certainly of intrinsic interest, we adopt one that requires a minimum of ordinal machinery. The starting point here is the binary predicate P_g introduced in Example 3.3.2, which depends on a given binary function g. From now on we write $x <_g y$ rather than $\mathsf{P}_g(x, y)$. This does not mean that $<_g$ is always an order relation, but note that $<_g$ is a transitive predicate (Theorem 3.3.4).

The predicate $<_g$ is determined by elementary functions, elementary predicates, and the function g. If g is $\mathscr{C}_\mathrm{d}$-hyperenumerable and $\mathscr{C}$ is closed under elementary operations, then $<_g$ is $\mathscr{C}_\mathrm{d}$-hyperenumerable (Corollary 5.2.4.1).

Next we introduce a set $\mathcal{O}_g$, which also depends on a given binary function g. We shall say that $\mathcal{O}_g$ is a *system of ordinal notations*, in a sense that is made precise later. The inductive specification of $\mathcal{O}_g$ is as follows:

$\mathcal{O}1$: If $y = 0$ then $y \in \mathcal{O}_g$.
$\mathcal{O}2$: If $\exists w$: $y = \mathsf{exp}(2, w) \wedge w \in \mathcal{O}_g$ then $y \in \mathcal{O}_g$.
$\mathcal{O}3$: If $\exists w$: $y = \mathsf{exp}(3, w + 1)$

$$\wedge\ \forall z \exists v \exists v' \colon \mathsf{G}_g(w, z, v) \wedge \mathsf{G}_g(w, z+1, v') \wedge v \in \mathcal{O}_g \wedge v <_g v'$$

then $y \in \mathcal{O}_g$.

The set $\mathcal{O}_g$ is determined by the function g, and in case g is $\mathscr{C}_\mathrm{d}$-hyperenumerable and $\mathscr{C}$ is closed under elementary operations, then $\mathcal{O}_g$ is also $\mathscr{C}_\mathrm{d}$-hyperenumerable (Corollary 5.2.4.1).

Theorem 5.4.1. *Let g be a binary function and $y \in \mathcal{O}_g$. Then*

(i) $y = 0 \vee 0 <_g y$.
(ii) *If* $x <_g y$ *then* $x \in \mathcal{O}_g$.
(iii) *If* $x <_g y$ *and* $x' <_g y$ *then* $x = x' \vee x <_g x' \vee x' <_g x$.

Proof. Each one of the assertions is proved by induction on the inductive specification of $\mathcal{O}_g$. To prove (i) we introduce a binary predicate P_1 such that

$$P_1(y) \equiv y = 0 \vee 0 <_g y,$$

and prove by induction that $y \in \mathcal{O}_g$ implies $P_1(y)$. Case $\mathcal{O}1$ is trivial. In case $\mathcal{O}2$ from the induction hypothesis it follows that there is w such that

$$y = \mathsf{exp}(2, w) \wedge w \in \mathcal{O}_g \wedge P_1(w).$$

From $P_1(w)$ and using $g1$ it follows immediately that $0 <_g y$. In case $\mathcal{O}3$ it follows from the induction hypothesis that there is a number w such that

$$y = \mathsf{exp}(3, w + 1)$$
$$\forall z \exists v \exists v'\colon \mathsf{G}_g(w, z, v) \wedge \mathsf{G}_g(w, z + 1, v') \wedge v \in \mathcal{O}_g \wedge P_1(v) \wedge v <_g v'.$$

Taking $z = 0$ we get a number v such that $\mathsf{G}_g(w, 0, v)$ and $P_1(v)$. From $P_1(v)$, $v <_g v'$, and $g2$ it follows that $0 <_g y$.

To prove part (ii) we introduce the unary predicate P_2 such that

$$P_2(y) \equiv \forall x\colon x \nless_g y \vee x \in \mathcal{O}_g,$$

and prove by induction that $y \in \mathcal{O}_g$ implies $P_2(y)$. The case $\mathcal{O}1$ is trivial, for if $y = 0$ then $x \nless_g y$ holds for all x. In case $\mathcal{O}2$ it follows from the induction hypothesis that there is a number w such that

$$y = \mathsf{exp}(2, w), \quad w \in \mathcal{O}_g, \quad P_2(w).$$

If $x <_g y$ then either $x = w$, hence $x \in \mathcal{O}_g$, or $x <_g w$ in which case $P_2(w)$ implies that $x \in \mathcal{O}_g$. In case $\mathcal{O}3$ the induction hypothesis implies the existence of a number w such that $y = \mathsf{exp}(3, w + 1)$, and

$$\forall z \exists v \exists v'\colon \mathsf{G}_g(w, z, v) \wedge \mathsf{G}_g(w, z + 1, v') \wedge y \in \mathcal{O}_g \wedge P_2(v) \wedge v <_g v'.$$

If we assume that $x <_g y$ it follows that there is a z and there is a v such that $\mathsf{G}_g(w, z, v)$ and $x <_g v$. On the other hand, from the induction hypothesis we have $P_2(v)$. It follows that $x <_g y$.

To prove part (iii) we introduce the unary predicate P_3 such that

$$P_3(y) \equiv \forall x \forall x'\colon x \nless_g y \vee x' \nless_g y \vee x = x' \vee x <_g x' \vee x' <_g x.$$

To prove by induction that $y \in \mathcal{O}_g$ implies $P_3(y)$ we note that case $\mathcal{O}1$ is trivial, and in case $\mathcal{O}2$ the induction hypothesis implies that there is a number w such that $y = \mathsf{exp}(2, w)$, $w \in \mathcal{O}_g$, and $P_3(w)$. If we assume that $x <_g y$ and $x' <_g y$ we have four combinations, depending on $x = w$ or $x <_g w$, and $x' = w$ or $x' <_g w$. The cases where $x = w$ or $x' = w$ are trivial. If $x <_g w$ and $x' <_g w$ we use $P_3(w)$ and get that $x = x'$, or $x <_g x'$, or $x' <_g x$. Finally, we consider case $\mathcal{O}3$, where from the induction hypothesis we get a number w such that $y = \mathsf{exp}(3, w + 1)$ and

$$\forall z \exists v \exists v'\colon \mathsf{G}_g(w, z, v) \wedge \mathsf{G}_g(w, z + 1, v') \wedge v \in \mathcal{O}_g \wedge P_3(v) \wedge v <_g v'.$$

If we assume that $x <_g y$ and $x' <_g y$ it follows that there are numbers z_1, z_2, v_1, and v_2 such that

$$\mathsf{G}_g(w, z_1, v_1) \wedge x <_g v_1$$
$$\mathsf{G}_g(w, z_2, v_2) \wedge x' <_g v_2.$$

If we take $z = \max\{z_1, z_2\}$ it follows that there is a v such that $G_g(w, z, v)$, $x <_g v$, and $x' <_g v$. Since we have $P_3(v)$ it follows that $x = x'$, or $x <_g x'$, or $x' <_g x$. □

EXAMPLE 5.4.1. Let $g = \mathsf{UD}_2$. The elements of $\mathcal{O}_g$ are determined by statements $\mathcal{O}1$ and $\mathcal{O}2$; $\mathcal{O}_g$ contains only the numbers 0, 2, 4, $\exp(2, 4)$, and so on, and the relation $<_g$ is a subset of the standard relation $<$.

EXAMPLE 5.4.2. Let g be a binary function such that for some numbers w and z we have $g(\mathsf{w}, \mathsf{z}) \simeq \exp(3, \mathsf{w} + 1)$. It follows that the number $\mathsf{y} = \exp(3, \mathsf{w} + 1)$ is not an element of $\mathcal{O}_g$. For, if we take $\mathcal{O}' = \mathcal{O}_g - \{\mathsf{y}\}$, it is easy to show that $\mathcal{O}'$ is a solution for the inductive specification of $\mathcal{O}_g$; hence $\mathcal{O}_g = \mathcal{O}'$.

For every number y and binary function g we define a set $\Sigma_g(y)$ such that

$$x \in \Sigma_g(y) \equiv x <_g y.$$

The properties of $\Sigma_g(y)$ are determined by the specification of $<_g$, and in case $y \in \mathcal{O}_g$ by the specification of $\mathcal{O}_g$. We note some of them that are relevant for the discussion below.

$\Sigma 1$: If $x \in \Sigma_g(y)$ then $\Sigma_g(x) \subseteq \Sigma_g(y)$ (Theorem 3.3.4).
$\Sigma 2$: $\Sigma_g(0) = \varnothing$.
$\Sigma 3$: If $y = \exp(2, w)$ then $\Sigma_g(y) = \Sigma_g(w) \cup \{w\}$ (from $g1$).
$\Sigma 4$: If $y = \exp(3, w + 1)$ then $x \in \Sigma_g(y)$ if and only if there are z and v such that $g(w, z) \simeq v$ and $x \in \Sigma_g(v)$ (from $g2$).
$\Sigma 5$: If $y \in \mathcal{O}_g$ then $\Sigma_g(y) \subseteq \mathcal{O}_g$ (Theorem 5.4.1 (ii)).
$\Sigma 6$: If $y \in \mathcal{O}_g$, $x \in \Sigma_g(y)$, and $x' \in \Sigma_g(y)$ then either $x = x'$, or $x <_g x'$, or $x' <_g x$ (Theorem 5.4.1 (iii)).
$\Sigma 7$: If $y \in \mathcal{O}_g$ and $y = \exp(3, w + 1)$ then for every z there are v and v' such that $g(w, z) \simeq v$, $g(w, z + 1) \simeq v'$, $v \in \Sigma_g(y)$, and $v <_g v'$ (from $g2$ and $\mathcal{O}3$).

A *descending chain* in $\Sigma_g(y)$ is an infinite sequence $y_0, y_1, \ldots, y_n, \ldots$ such that $y_0 \in \Sigma_g(y)$ and $y_{n+1} <_g y_n$ for all $n \geq 0$. It follows that $y_n \in \Sigma_g(y)$ for all $n \geq 0$.

Theorem 5.4.2. *If $y \in \mathcal{O}_g$ then there is no descending chain in $\Sigma_g(y)$.*

PROOF. By induction in the specification of $\mathcal{O}_g$. The case $\mathcal{O}1$ is trivial for $\Sigma_g(0) = \varnothing$. In case $\mathcal{O}2$ we have $y = \exp(2, w)$ and there is no descending chain in $\Sigma_g(w)$. It follows that the same is true for $\Sigma_g(y)$, for if $y_0, y_1, y_2, \ldots, y_n, \ldots$ is a descending chain in $\Sigma_g(y)$ then $y_2, y_3, \ldots, y_n, \ldots$ is a descending chain in $\Sigma_g(w)$. In case $\mathcal{O}3$ we note that a descending chain in $\Sigma_g(y)$ is also a descending chain in $\Sigma_g(v)$, where for some z we have $g(w, z) \simeq v$, contradicting the induction hypothesis on $\Sigma_g(v)$. □

Corollary 5.4.2.1. *Let $y \in \mathcal{O}_g$. Then $y \not<_g y$.*

PROOF. If $y <_g y$ then there is a descending chain in $\Sigma_g(y)$. □

From the preceding results it follows that whenever $y \in \mathcal{O}_g$ the set $\Sigma_g(y)$ is in fact a well-ordering under the relation $<_g$. This is because $<_g$ is transitive (Theorem 3.3.4), strict (Corollary 5.4.2.1), and connected (Theorem 5.4.1 (iii)). Furthermore, $<_g$ is well-founded in $\Sigma_g(y)$ by Theorem 5.4.2. The order type of $\Sigma_g(y)$ under $<_g$ is an ordinal, which we denote by $|y|_g$. Note that $|y|_g$ is undefined when $y \notin \mathcal{O}_g$.

The notational power of g (or rather of $\mathcal{O}_g$) depends on the number of ordinals that have notation in $\mathcal{O}_g$. We note that the following relations hold whenever $y \in \mathcal{O}_g$.

ON 1: $|0|_g = 0$ ($\Sigma 2$).
ON 2: If $y = \mathsf{exp}(2, w)$ and $|w|_g = \mu$ then $|y|_g = \mu + 1$ ($\Sigma 3$).
ON 3: If $y = \mathsf{exp}(3, w + 1)$ then $|y|_g = \lim_{n\to\infty} |v_n|_g$, where for each $n \in \mathbb{N}$, $g(w, n) \simeq v_n$ ($\Sigma 4$ and $\Sigma 7$).
ON 4: If $x \in \Sigma_g(y)$ then $|x|_g < |y|_g$ ($\Sigma 1$).
ON 5: If $|y|_g = \mu$ and $\nu < \mu$ then there is $x \in \Sigma_g(y)$ such that $|x|_g = \nu$.

From these properties we infer that the ordinals with notation in $\mathcal{O}_g$ are a segment of ordinals; hence there is a least countable ordinal μ such that μ has no notation in $\mathcal{O}_g$ and every ordinal less than μ has a notation in $\mathcal{O}_g$. We put $|g| = \mu$. It is clear that $|g|$ is a limit ordinal.

EXAMPLE 5.4.3 Let g be a binary function such that $g(0, 0) \simeq 0$, $g(0, y + 1) \simeq \mathsf{exp}(2, g(0, y))$, $g(1, 0) \simeq 0$, and $g(1, y + 1) \simeq \mathsf{exp}(2, \mathsf{exp}(2, g(1, y)))$. It is clear that both 3 and 9 are elements of $\mathcal{O}_g$, and $|3|_g = |9|_g = \omega$, where ω is the first infinite ordinal.

Now we assume two binary functions g and g', which determine systems of notations $\mathcal{O}_g$ and $\mathcal{O}_{g'}$. If $y \in \mathcal{O}_g$ and $x \in \mathcal{O}_{g'}$ then there are two possibilities: $|x|_{g'} < |y|_g$ or $|x|_{g'} \geq |y|_g$. The relation $|x|_{g'} < |y|_g$ can be given a numerical characterization if y is restricted to $\mathcal{O}_g$ and the relation is assumed to be false when $x \notin \mathcal{O}_{g'}$. We formalize this via two binary predicates $\mathfrak{D}_g^{g'}$ and $\mathfrak{E}_g^{g'}$, which are introduced using recursive specifications. We consider first the specification of $\mathfrak{D}_g^{g'}$, which involves elementary functions, elementary predicates, and the predicates G_g, $\mathsf{G}_{g'}$, $<_g$, and $<_{g'}$. The specification statements are as follows:

D 1: If $x = 0 \wedge y \neq 0$ then $\mathfrak{D}_g^{g'}(x, y)$.
D 2: If $\exists w \exists z \exists v$: $y = \mathsf{exp}(3, w + 1) \wedge \mathsf{G}_g(w, z, v) \wedge \mathfrak{D}_g^{g'}(x, v)$ then $\mathfrak{D}_g^{g'}(x, y)$.
D 3: If $\exists w \exists w' \forall z \exists v \exists v'$: $y = \mathsf{exp}(2, w) \wedge x = \mathsf{exp}(3, w' + 1) \wedge \mathsf{G}_{g'}(w', z, v) \wedge \mathsf{G}_{g'}(w', z + 1, v') \wedge v <_{g'} v' \wedge \mathfrak{D}_g^{g'}(v, w)$ then $\mathfrak{D}_g^{g'}(x, y)$.
D 4: If $\exists w \exists w'$: $y = \mathsf{exp}(2, w) \wedge x = \mathsf{exp}(2, w') \wedge \mathfrak{D}_g^{g'}(w', w)$ then $\mathfrak{D}_g^{g'}(x, y)$.

Theorem 5.4.3. *Let g and g' be binary functions and $y \in \mathcal{O}_g$. Then*

(i) *If $x \in \mathcal{O}_{g'}$ and $|x|_{g'} < |y|_g$ then $\mathfrak{D}_g^{g'}(x, y)$.*
(ii) *If $\mathfrak{D}_g^{g'}(x, y)$ then $x \in \mathcal{O}_{g'}$ and $|x|_{g'} < |y|_g$.*

Proof. To prove (i) we introduce the unary predicate P_1 such that

$$P_1(y) \equiv \forall x\colon x \notin \mathcal{O}_{g'} \vee (x \in \mathcal{O}_{g'} \wedge y \in \mathcal{O}_g \wedge |x|_{g'} \geq |y|_g) \vee \mathfrak{D}_g^{g'}(x, y).$$

We prove by induction on $\mathcal{O}_g$ that $y \in \mathcal{O}_g$ implies $P_1(y)$. There are three cases in the induction. Case $\mathcal{O}1$ is trivial. In case $\mathcal{O}2$ we have $y = \mathsf{exp}(2, w)$ and by the induction hypothesis $w \in \mathcal{O}_g$ and $P_1(w)$ holds. If $x \in \mathcal{O}_{g'}$ and $|x|_{g'} < |y|_g$, there are three possibilities: $x = 0$ and $\mathfrak{D}_g^{g'}(x, y)$ follows by D1, or $x = \mathsf{exp}(2, w')$ and $\mathfrak{D}_g^{g'}(x, y)$ follows from the induction hypothesis and D4, or $x = \mathsf{exp}(3, w' + 1)$ and $\mathfrak{D}_g^{g'}(x, y)$ follows from the induction hypothesis and D3. In case $\mathcal{O}3$ $y = \mathsf{exp}(3, w + 1)$. If $x \in \mathcal{O}_{g'}$ and $|x|_{g'} < |y|_g$, it follows that there are z and v such that

$$\mathsf{G}_g(w, z, v) \wedge v \in \mathcal{O}_g \wedge |x|_{g'} < |v|_g.$$

From the induction hypothesis we have $P_1(v)$; hence $\mathfrak{D}_g^{g'}(x, v)$ holds. From D2 it follows that $\mathfrak{D}_g^{g'}(x, y)$ holds.

To prove (ii) we introduce the binary predicate P_2 such that

$$P_2(x, y) \equiv y \notin \mathcal{O}_g \vee (x \in \mathcal{O}_{g'} \wedge y \in \mathcal{O}_g \wedge |x|_{g'} < |y|_g),$$

and we prove by induction on $\mathfrak{D}_g^{g'}$ that $\mathfrak{D}_g^{g'}(x, y)$ implies $P_2(x, y)$. There are four cases in the induction, and in each case we assume that $y \in \mathcal{O}_g$ and we prove that $x \in \mathcal{O}_{g'}$ and $|x|_{g'} < |y|_g$. Case D1 is trivial. In case D2 we know there are w, z, and v such that

$$y = \mathsf{exp}(3, w + 1), \quad \mathsf{G}_g(w, z, v), \quad \mathfrak{D}_g^{g'}(x, v), \quad \text{and} \quad P_2(x, v).$$

Since $v \in \mathcal{O}_g$ it follows from $P_2(x, v)$ that $x \in \mathcal{O}_{g'}$ and $|x|_{g'} < |v|_g < |y|_g$. In case D3 there are w and w' such that $y = \mathsf{exp}(2, w)$, $x = \mathsf{exp}(3, w' + 1)$, and

$$\forall z \exists v \exists v'\colon \mathsf{G}_{g'}(w', z, v) \wedge \mathsf{G}_{g'}(w', z + 1, v') \wedge \mathfrak{D}_g^{g'}(v, w) \wedge P_2(v, w)$$
$$\wedge\, v <_{g'} v'.$$

From $P_2(v, w)$ it follows that $v \in \mathcal{O}_{g'}$ and $|v|_{g'} < |w|_g$; hence $x \in \mathcal{O}_{g'}$ and $|x|_{g'} \leq |w|_g < |y|_g$. Case D4 is immediate from the induction hypothesis. □

Corollary 5.4.3.1. *Let g and g' be binary functions, and assume $y \in \mathcal{O}_g$. Then $\mathfrak{D}_g^{g'}(x, y)$ holds if and only if $x \in \mathcal{O}_{g'}$ and $|x|_{g'} < |y|_g$.*

Proof. Immediate from Theorem 5.4.3. □

Theorem 5.4.4. *Let $\mathscr{C}$ be a class of functions, and assume g and g' are binary functions. Then*

(i) *If $\mathscr{C}$ is closed under elementary operations and g and g' are $\mathscr{C}_{\mathsf{d}}$-hyperenumerable, then $\mathfrak{D}_g^{g'}$ is $\mathscr{C}_{\mathsf{d}}$-hyperenumerable.*
(ii) *If $\mathscr{C}$ is closed under recursive operations and g and g' are $\mathscr{C}_{\mathsf{pd}}$-hyperenumerable, then $\mathfrak{D}_g^{g'}$ is $\mathscr{C}_{\mathsf{pd}}$-hyperenumerable.*

PROOF. Part (i) follows from Corollary 5.2.4.1, and part (ii) from Corollary 5.2.4.2. □

If g is a binary function we put $\mathfrak{D}_g = \mathfrak{D}_g^g$. Note that in case $y \in \mathcal{O}_g$ then $\mathfrak{D}_g(x, y)$ holds if and only if $x \in \mathcal{O}_g$ and $|x|_g < |y|_g$.

We proceed now to give the specification of the binary predicate $\mathfrak{E}_g^{g'}$, where g and g' are binary functions. We introduce first a unary predicate $\mathsf{R}_{g'}$ such that $\mathsf{R}_{g'}(x)$ implies $x \notin \mathcal{O}_{g'}$:

$$\begin{aligned}\mathsf{R}_{g'}(x) \equiv x \neq 0 \wedge \forall w\colon x \neq \mathsf{exp}(2, w) \wedge x \neq \mathsf{exp}(3, w+1) \\ \vee \exists w \exists z \forall v\colon x = \mathsf{exp}(3, w+1) \wedge \sim \mathsf{G}_{g'}(w, z, v) \\ \vee \exists w \exists z \exists v \exists v'\colon x = \mathsf{exp}(3, w+1) \wedge \mathsf{G}_{g'}(w, z, v) \wedge \mathsf{G}_{g'}(w, z+1, v') \\ \wedge\, v \nless_{g'} v'.\end{aligned}$$

Note that the specification of $\mathsf{R}_{g'}$ involves the negation of $\mathsf{G}_{g'}$ and $<_{g'}$. The predicate $\mathfrak{E}_g^{g'}$ is given by the following inductive specification:

E1: If $y = 0$ then $\mathfrak{E}_g^{g'}(x, y)$.
E2: If $\mathsf{R}_{g'}(x)$ then $\mathfrak{E}_g^{g'}(x, y)$.
E3: If $\exists w \forall z \exists v\colon y = \mathsf{exp}(3, w+1) \wedge \mathsf{G}_g(w, z, v) \wedge \mathfrak{E}_g^{g'}(x, v)$ then $\mathfrak{E}_g^{g'}(x, y)$.
E4: If $\exists w \exists w' \exists z \exists v\colon y = \mathsf{exp}(2, w) \wedge x = \mathsf{exp}(3, w'+1) \wedge \mathsf{G}_{g'}(w', z, v) \wedge \mathfrak{E}_g^{g'}(v, w)$ then $\mathfrak{E}_g^{g'}(x, y)$.
E5: If $\exists w \exists w'\colon y = \mathsf{exp}(2, w) \wedge x = \mathsf{exp}(2, w') \wedge \mathfrak{E}_g^{g'}(w', w)$ then $\mathfrak{E}_g^{g'}(x, y)$.

Theorem 5.4.5. *Let g and g' be binary functions and $y \in \mathcal{O}_g$. Then*

(i) *If $x \notin \mathcal{O}_{g'}$, or $x \in \mathcal{O}_{g'}$ and $|x|_{g'} \geq |y|_g$, then $\mathfrak{E}_g^{g'}(x, y)$.*
(ii) *If $\mathfrak{E}_g^{g'}(x, y)$, and $x \in \mathcal{O}_{g'}$, then $|x|_{g'} \geq |y|_g$.*

PROOF. To prove (i) we introduce the unary predicate P_1 such that

$$P_1(y) \equiv \forall x\colon (x \in \mathcal{O}_{g'} \wedge y \in \mathcal{O}_g \wedge |x|_{g'} < |y|_g) \vee \mathfrak{E}_g^{g'}(x, y).$$

We prove by induction on $\mathcal{O}_g$ that $y \in \mathcal{O}_g$ implies $P_1(y)$. There are three cases in the induction and case $\mathcal{O}1$ is trivial by E1. In the remaining cases we assume that $\mathsf{R}_{g'}(x) \equiv \mathsf{F}$, for otherwise $\mathfrak{E}_g^{g'}(x, y)$ follows by E2. Note that from the assumptions $\mathsf{R}_{g'}(x) \equiv \mathsf{F}$ and $x \notin \mathcal{O}_{g'}$ it follows that either $x = \mathsf{exp}(2, w)$ and $w \notin \mathcal{O}_{g'}$, or $x = \mathsf{exp}(3, w+1)$ and there are z and v such that $\mathsf{G}_{g'}(w, z, v)$ and $v \notin \mathcal{O}_{g'}$.

In case $\mathcal{O}2$ we have $y = \mathsf{exp}(2, w)$. If $x \notin \mathcal{O}_{g'}$, or $x \in \mathcal{O}_{g'}$ and $|x|_{g'} \geq |y|_g$, it follows that either $x = \mathsf{exp}(2, w')$ and $\mathfrak{E}_g^{g'}(w', w)$ holds, hence $\mathfrak{E}_g^{g'}(x, y)$ holds by E5, or $x = \mathsf{exp}(3, w' + 1)$ and there are z and v such that $\mathsf{G}_{g'}(w', z, v)$ and $\mathfrak{E}_g^{g'}(v, w)$, hence $\mathfrak{E}_g^{g'}(x, y)$ holds by E4.

In case $\mathcal{O}3$ $y = \mathsf{exp}(3, w + 1)$. If $x \notin \mathcal{O}_{g'}$, or $x \in \mathcal{O}_{g'}$ and $|x|_{g'} \geq |y|_g$, then for every z there is v such that $\mathsf{G}_g(w, z, v)$ and $\mathfrak{E}_g^{g'}(x, v)$ hold, hence $\mathfrak{E}_g^{g'}(x, y)$ holds by E3. This completes the proof of part (i).

To prove (ii) we introduce a binary predicate P_2 such that

$$P_2(x, y) \equiv x \notin \mathcal{O}_{g'} \vee y \notin \mathcal{O}_g \vee (x \in \mathcal{O}_{g'} \wedge y \in \mathcal{O}_g \wedge |x|_{g'} \geq |y|_g)$$

and prove that $\mathfrak{E}_g^{g'}(x, y)$ implies $P_2(x, y)$. There are five cases in the induction. Cases E1 and E2 are trivial. In the remaining cases we assume $x \in \mathcal{O}_{g'}$, $y \in \mathcal{O}_g$, and prove $|x|_{g'} \geq |y|_g$.

In case E3 we have $y = \mathsf{exp}(3, w + 1)$, and for every z there is v such that $\mathsf{G}_g(w, z, v)$ and $P_2(x, v)$. Since for each such v we have $v \in \mathcal{O}_g$ it follows that $|x|_{g'} \geq |v|_g$, hence $|x|_{g'} \geq |y|_g$.

In case E4 we have $y = \mathsf{exp}(2, w)$ and $x = \mathsf{exp}(3, w' + 1)$, and there are z and v such that $\mathsf{G}_{g'}(w', z, v)$ and $P_2(v, w)$ hold. Since $v \in \mathcal{O}_{g'}$ this means that $|v|_{g'} \geq |w|_g$, hence $|x|_{g'} \geq |y|_g$. Case E5 is straightforward. □

Corollary 5.4.5.1. *Let g and g' be binary functions, and assume $y \in \mathcal{O}_g$. The following conditions are equivalent*:

(i) $\mathfrak{E}_g^{g'}(x, y)$.
(ii) $x \notin \mathcal{O}_{g'} \vee (x \in \mathcal{O}_{g'} \wedge |x|_{g'} \geq |y|_g)$.
(iii) $\sim\mathfrak{D}_g^{g'}(x, y)$.

Proof. The implication from (i) to (ii) follows from Theorem 5.4.5 (ii). The implication from (ii) to (iii) follows from Theorem 5.4.3 (ii). The implication from (iii) to (i) follows from Theorem 5.4.3 (i) and Theorem 5.4.5 (i). □

Theorem 5.4.6. *Let $\mathscr{C}$ be a class of functions, and assume g and g' are binary functions. Then*

(i) *If $\mathscr{C}$ is closed under elementary operations, g is $\mathscr{C}_{\mathrm{d}}$-hyperenumerable, and g' is $\mathscr{C}$-hyperarithmetic, then $\mathfrak{E}_g^{g'}$ is $\mathscr{C}_{\mathrm{d}}$-hyperenumerable.*
(ii) *If $\mathscr{C}$ is closed under recursive operations, g is $\mathscr{C}_{\mathrm{pd}}$-hyperenumerable, and g' is $\mathscr{C}$-hyperarithmetic, then $\mathfrak{E}_g^{g'}$ is $\mathscr{C}_{\mathrm{pd}}$-hyperenumerable.*

Proof. Part (i) follows from Corollary 5.2.4.1 noting that in case g' is $\mathscr{C}$-hyperarithmetic then both $\mathsf{G}_{g'}$ and $\bar{\mathsf{G}}_{g'}$ are $\mathscr{C}_{\mathrm{d}}$-hyperenumerable, and furthermore $<_{g'}$ and $\not<_{g'}$ are also $\mathscr{C}_{\mathrm{d}}$-hyperenumerable. Part (ii) follows in a similar way from Corollary 5.2.4.2. □

If g is a binary function we put $\mathfrak{E}_g = \mathfrak{E}_g^g$. Note that in case $y \in \mathcal{O}_g$ then $\mathfrak{E}_g(x, y)$ holds if and only if $x \notin \mathcal{O}_g$ or $x \in \mathcal{O}_g$ and $|x|_g \geq |y|_g$.

The preceding results suggest the following notation. If g and g' are binary functions and y is a number, then $\mathcal{O}^{g'}_{g,y}$ is the set such that

$$x \in \mathcal{O}^{g'}_{g,y} \equiv \mathfrak{D}^{g'}_{g}(x, y).$$

If $g = g'$ we set $\mathcal{O}_{g,y} = \mathcal{O}^{g}_{g,y}$. Note that if $y \in \mathcal{O}_g$ then $x \in \mathcal{O}^{g'}_{g,y}$ if and only if $x \in \mathcal{O}_{g'}$ and $|x|_{g'} < |y|_g$.

Theorem 5.4.7. *Let $\mathscr{C}$ be a class of functions closed under elementary operations, let g and g' be binary functions such that g is $\mathscr{C}_{\mathrm{d}}$-hyperenumerable, and let g' be $\mathscr{C}$-hyperarithmetic. Then*

(i) *If $y \in \mathcal{O}_g$ then $\mathcal{O}^{g'}_{g,y}$ is $\mathscr{C}$-hyperarithmetic.*
(ii) *If $|g'| < |g|$ then $\mathcal{O}_{g'}$ is $\mathscr{C}$-hyperarithmetic.*

Proof. It is clear that $\mathcal{O}^{g'}_{g,y}$ is $\mathscr{C}_{\mathrm{d}}$-hyperenumerable. If $y \in \mathcal{O}_g$ then from Corollary 5.4.5.1 it follows that

$$x \notin \mathcal{O}^{g'}_{g,y} \equiv \mathfrak{E}^{g'}_{g}(x, y);$$

hence the complement of $\mathcal{O}^{g'}_{g,y}$ is also $\mathscr{C}_{\mathrm{d}}$-hyperenumerable and $\mathcal{O}^{g'}_{g,y}$ is $\mathscr{C}$-hyperarithmetic. To prove (ii) we note that if $|g'| < |g|$ then there is $y \in \mathcal{O}_g$ such that $|g'| = |y|_g$; hence $\mathcal{O}_{g'} = \mathcal{O}^{g'}_{g,y}$. □

Theorem 5.4.8. *Let $\mathscr{C}$ be a class of functions closed under elementary operations and g a binary $\mathscr{C}_{\mathrm{d}}$-hyperenumerable function such that $\mathcal{O}_g$ is $\mathscr{C}$-hyperarithmetic. Then*

(i) *There is a binary $\mathscr{C}$-hyperarithmetic function g' such that $\mathcal{O}_g = \mathcal{O}_{g'}$, $|y|_g = |y|_{g'}$ whenever $y \in \mathcal{O}_g$, and $|g| = |g'|$.*
(ii) *There is a total unary $\mathscr{C}$-hyperarithmetic function f such that $f(x) \in \mathcal{O}_g$ and $|f(x)|_g < |f(x+1)|_g$ hold for all x, and $\lim_{n\to\infty} |f(n)|_g = |g|$.*

Proof. To prove (i) we introduce a binary function g' such that

$$g'(w, x) \simeq v \equiv \mathsf{exp}(3, w+1) \in \mathcal{O}_g \wedge \mathsf{G}_g(w, x, v).$$

It is clear that g' is $\mathscr{C}_{\mathrm{d}}$-hyperenumerable, and furthermore

$$\sim\mathsf{G}_{g'}(w, x, v) \equiv \mathsf{exp}(3, w+1) \notin \mathcal{O}_g \vee \exists v' \colon \mathsf{G}_g(w, x, v') \wedge v' \neq v.$$

It follows that g' is $\mathscr{C}$-hyperarithmetic. Since $g' \subseteq_2 g$ it follows that $\mathcal{O}_{g'} \subseteq \mathcal{O}_g$ and whenever $y \in \mathcal{O}_{g'}$ then $|y|_{g'} = |y|_g$. In the other direction we prove by induction on $\mathcal{O}_g$ that whenever $y \in \mathcal{O}_g$ then $y \in \mathcal{O}_{g'}$ and $\Sigma_g(y) \subseteq \Sigma_{g'}(y)$. It follows that $\mathcal{O}_g = \mathcal{O}_{g'}$ and $|g| = |g'|$.

To prove (ii) we introduce the binary predicate P such that

$$P(x, y) \equiv y \in \mathcal{O}_g \wedge \mathfrak{D}_g(x, y).$$

It is clear that P is $\mathscr{C}_{\mathrm{d}}$-hyperenumerable. We can also write

$$\sim P(x, y) \equiv y \notin \mathcal{O}_g \vee \mathfrak{E}_{g'}(x, y),$$

where g' is the function introduced in part (i). It follows that P is $\mathscr{C}$-hyperarithmetic. The unary function f is introduced using primitive recursion and unbounded minimalization:

$$f(0) \simeq 0$$

$$f(x+1) \simeq \mu y P(f(x), y).$$

Since $\mathscr{C}_{\mathsf{ha}\#}$ is closed under recursive operations, it follows that f is $\mathscr{C}$-hyperarithmetic. It is easy to check that f is total, $f(x) \in \mathcal{O}_g$, and $|f(x)|_g < |f(x+1)|_g$ hold for all x. To prove that $\lim_{n\to\infty} |f(n)|_g = |g|$ we note that f is 1-1; hence if $y \in \mathcal{O}_g$ there is n such that $y \leq f(n)$. We conclude that either $|y|_g \leq |f(n)|_g$ or $f(n+1) = y$, so in either case there is n such that $|y|_g \leq |f(n)|_g$. □

Corollary 5.4.8.1. *If g is a binary recursively hyperenumerable function such that $\mathcal{O}_g$ is hyperarithmetic, then there is a total unary hyperarithmetic function f such that $f(x) \in \mathcal{O}_g$ and $|f(x)|_g < |f(x+1)|_g$ hold for all x, and* $\lim_{n\to\infty} |f(n)|_g = |g|$.

PROOF. Immediate from Theorem 5.4.8 with $\mathscr{C} = \mathsf{RC}$. □

The preceding results have been obtained from strictly closure assumptions. Now we extend our approach and consider reflexive structures.

Theorem 5.4.9. *Let $\mathscr{C}$ be a* FRS *that is closed under quantification. Assume g and g' are binary $\mathscr{C}$-computable functions, and furthermore the predicate $\mathsf{G}_{g'}$ is $\mathscr{C}$-decidable. If $y \in \mathcal{O}_g$ then $\mathcal{O}_{g,y}^{g'}$ is $\mathscr{C}$-decidable.*

PROOF. Let (ϕ, S) be an internal interpreter for $\mathscr{C}$. We introduce by partial definition by cases a 3-ary function f such that

(1)	$f(z,x,y) \simeq 1$	if $y = 0 \wedge \mathsf{R}_{g'}(x)$
(2)	$\simeq 0$	if $x = 0$
(3)	$\simeq \phi^2(z,(x \dot{-} 1)_0,(y \dot{-} 1)_0)$	if $P_1(z,x,y)$
(4)	$\simeq \forall v \phi^2(z, g'((x \dot{-} 1)_1 \dot{-} 1, v),(y \dot{-} 1)_0)$	if $P_2(z,x,y)$
(5)	$\simeq \exists v \phi^2(z,x,g((y \dot{-} 1)_1 \dot{-} 1, v))$	if $\exists w$: $\mathsf{exp}(3, w+1) = y$,

where P_1 and P_2 are elementary predicates such that

$$P_1(z,x,y) \equiv \exists w \exists w'\colon \mathsf{exp}(2,w) = y \wedge \mathsf{exp}(2,w') = x$$

$$P_2(z,x,y) \equiv \exists w \exists w'\colon \mathsf{exp}(2,w) = y \wedge \mathsf{exp}(2,w'+1) = x.$$

The function f is $\mathscr{C}$-computable, and we take a number z given by the RT such that

$$f(\mathsf{z},x,y) \simeq \phi^2(\mathsf{z},x,y).$$

We shall prove that whenever $y \in \mathcal{O}_g$ then $f(z, x, y)$ as a function of x is the characteristic function of $\mathcal{O}^{g'}_{g,y}$. Let P be the unary predicate such that

$$P(y) \equiv \forall x: (\mathfrak{D}^{g'}_g(x, y) \wedge f(z, x, y) \simeq 0) \vee (\mathfrak{E}^{g'}_g(x, y) \wedge f(z, x, y) \simeq 1).$$

We prove by induction that $y \in \mathcal{O}_g$ implies $P(y)$. The case $\mathcal{O}1$ means that $y = 0$, and we have $\mathfrak{E}^{g'}_g(x, y)$ and $f(z, x, y) \simeq 1$ by equation (1).

In case $\mathcal{O}2$ we have $y = \mathsf{exp}(2, w)$, and $P(w)$ holds by the induction hypothesis. We assume first that $\mathfrak{D}^{g'}_g(x, y)$ holds to prove $f(z, x, y) \simeq 0$. By inversion one of rules D1, D3, or D4 applies. In rule D1 $f(z, x, y) \simeq 0$ follows by equation (2) (note that $\mathsf{R}_{g'}(x)$ fails whenever $x = 0$). In rule D3 we have $x = \mathsf{exp}(3, w' + 1)$ and for every v there is v' such that $g'(w', v) \simeq v'$ and $\mathfrak{D}^{g'}_g(v', w)$ holds; hence $f(z, v', w) \simeq 0$. Since $\mathsf{R}_{g'}(x)$ fails it follows that equation (4) applies and $f(z, x, y) \simeq 0$. In rule D4 equation (3) applies and $f(z, x, y) \simeq 0$.

We continue with case $\mathcal{O}2$ but now we assume $\mathfrak{E}^{g'}_g(x, y)$ holds to prove $f(z, x, y) \simeq 1$. By inversion one of the rules E2, E4, or E5 applies. Rule E2 is trivial by equation (1), and we may assume $\mathsf{R}_{g'}(x)$ fails in the other rules. In rule E4 $x = \mathsf{exp}(3, w' + 1)$ and since $\mathsf{R}_{g'}(x)$ fails it follows that $g'(w', v)$ is defined for all v. Furthermore, there is v such that $g(w', v) \simeq v'$ and $\mathfrak{E}^{g'}_g(v', w)$ holds; hence $f(z, v', w) \simeq 1$. From equation (4) it follows that $f(z, x, y) \simeq 1$. If rule E5 applies then $f(z, x, y) \simeq 1$ follows by equation (3).

Now we consider case $\mathcal{O}3$ where $y = \mathsf{exp}(3, w + 1)$ and $g(w, v)$ is defined for all v. If $\mathfrak{D}^{g'}_g(x, y)$ holds then by inversion one of the rules D1 or D2 applies. Rule D1 is trivial by equation (2). In rule D2 there is v such that $g(w, v) \simeq v'$ and $\mathfrak{D}^{g'}_g(x, v')$ holds. It follows that $x \in \mathcal{O}_{g'}$ and $P(v')$ holds; hence $f(z, x, v') \simeq 0$. From equation (5) it follows that $f(z, x, y) \simeq 1$. If $\mathfrak{E}^{g'}_g(x, y)$ holds then one of the rules E2 or E3 applies. Rule E2 is trivial, and in rule E3 for every v there is v' such that $g(w, v) \simeq v'$ and $\mathfrak{E}^{g'}_g(x, v')$ holds; hence $f(z, x, v') \simeq 1$. Since $x \neq 0$ equation (5) applies and $f(z, x, y) \simeq 1$. □

Corollary 5.4.9.1. *Let c be a unary function and g a binary function such that G_g is in $\mathsf{HC_d}(c)$. If $y \in \mathcal{O}_g$ then $\mathcal{O}_{g,y}$ is also in $\mathsf{HC_d}(c)$.*

PROOF. Immediate from Theorem 5.4.9 and Theorem 5.3.4 (v). □

EXERCISES

5.4.1. Let g and g' be binary functions such that $g \subseteq_2 g'$. Prove the following:
(a) $<_g \subseteq_2 <_{g'}$ and $\mathcal{O}_g \subseteq \mathcal{O}_{g'}$.
(b) If $y \in \mathcal{O}_g$ then $|y|_g = |y|_{g'}$.

5.4.2. Prove that the set $\mathcal{O}'$ in Example 5.4.2 is a solution for the inductive specification of $\mathcal{O}_g$.

5.4.3. Let g be a total binary function given by the recursion

$$g(w, 0) = \mathsf{exp}(3, w \dot{-} 1)$$

$$g(w, y + 1) = \mathsf{exp}(2, g(w, y)).$$

Evaluate $|g|$.

5.4.4. Let g be the function of Example 5.4.3 and f a total unary function such that $f(x) \in \mathcal{O}_g$ and $f(x) <_g f(x+1)$ hold for all x. Prove there is $y \in \mathcal{O}_g$ such that $y \nless_g f(x)$ for all x.

5.4.5. Let g be a binary function. Prove that the following conditions are equivalent:
(a) Whenever $y \in \mathcal{O}_g$, $y' \in \mathcal{O}_g$, and $|y|_g = |y'|_g$, then $y = y'$.
(b) Whenever $y \in \mathcal{O}_g$ and $y' \in \mathcal{O}_g$, then either $y = y'$, or $y <_g y'$, or $y' <_g y$.
(c) There is a unary total function f such that $f(x) \in \mathcal{O}_g$ and $f(x) <_g f(x+1)$ hold for all x, and furthermore if $y \in \mathcal{O}_g$ there is some x such that $y <_g f(x)$.

5.4.6. Give an example of a binary function g such that $\mathcal{O}_g$ satisfies the conditions in Exercise 5.4.5, $\omega < |g|$, and whenever $\mu < |g|$ then $\mu + \omega < |g|$.

5.4.7. Let $\mathscr{C}$ be closed under elementary operations, and assume $\mathscr{C}_{\mathrm{pd}} \subseteq \mathscr{C}_{\mathrm{ha}}$. Let g be a binary $\mathscr{C}$-computable function such that $\mathcal{O}_g$ is not $\mathscr{C}$-hyperarithmetic. Prove that if g' is a binary $\mathscr{C}_{\mathrm{d}}$-hyperenumerable function then $|g'| \leq |g|$.

5.4.8. Let g be a recursive function such that $\mathcal{O}_g$ is not hyperarithmetic. Prove that if g' is a binary recursively hyperenumerable function then $|g'| \leq |g|$.

5.4.9. Let g be the binary recursive function in Exercise 5.4.3. Prove that $\mathcal{O}_g$ is recursively enumerable.

5.4.10. Let $\mathscr{C}$ be a class closed under elementary operations. Assume g is a binary $\mathscr{C}$-hyperarithmetic function, and f is a total unary $\mathscr{C}_{\mathrm{d}}$-hyperenumerable function such that $f(x) \in \mathcal{O}_g$ for all x, and whenever $y \in \mathcal{O}_g$ there is x such that $|y|_g < |f(x)|_g$. Prove that $\mathcal{O}_g$ is $\mathscr{C}$-hyperarithmetic.

Notes

The critical relation between hyperenumeration and ordinals is discussed in the introduction to this section, where we mention different possibilities to exploit the connection. The one we follow, via ordinal notation, originates with Kleene [1] and has been discussed in many places. We find it particularly attractive, because it contributes to the clarification of the notion of ordinal, which is crucial in the foundations of mathematics. Furthermore, it is simpler to handle than the alternative approach via recursive well-orderings, as in Hinman [8], for example.

In this section we operate assuming only closure considerations. The more fruitful approach via reflexive structures is discussed in the next section.

§5. Reflexive Systems

The theory presented in the preceding section applies to any notation system $\mathcal{O}_g$, induced by a binary function g. No complete examples were described, because in order to be interesting such a system must satisfy closure properties, which require special techniques. For example, we would like that whenever f is a total function such that $f(x) \in \mathcal{O}_g$ and $|f(x)|_g < |f(x+1)|_g$ for every x, and $\lim_{n \to \infty} |f(n)|_g = \mu$, then there is a notation $y \in \mathcal{O}_g$ such that $|y|_g = \mu$. While this condition cannot be satisfied for arbitrary function f, it is desirable that it holds under some restrictions, for example, when f is recursive in g. Systems of notations with such a property are said to be *reflexive* and can be constructed using the theory of functional reflexive structures of Chapter 4.

Let $\mathscr{C}$ be a FRS with internal interpreter (ϕ, S). We introduce a system of notations $\mathcal{O}_g$, where $g = \phi^1$ (i.e., $\phi^1(w, v) \simeq \phi(\mathsf{S}(w, v))$). Such a system will be denoted by $\mathcal{O}_\phi$; similarly, we write $<_\phi$ instead of $<_g$, and $|y|_\phi$ instead $|y|_g$. The notation $|\phi|$ denotes the least ordinal with no notation in $\mathcal{O}_\phi$. We say that the system $\mathcal{O}_\phi$ is *induced* by the FRS $\mathscr{C}$ with internal interpreter (ϕ, S). Different interpreters for $\mathscr{C}$ induce different systems of notation, but we shall show that they are all equivalent.

Let $\mathscr{C}$ be a FRS with internal interpreter (ϕ, S). Let g be the total 3-ary function given by the abstraction property such that

$$\phi^1(g(z, x, w), v) \simeq \phi^2(z, x, \phi^1(w, v)).$$

Using course-of-values recursion we introduce a total 3-ary function f;

$$\begin{aligned} f(z, x, 0) &= x \\ f(z, x, y + 1) &= \mathsf{exp}(2, f(z, x, (y)_0)) && \text{if } \exists w \le y\colon y + 1 = \mathsf{exp}(2, w) \\ &= \mathsf{exp}(3, g(z, x, (y)_1 \dot{-} 1) + 1) && \text{if } \exists w \le y\colon y + 1 = \mathsf{exp}(3, w + 1) \\ &= 5 && \text{otherwise.} \end{aligned}$$

It follows that f is a total function primitive recursive in S; hence it is $\mathscr{C}$-computable. From the RT we know there is a number z such that

$$f(\mathsf{z}, x, y) = \phi^2(\mathsf{z}, x, y).$$

The binary function $+_\phi$ (notation infixed) is given by $x +_\phi y = f(\mathsf{z}, x, y)$. With the same z we set $d(x, w) = g(\mathsf{z}, x, w)$. From the above specification for the function f we get the following relations:

$$x +_\phi 0 = x$$

$$x +_\phi \mathsf{exp}(2, w) = \mathsf{exp}(2, x +_\phi w)$$

$$x +_\phi \mathsf{exp}(3, w + 1) = \mathsf{exp}(3, d(x, w) + 1),$$

where d is a total binary $\mathscr{C}$-computable function such that

$$\phi^1(d(x, w), v) \simeq x +_\phi \phi^1(w, v).$$

Theorem 5.5.1. *Let $\mathscr{C}$ be a* FRS *with internal interpreter (ϕ, S). If $y <_\phi y'$ then $x +_\phi y <_\phi x +_\phi y'$.*

PROOF. We use induction on the specification of $y <_\phi y'$ with x fixed. There are two cases, and both follow immediately from the above relations. □

Corollary 5.5.1.1. *Let $\mathscr{C}$ be a* FRS *with internal interpreter (ϕ, S). If $y \in \mathcal{O}_\phi$ and $y \neq 0$ then $x <_\phi x +_\phi y$ for any number x.*

PROOF. From Theorem 5.4.1 (i) it follows that $0 <_\phi y$; hence $x = x +_\phi 0 <_\phi x +_\phi y$. □

Theorem 5.5.2. *Let $\mathscr{C}$ be a FRS with internal interpreter (ϕ, S). If $x \in \mathscr{O}_\phi$ and $y \in \mathscr{O}_\phi$ then $x +_\phi y \in \mathscr{O}_\phi$ and $|x +_\phi y|_\phi = |x|_\phi + |y|_\phi$.*

PROOF. By induction on the specification of $y \in \mathscr{O}_\phi$, with x fixed. Cases $\mathscr{O}1$ and $\mathscr{O}2$ are straightforward. In case $\mathscr{O}3$ $y = \mathsf{exp}(3, w + 1)$, and for every n we have $\phi^1(w, n) \simeq v_n$, $v_n <_\phi v_{n+1}$, $v_n \in \mathscr{O}_\phi$, $x +_\phi v_n \in \mathscr{O}_\phi$, and $|x +_\phi v_n|_\phi = |x|_\phi + |v_n|_\phi$. From Theorem 5.5.1 we know that $x +_\phi v_n <_\phi x +_\phi v_{n+1}$. Since $x +_\phi y = \mathsf{exp}(3, d(x, w) + 1)$, where for every n $\phi^1(d(x, w), n) \simeq x +_\phi v_n$, it follows that $x +_\phi y \in \mathscr{O}_\phi$. Furthermore, $|x +_\phi y|_\phi = \lim_{n\to\infty} |x +_\phi v_n|_\phi = \lim_{n\to\infty} |x|_\phi + |v_n|_\phi = |x|_\phi + |y|_\phi$, since $\lim_{n\to\infty} |v_n|_\phi = |y|_\phi$. □

Theorem 5.5.3. *Let $\mathscr{C}$ be a FRS with internal interpreter (ϕ, S), and let f be a unary total $\mathscr{C}$-computable function such that $f(x) \in \mathscr{O}_\phi$ holds for all x. There is $y \in \mathscr{O}_\phi$ such that $|f(x)|_\phi < |y|_\phi$ holds for all x.*

PROOF. We set $h(0) = f(0)$ and $h(x + 1) = h(x) +_\phi \mathsf{exp}(2, f(x + 1))$. It follows that h is a total $\mathscr{C}$-computable function, $h(x) \in \mathscr{O}_\phi$, $|f(x)|_\phi \le |h(x)|_\phi$, and $h(x) <_\phi h(x + 1)$ hold for all x. If we take w such that $h(x) = \phi^1(w, x)$ and $y = \mathsf{exp}(3, w + 1)$, it follows that $|f(x)|_\phi < |y|_\phi$ for all x. □

Theorem 5.5.4. *Let $\mathscr{C}$ be a FRS with internal interpreter (ϕ, S), and let g be a binary $\mathscr{C}$-computable function. There is a total unary $\mathscr{C}$-computable function h such that whenever $y \in \mathscr{O}_g$ then $h(y) \in \mathscr{O}_\phi$ and $|y|_g \le |h(y)|_\phi$.*

PROOF. We introduce by primitive recursion a 3-ary $\mathscr{C}$-computable function q such that

$$q(z, w, 0) \simeq \phi^1(z, g(w, 0))$$

$$q(z, w, v + 1) \simeq q(z, w, v) +_\phi \phi^1(z, g(w, v + 1)).$$

Let $\phi^1(p(z, w), v) \simeq q(z, w, v)$, where p is given by the abstraction property. Using course-of-values recursion we introduce a total $\mathscr{C}$-computable function f:

$$f(z, 0) \simeq 1$$

$$\begin{aligned} f(z, y + 1) &\simeq \mathsf{exp}(2, f(z, (y)_0)) && \text{if } (y)_1 = 0 \\ &\simeq \mathsf{exp}(3, p(z, (y)_1 \dot- 1) + 1) && \text{otherwise.} \end{aligned}$$

Finally, we take z a number given by the RT such that $f(\mathsf{z}, y) = \phi^1(\mathsf{z}, y)$, and we set $h(y) = f(\mathsf{z}, y)$. With the same z we set $q'(w, v) \simeq q(\mathsf{z}, w, v)$ and $d(w) = p(\mathsf{z}, w)$. Note the following relations derived from the above specifications:

$$h(0) = 1$$

$$h(\mathsf{exp}(2, w)) = \mathsf{exp}(2, h(w))$$

$$h(\mathsf{exp}(3, w + 1)) = \mathsf{exp}(3, d(w) + 1),$$

where d is a total $\mathscr{C}$-computable function such that $\phi^1(d(w), v) \simeq q'(w, v)$. Note that the function q' is total and satisfies the following relations:

$$q'(w, 0) = h(g(w, 0))$$
$$q'(w, v + 1) = q'(w, v) +_\phi h(g(w, v + 1)).$$

We prove now by induction on the specification of $\mathcal{O}_g$ that whenever $y \in \mathcal{O}_g$ then $h(y) \in \mathcal{O}_\phi$, and $|y|_g \leq |h(y)|_\phi$. There are three cases in the induction, and cases $\mathcal{O}1$ and $\mathcal{O}2$ follow immediately from the specification of h given by the above equations. In case $\mathcal{O}3$ we have $y = \mathsf{exp}(3, w + 1)$, and for every n there is v_n such that $g(w, n) \simeq v_n$, $v_n \in \mathcal{O}_g$, $h(v_n) \in \mathcal{O}_\phi$, and $|v_n|_g \leq |h(v_n)|_\phi$. In this case $h(y) = \mathsf{exp}(3, d(w) + 1)$, where $\phi^1(d(w), n) \simeq q'(w, n)$. It follows easily by induction on n that $q'(w, n) \in \mathcal{O}_\phi$, and furthermore $|h(v_n)|_\phi \leq |q'(w, n)|_\phi$, and $q'(w, n) <_\phi q'(w, n + 1)$. Hence $h(y) \in \mathcal{O}_\phi$, and

$$|y|_g = \lim_{n \to \infty} |v_n|_g \leq \lim_{n \to \infty} |h(v_n)|_\phi \leq \lim_{n \to \infty} |q'(w, n)|_\phi = |h(y)|_\phi.$$

This completes the proof. □

EXAMPLE 5.5.1. Consider the function f introduced in Example 5.4.4. Assume that $\mathcal{O}_g = \mathcal{O}_\phi$ and that (ϕ, S) is an internal interpreter for a FRS $\mathscr{C}$. It follows from Theorem 5.5.3 that the function f is not $\mathscr{C}$-computable.

The preceding results hold for any internal interpreter (ϕ, S) of the FRS $\mathscr{C}$. Note that if S is primitive recursive (as in the interpreter $(\phi_c, \mathsf{S}_\delta)$) then $+_\phi$ is also primitive recursive, and similarly for the function h of Theorem 5.5.4.

We recall now that for every FRS $\mathscr{C}$ there is an internal interpreter (ϕ, S), where the function S is strict (e.g., if $\mathsf{S} = \mathsf{S}_\delta$ where S_δ is the elementary function introduced in Chapter 2). It follows from this that a function obtained via the abstraction property is always strict. In particular, the function d that appears in the specification of the function $+_\phi$ is strict.

Theorem 5.5.5. *Let $\mathscr{C}$ be a* FRS *with internal interpreter (ϕ, S), where the function* S *is strict. Assume $y \in \mathcal{O}_\phi$. Then*

(i) *If $x +_\phi u = y$ then $u = 0$ or $x < y$.*
(ii) *If $x +_\phi u' = x +_\phi u = y$ then $u' = u$.*
(iii) *If $x \in \mathcal{O}_\phi$, $u' \in \mathcal{O}_\phi$, and $x +_\phi u' <_\phi x +_\phi u = y$, then $u' <_\phi u$.*
(iv) *If $x \in \mathcal{O}_\phi$ and $x +_\phi u = y$, then $u \in \mathcal{O}_\phi$.*

PROOF. Each one of these properties is proved by induction on the specification of $y \in \mathcal{O}_\phi$. We note that in case $x +_\phi u = y$ and $u \neq 0$ then the form of u is determined by the form of y. More precisely, either $y = \mathsf{exp}(2, w)$, in which case $u = \mathsf{exp}(2, w')$ and $w = x +_\phi w'$, or $y = \mathsf{exp}(3, w + 1)$, in which case $u = \mathsf{exp}(3, w' + 1)$ and $w = d(x, w')$, where d is the strict function that appears in the specification of $+_\phi$.

We assume that x is a fixed number that remains unchanged in all cases of each induction.

To prove (i) we introduce the unary predicate P_1, such that

$$P_1(y) \equiv \forall u\colon x +_\phi u \neq y \vee u = 0 \vee x < y,$$

and prove that $y \in \mathcal{O}_\phi$ implies $P_1(y)$. Case $\mathcal{O}1$ is trivial, and case $\mathcal{O}2$ follows easily from the induction hypothesis. In case $\mathcal{O}3$ we use the fact that d is strict; hence $x < d(x, w') < y$.

To prove (ii) we introduce the unary predicate P_2, such that

$$P_2(y) \equiv \forall u \forall u'\colon x +_\phi u' \neq y \vee x +_\phi u \neq y \vee u' = u,$$

and prove that $y \in \mathcal{O}_\phi$ implies $P_2(y)$. In each case we assume that $x +_\phi u' = x +_\phi u = y$ and prove $u' = u$. Note that under such an assumption if $u = 0$ then $x = y$ and from part (i) it follows that $u' = 0$. Similarly, if $u' = 0$ it follows that $u = 0$. So we assume in each case that $u \neq 0$ and $u' \neq 0$. We conclude from this that case $\mathcal{O}1$ is impossible, case $\mathcal{O}2$ follows easily from the induction hypothesis, and case $\mathcal{O}3$ follows using the fact that the function d is 1-1.

In the proof of (iii) the number u' is assumed fixed throughout the induction, and both x and u' are assumed to be elements of $\mathcal{O}_\phi$. The predicate P_3 is given by

$$P_3(y) \equiv \forall u\colon x +_\phi u' \not<_\phi x +_\phi u \vee x +_\phi u \neq y \vee u' <_\phi u.$$

The assumption about x and u' is used in the following preliminary argument. Assume that $x +_\phi u' <_\phi x +_\phi u = y$ and $u = 0$. This means that $x +_\phi u' <_\phi x$ and from Corollary 5.5.1.1 it follows that $x <_\phi x$, which contradicts Corollary 5.4.2.1. So in the induction we assume that $u \neq 0$. Case $\mathcal{O}1$ is impossible, and case $\mathcal{O}2$ follows using part (ii) with the induction hypothesis. Case $\mathcal{O}3$ follows immediately from the induction hypothesis.

To prove (iv) the number x is fixed and assumed to be an element of $\mathcal{O}_\phi$. The predicate P_4 is given by

$$P_4(y) \equiv \forall u\colon x +_\phi u \neq y \vee u \in \mathcal{O}_\phi.$$

If $u = 0$ it is clear that $u \in \mathcal{O}_\phi$, so we assume in the induction that $u \neq 0$. Case $\mathcal{O}1$ is trivial, and case $\mathcal{O}2$ is straightforward from the induction hypothesis. In case $\mathcal{O}3$ $y = \mathsf{exp}(3, w + 1)$; hence $u = \mathsf{exp}(3, w' + 1)$ and $w = d(x, w')$. It follows that for every n there is v_n such that $\phi^1(w', n) \simeq v_n$, $x +_\phi v_n \in \mathcal{O}_\phi$, and $x +_\phi v_n <_\phi x +_\phi v_{n+1}$. By the induction hypothesis we have $v_n \in \mathcal{O}_\phi$, and from part (iii) we have $v_n <_\phi v_{n+1}$, for every n. It follows that $u \in \mathcal{O}_\phi$. □

Theorem 5.5.6. *Let $\mathscr{C}$ be a* FRS *with internal interpreter* (ϕ, S), *where the function* S *is strict. If P is a k-ary $\mathscr{C}_{\mathrm{d}}$-hyperenumerable predicate there is a total k-ary $\mathscr{C}$-computable function h such that*

$$P(\mathbf{x}) \equiv h(\mathbf{x}) \in \mathcal{O}_\phi.$$

Proof. We assume that $P(\mathbf{x}) \equiv \forall \alpha \exists v Q(\mathbf{x}, \bar{\alpha}(v))$, where the predicate Q is $\mathscr{C}$-decidable. Using primitive recursion we introduce a $(k + 3)$-ary $\mathscr{C}$-computable function q such that

$$q(z, \mathbf{x}, w, 0) \simeq \phi^{k+1}(z, \mathbf{x}, w \square \langle 0 \rangle)$$

$$q(z, \mathbf{x}, w, v + 1) \simeq q(z, \mathbf{x}, w, v) +_\phi \phi^{k+1}(z, \mathbf{x}, w \square \langle v + 1 \rangle).$$

Let g be the $(k + 2)$-ary total $\mathscr{C}$-computable function given by the abstraction property such that

$$q(z, \mathbf{x}, w, v) \simeq \phi^1(g(z, \mathbf{x}, w), v).$$

Using total definition by cases we introduce a $(k + 2)$-ary total $\mathscr{C}$-computable function f such that

$$\begin{aligned} f(z, \mathbf{x}, w) &= 1 && \text{if } Q(\mathbf{x}, w) \\ &= \mathsf{exp}(3, g(z, \mathbf{x}, w) + 1) && \text{otherwise.} \end{aligned}$$

Finally, we fix z a number given by the RT such that

$$f(\mathsf{z}, \mathbf{x}, w) = \phi^{k+1}(\mathsf{z}, \mathbf{x}, w),$$

and we set $h'(\mathbf{x}, w) = f(\mathsf{z}, \mathbf{x}, w)$, $q'(\mathbf{x}, w, v) \simeq q(\mathsf{z}, \mathbf{x}, w, v)$, and $d(\mathbf{x}, w) = g(\mathsf{z}, \mathbf{x}, w)$. From the above specifications we derive the following relations:

$$q'(\mathbf{x}, w, 0) \simeq h'(\mathbf{x}, w \square \langle 0 \rangle)$$

$$q'(\mathbf{x}, w, v + 1) \simeq q'(\mathbf{x}, w, v) +_\phi h'(\mathbf{x}, w \square \langle v + 1 \rangle),$$

which implies that the function q' is total. Furthermore,

$$\begin{aligned} h'(\mathbf{x}, w) &= 1 && \text{if } Q(\mathbf{x}, w) \\ &= \mathsf{exp}(3, d(\mathbf{x}, w) + 1) && \text{otherwise,} \end{aligned}$$

where the function d is such that

$$\phi^1(d(\mathbf{x}, w), v) = q'(\mathbf{x}, w, v).$$

We take h the function such that $h(\mathbf{x}) = h'(\mathbf{x}, 0)$ and proceed to prove that

$$P(\mathbf{x}) \equiv h(\mathbf{x}) \in \mathcal{O}_\phi.$$

(a) First we prove that if $Q(\mathbf{x}, w) \equiv \mathsf{F}$ and $h'(\mathbf{x}, w) \in \mathcal{O}_\phi$, then $h'(\mathbf{x}, w \square \langle v \rangle) \in \mathcal{O}_\phi$ and $|h'(\mathbf{x}, w \square \langle v \rangle)|_\phi < |h'(\mathbf{x}, w)|_\phi$ for every v. Note that in this case $h'(\mathbf{x}, w) = \mathsf{exp}(3, d(\mathbf{x}, w) + 1)$, where $\phi^1(d(\mathbf{x}, w), v) = q'(\mathbf{x}, w, v)$. It follows that $q'(\mathbf{x}, w, v) \in \mathcal{O}_\phi$, and $|q'(\mathbf{x}, w, v)|_\phi < |h'(\mathbf{x}, w)|_\phi$ for every v. So we prove for every v that $h'(\mathbf{x}, w \square \langle v \rangle) \in \mathcal{O}_\phi$ and $|h'(\mathbf{x}, w \square \langle v \rangle)|_\phi \leq |q'(\mathbf{x}, w, v)|_\phi$. The case $v = 0$ is trivial from the specification of q'. The case $v + 1$ follows immediately from the specification of q', Theorem 5.5.5 (iv), and Theorem 5.5.2.

(b) From (a) it follows that in case $h'(\mathbf{x}, 0) \in \mathcal{O}_\phi$ then $\forall \alpha \exists v Q(\mathbf{x}, \bar{\alpha}(v))$. For let α be a total unary function. If $\forall v \sim Q(\mathbf{x}, \bar{\alpha}(v))$ it follows that for every v $h'(\mathbf{x}, \bar{\alpha}(v)) \in \mathcal{O}_\phi$ and $|h'(\mathbf{x}, \bar{\alpha}(v + 1))|_\phi < |h'(\mathbf{x}, \bar{\alpha}(v))|_\phi$, which is impossible.

(c) Here we prove the converse of (a). We assume $Q(\mathbf{x}, w) \equiv \mathsf{F}$ and $h'(\mathbf{x}, w \square \langle v \rangle) \in \mathcal{O}_\phi$ for every v, and prove that $h'(\mathbf{x}, w) \in \mathcal{O}_\phi$. Clearly, it is sufficient to prove that $q'(\mathbf{x}, w, v) \in \mathcal{O}_\phi$ and $q'(\mathbf{x}, w, v) <_\phi q'(\mathbf{x}, w, v + 1)$ hold for all v. The condition $q'(\mathbf{x}, w, v) \in \mathcal{O}_\phi$ is proved by induction on v. The case $v = 0$ is trivial. The case $v + 1$ follows from the induction hypothesis and Theorem 5.5.2. The relation $q'(\mathbf{x}, w, v) <_\phi q'(\mathbf{x}, w, v + 1)$ is clear from the specification of q', Corollary 5.5.1.1, noting that $h'(\mathbf{x}, w)$ is always different from 0.

(d) We use (c) to prove that if $h'(\mathbf{x}, 0) \notin \mathcal{O}_\phi$ then $\sim P(\mathbf{x})$. Such an assumption and (c) imply that there is a total function α such that $h'(\mathbf{x}, \bar{\alpha}(v)) \notin \mathcal{O}_\phi$ and $Q(\mathbf{x}, \bar{\alpha}(v)) \equiv \mathsf{F}$ for all v; hence $\sim \forall\alpha\exists v Q(\mathbf{x}, \bar{\alpha}(v))$.

From (b) and (d) it follows that

$$P(\mathbf{x}) \equiv h(\mathbf{x}) \in \mathcal{O}_\phi. \qquad \square$$

Corollary 5.5.6.1. *Let $\mathscr{C}$ be a* FRS, *and assume $\mathscr{C} \subseteq \mathscr{C}_{\mathsf{ha}\#}$. Then $\mathscr{C}_{\mathsf{dh}\#}$ has the selector property relative to $\mathscr{C}_{\mathsf{dhe}}$.*

Proof. Let P be a $(k + 1)$-ary $\mathscr{C}_{\mathsf{d}}$-hyperenumerable predicate, and let (ϕ, S) be an internal interpreter for $\mathscr{C}$ where S is strict. From the preceding theorem there is a total $\mathscr{C}$-computable function h such that

$$P(\mathbf{x}, y) \equiv h(\mathbf{x}, y) \in \mathcal{O}_\phi.$$

We introduce a $(k + 1)$-ary predicate P' such that

$$P'(\mathbf{x}, y) \equiv P(\mathbf{x}, y) \wedge \forall u\colon u < y \wedge \mathfrak{E}_\phi(h(\mathbf{x}, u), \mathsf{exp}(2, h(\mathbf{x}, y)))$$
$$\vee\ u \geq y \wedge \mathfrak{E}_\phi(h(\mathbf{x}, u), h(\mathbf{x}, y)).$$

We prove first that the predicate P' is single-valued. Assume that $P'(\mathbf{x}, y)$ and $P'(\mathbf{x}, y')$ both hold. It is sufficient to derive a contradiction from the assumption that $y' < y$. Note that $h(\mathbf{x}, y) \in \mathcal{O}_\phi$ and $h(\mathbf{x}, y') \in \mathcal{O}_\phi$. From $P'(\mathbf{x}, y)$ we get

$$\mathfrak{E}_\phi(h(\mathbf{x}, y'), \mathsf{exp}(2, h(\mathbf{x}, y))),$$

which means that $|h(\mathbf{x}, y')|_\phi > |h(\mathbf{x}, y)|_\phi$. From $P'(\mathbf{x}, y')$ we get

$$\mathfrak{E}_\phi(h(\mathbf{x}, y), h(\mathbf{x}, y')),$$

which means that $|h(\mathbf{x}, y)|_\phi \geq |h(\mathbf{x}, y')|_\phi$, so we have a contradiction.

Since P' is single-valued it follows there is a k-ary function f such that

$$\mathsf{G}_f(\mathbf{x}, y) \equiv P'(\mathbf{x}, y).$$

Furthermore, from the assumption that $\mathscr{C} \subseteq \mathscr{C}_{\mathsf{ha}\#}$ it follows that ϕ is $\mathscr{C}$-hyperarithmetic; hence $\mathfrak{E}_\phi$ is $\mathscr{C}_{\mathsf{d}}$-hyperenumerable, and P' is also $\mathscr{C}_{\mathsf{d}}$-hyperenumerable. We conclude that f is $\mathscr{C}_{\mathsf{d}}$-hyperenumerable.

We must prove that f is a selector function for the predicate P. It is clear that in case $f(\mathbf{x}) \simeq y$ then $P(\mathbf{x}, y)$ holds. Conversely, assume that there is y such that $P(\mathbf{x}, y)$ holds. This means $h(\mathbf{x}, y) \in \mathcal{O}_\phi$ and $|h(\mathbf{x}, y)|_\phi = \mu$. We can find a y such that μ is minimal. If we take y the least (in the order of the natural numbers) such that μ is minimal, it follows that $P'(\mathbf{x}, y)$ holds; hence $f(\mathbf{x}) \simeq y$. $\square$

Corollary 5.5.6.2. *Let $\mathscr{C}$ be a* FRS, *and assume $\mathscr{C} \subseteq \mathscr{C}_{\mathsf{ha}\#}$. Then $\mathscr{C}_{\mathsf{dh}\#}$ is also a* FRS.

Proof. Let (ϕ, S) be an internal interpreter for $\mathscr{C}$, and let P be the binary predicate such that

$$P(z, y) \equiv \forall\alpha\exists v \mathsf{G}_\phi(\mathsf{S}^2(z, y, \bar{\alpha}(v)), 0).$$

Since G_ϕ is $\mathscr{C}$-hyperarithmetic it follows that P is $\mathscr{C}_{\mathsf{d}}$-hyperenumerable. We take ϕ' such that

$$\phi'(z) \simeq \sigma y P(z, y).$$

To prove (ϕ', S) is an internal interpreter for $\mathscr{C}_{\mathsf{dh}\#}$ we take a k-ary function h that is $\mathscr{C}_{\mathsf{d}}$-hyperenumerable; hence

$$\mathsf{G}_h(\mathbf{x}, y) \equiv \forall\alpha\exists v Q(\mathbf{x}, y, \bar{\alpha}(v)),$$

where the predicate Q is $\mathscr{C}$-decidable. Let z be a (ϕ, S)-index for χ_Q; hence

$$\begin{aligned} Q(\mathbf{x}, y, v) &\equiv \mathsf{G}_\phi(\mathsf{S}^2(\mathsf{S}^k(\mathsf{z}, \mathbf{x}), y, v), 0) \\ \mathsf{G}_h(\mathbf{x}, y) &\equiv P(\mathsf{S}^k(\mathsf{z}, \mathbf{x}), y) \\ h(\mathbf{x}) &\simeq \sigma y P(\mathsf{S}^k(\mathsf{z}, \mathbf{x}), y) \\ &\simeq \phi'(\mathsf{S}^k(\mathsf{z}, \mathbf{x})). \end{aligned}$$

□

Corollary 5.5.6.3. *Let $\mathscr{C}$ be a class closed under recursive operations. Then $\mathscr{C}_{\mathsf{dh}\#}$ has the selector property relative to $\mathscr{C}_{\mathsf{dhe}}$.*

Proof. Let P be a $(k + 1)$-ary predicate that is $\mathscr{C}_{\mathsf{d}}$-hyperenumerable; hence

$$P(\mathbf{x}, y) \equiv \forall\alpha\exists v Q(\mathbf{x}, y, \bar{\alpha}(v)),$$

where Q is $\mathscr{C}$-decidable. We take $\mathscr{C}' = \mathsf{RC}(Q) = \mathsf{RC}(\chi_Q)$; hence $\mathscr{C}' \subseteq \mathscr{C}'_{\mathsf{ha}\#}$, $\mathscr{C}' \subseteq \mathscr{C}$, and $\mathscr{C}'_{\mathsf{dhe}} \subseteq \mathscr{C}_{\mathsf{dhe}}$. Since P is $\mathscr{C}'_{\mathsf{d}}$-hyperenumerable, it has a selector function in $\mathscr{C}'_{\mathsf{dh}\#} \subseteq \mathscr{C}_{\mathsf{dh}\#}$. □

Let $\mathscr{C}$ be a FRS with internal interpreter (ϕ, S). From Theorem 5.5.3 it follows that $|\phi|$ is maximal in $\mathscr{C}$. With this understanding we define $|\mathscr{C}| = |\phi| =$ the least ordinal, which has no notation in a system $\mathcal{O}_g$ whenever g is $\mathscr{C}$-computable.

Theorem 5.5.7. *Let $\mathscr{C}$ be a* FRS *with internal interpreter (ϕ, S), where* S *is strict and $\mathscr{C} \subseteq \mathscr{C}_{\mathsf{ha}\#}$. Then*

(i) *$\mathcal{O}_\phi$ is $\mathscr{C}$-creative in the* RRS *$\mathscr{C}_{\mathsf{dhe}}$.*
(ii) *$\overline{\mathcal{O}_\phi}$ is not $\mathscr{C}_{\mathsf{d}}$-hyperenumerable and $\mathcal{O}_\phi$ is not $\mathscr{C}$-hyperarithmetic.*
(iii) *If P is a k-ary predicate where $\bar{P}$ is $\mathscr{C}_{\mathsf{d}}$-hyperenumerable, and h is a total k-ary $\mathscr{C}_{\mathsf{d}}$-hyperenumerable function such that $h(P) \subseteq \mathcal{O}_\phi$, there is $y \in \mathcal{O}_\phi$ such that $h(P) \subseteq \mathcal{O}_{\phi, y}$.*

PROOF. The class $\mathscr{C}_{\mathsf{dhe}}$ is a RRS by Corollary 5.1.13.1 and (i) follows from Theorem 4.5.7 (iii) and Theorem 5.5.6. Part (ii) is immediate from (i). To prove (iii) we derive a contradiction from the assumption that for every $y \in \mathcal{O}_\phi$ there is $(\mathbf{x}) \in \mathbb{N}^k$ such that $P(\mathbf{x})$ holds and $|y|_\phi \leq |h(\mathbf{x})|_\phi$. In fact if this is the case it follows that

$$y \notin \mathcal{O}_\phi \equiv \forall \mathbf{x}\colon \bar{P}(\mathbf{x}) \vee \mathfrak{E}_\phi(y, \mathsf{exp}(2, h(\mathbf{x}))),$$

which means that $\bar{\mathcal{O}}_\phi$ is $\mathscr{C}_{\mathsf{d}}$-hyperenumerable and contradicts (ii). □

Corollary 5.5.7.1. *Let $\mathscr{C}$ be a* FRS *where $\mathscr{C} \subseteq \mathscr{C}_{\mathsf{ha}\#}$. Then $|\mathscr{C}| = |\mathscr{C}_{\mathsf{dh}\#}|$.*

PROOF. Let (ϕ, S) be an internal interpreter for $\mathscr{C}$, where the function S is strict. If $|\phi| < |\mathscr{C}_{\mathsf{dh}\#}|$ we note that ϕ is $\mathscr{C}$-hyperarithmetic; hence from Theorem 5.4.7 (ii) it follows that $\mathcal{O}_\phi$ is $\mathscr{C}$-hyperarithmetic, contradicting Theorem 5.5.7 (ii).

□

Corollary 5.5.7.2. *Let $\mathscr{C}$ be a* FRS *with internal interpreter (ϕ, S), where* S *is strict and $\mathscr{C} \subseteq \mathscr{C}_{\mathsf{ha}\#}$. Assume P is a k-ary predicate. The following conditions are equivalent:*

(i) *P is $\mathscr{C}$-hyperarithmetic.*
(ii) *There is a total k-ary $\mathscr{C}$-computable function h, and $y \in \mathcal{O}_\phi$, such that*

$$P(\mathbf{x}) \equiv h(\mathbf{x}) \in \mathcal{O}_{\phi,y}.$$

(iii) *There is a total k-ary $\mathscr{C}$-hyperarithmetic function h, and $y \in \mathcal{O}_\phi$, such that*

$$P(\mathbf{x}) \equiv h(\mathbf{x}) \in \mathcal{O}_{\phi,y}.$$

PROOF. Assume (i) holds. From Theorem 5.5.6 it follows that there is a total k-ary $\mathscr{C}$-computable function h such that

$$P(\mathbf{x}) \equiv h(\mathbf{x}) \in \mathcal{O}_\phi,$$

and from Theorem 5.5.7 there is $y \in \mathcal{O}_\phi$ such that $h(P) \subseteq \mathcal{O}_{\phi,y}$. The implication from (ii) to (iii) is trivial, and (i) follows from (iii) by Theorem 5.4.7 (i). □

Let $\mathscr{C}$ be a FRS and g a binary $\mathscr{C}$-hyperarithmetic function. From Theorem 5.4.7 (ii) it follows that in case $|g| < |\mathscr{C}|$ and $\mathscr{C} \subseteq \mathscr{C}_{\mathsf{dh}\#}$ then $\mathcal{O}_g$ is $\mathscr{C}$-hyperarithmetic. We now prove, under some restrictions, the converse of this relation.

Theorem 5.5.8. *Let $\mathscr{C}$ be a* FRS *such that $\mathscr{C} \subseteq \mathscr{C}_{\mathsf{ha}\#}$ and g is a binary $\mathscr{C}_{\mathsf{d}}$-hyperenumerable function. If $\mathcal{O}_g$ is $\mathscr{C}$-hyperarithmetic then $|g| < |\mathscr{C}|$.*

PROOF. Let $\mathscr{C}' = \mathscr{C}_{\mathsf{dh}\#}$. Since $|\mathscr{C}| = |\mathscr{C}'|$ it is sufficient to prove that $|g| < |\mathscr{C}'|$. From Theorem 5.4.8 it follows that there is a total unary $\mathscr{C}$-hyperarithmetic function f such that $f(x) \in \mathcal{O}_g$ and $|f(x)|_g < |f(x+1)|_g$ hold for all x, and $\lim_{n\to\infty} |f(n)|_g = |g|$. Let (ϕ, S) be an internal interpreter for $\mathscr{C}'$. From Theorem 5.5.4 it follows that there is a total unary $\mathscr{C}'$-computable function h such that $h(y) \in \mathcal{O}_\phi$ whenever $y \in \mathcal{O}_g$ and $|y|_g \leq |h(y)|_\phi$. If we set $f' = h \circ f$ it follows that f' is a total unary $\mathscr{C}'$-computable function such that $f'(x) \in \mathcal{O}_\phi$ for all x; hence by Theorem 5.5.3 there is $y \in \mathcal{O}_\phi$ such that $|f'(x)|_\phi < |y|_\phi$ for all x. Furthermore, if $v \in \mathcal{O}_g$ then by Theorem 5.4.8 there is x such that $|v|_g \leq |f(x)|_g \leq |f'(x)|_\phi < |y|_\phi$; hence $|g| < |\phi| = |\mathscr{C}'|$. □

Theorem 5.5.9. *Let $\mathscr{C}$ and $\mathscr{C}'$ be* FRSs *such that $\mathscr{C} \subseteq \mathscr{C}_{\mathsf{ha}\#}$, $\mathscr{C}' \subseteq \mathscr{C}'_{\mathsf{ha}\#}$, and $\mathscr{C} \subseteq \mathscr{C}'$. Assume (ϕ, S) is an internal interpreter for $\mathscr{C}$, where the function* S *is strict. The following conditions are equivalent:*

(i) $|\mathscr{C}| < |\mathscr{C}'|$.
(ii) *There is a set A in the class $(\mathscr{C}_{\mathsf{dhe}} - \mathscr{C}_{\mathsf{ha}}) \cap \mathscr{C}'_{\mathsf{ha}}$.*
(iii) *The set $\mathcal{O}_\phi$ is $\mathscr{C}'$-hyperarithmetic.*
(iv) $\mathscr{C}_{\mathsf{dhe}} \subseteq \mathscr{C}'_{\mathsf{ha}}$.

Proof. To prove (ii) from (i) note that $|\phi| < |\mathscr{C}'|$; hence $\mathcal{O}_\phi$ is in the class $(\mathscr{C}_{\mathsf{dhe}} - \mathscr{C}_{\mathsf{ha}}) \cap \mathscr{C}'_{\mathsf{ha}}$ by Theorem 5.4.7 (ii) and Theorem 5.5.7. So we take $A = \mathcal{O}_\phi$. To prove (iii) from (ii) let (Φ, S') be an internal interpreter for $\mathscr{C}'$, where S' is strict. From Theorem 5.5.4 there is a total unary $\mathscr{C}'$-computable function f such that if $y \in \mathcal{O}_\phi$ then $f(y) \in \mathcal{O}_\Phi$ and $|y|_\Phi \leq |f(y)|_\Phi$, and furthermore whenever $y' \in \mathcal{O}_\phi$ and $|y|_\phi \leq |y'|_\phi$ then $|f(y)|_\Phi \leq |f(y')|_\Phi$ (see Exercise 5.5.8). From Theorem 5.5.6 it follows that there is a total unary $\mathscr{C}$-computable function h such that

$$x \in A \equiv h(x) \in \mathcal{O}_\phi.$$

We take $f' = f \circ h$, so it follows that f' is $\mathscr{C}'$-computable and $f'(A) \subseteq \mathcal{O}_\Phi$; hence from Theorem 5.5.7 (iii) it follows that there is $v \in \mathcal{O}_\Phi$ such that $f'(A) \subseteq \mathcal{O}_{\Phi,v}$. From this it follows that $f(\mathcal{O}_\phi) \subseteq \mathcal{O}_{\Phi,v}$, hence $|\phi| < |v|_\Phi < |\Phi|$, so from Theorem 5.4.7 (ii) it follows that $\mathcal{O}_\phi$ is $\mathscr{C}'$-hyperarithmetic.

The implication from (iii) to (iv) is immediate from Theorem 5.5.6, and the implication from (iv) to (i) follows from Theorem 5.5.8. □

Corollary 5.5.9.1. *Let A be a recursively hyperenumerable set, and let $\mathscr{C}' =$* RC(A). *The following conditions are equivalent:*

(i) $|\mathsf{RC}| < |\mathscr{C}'|$.
(ii) *A is not hyperarithmetic.*
(iii) $\mathsf{RC}_{\mathsf{dhe}} \subseteq \mathscr{C}'_{\mathsf{ha}}$.

Proof. Assume (i) and also that A is hyperarithmetic. It follows that $\mathscr{C}' \subseteq \mathsf{RC}_{\mathsf{ha}\#} \subseteq \mathsf{RC}_{\mathsf{dh}\#}$, hence $|\mathscr{C}'| \leq |\mathsf{RC}_{\mathsf{dh}\#}| = |\mathsf{RC}|$, which contradicts (i). To prove (iii) from (ii) note that if A is not hyperarithmetic then A is in the class

$(\mathsf{RC}_{\mathsf{dhe}} - \mathsf{RC}_{\mathsf{ha}}) \cap \mathscr{C}'_{\mathsf{ha}}$; hence (iii) follows from Theorem 5.5.9 (ii) and (iv). The implication from (iii) to (i) is immediate from Theorem 5.5.9. □

EXERCISES

5.5.1. Let $\mathscr{C}$ be a FRS with internal interpreter (ϕ, S). Assume $x \in \mathcal{O}_\phi$, $y \in \mathcal{O}_\phi$, and $y <_\phi x +_\phi y$. Prove that either $x = y$, or $x <_\phi y$, or $y <_\phi x$.

5.5.2. Let $\mathscr{C}$ be a FRS with internal interpreter (ϕ, S). Prove that there are elements $x \in \mathcal{O}_\phi$, $y \in \mathcal{O}_\phi$, such that $x \neq y$, $x \not<_\phi y$, and $y \not<_\phi x$.

5.5.3. Let $\mathscr{C}$ be a FRS with internal interpret (ϕ, S). Assume f is a unary total function such that $f(x) \in \mathcal{O}_\phi$, and $f(x) <_\phi f(x + 1)$ for all x. Let h be the unary function such that

$$h(0) \simeq 0$$

$$h(x + 1) \simeq \mu y\colon y \in \mathcal{O}_\phi \wedge y \not<_\phi f(x).$$

Prove that h is total, $h(x) \leq h(x + 1)$ for all x, and there is y such that $h(x) < y$ for all x.

5.5.4. Let $\mathscr{C}$ be a class of functions closed under elementary operations. Assume A is an infinite $\mathscr{C}_{\mathsf{d}}$-hyperenumerable set. Prove there is an infinite $\mathscr{C}$-hyperarithmetic set B such that $B \subseteq A$.

5.5.5. Let $\mathscr{C}$ be a class of functions closed under recursive operations. Assume A and B are $\mathscr{C}_{\mathsf{d}}$-hyperenumerable sets. Prove there are $\mathscr{C}_{\mathsf{d}}$-hyperenumerable sets A' and B' such that $A' \subseteq A$, $B' \subseteq B$, $A' \cup B' = A \cup B$, and $A' \cap B' = \varnothing$.

5.5.6. Let $\mathscr{C}$ be a FRS with internal interpreter (ϕ, S) where S is recursive and strict. Assume that $\mathscr{C}$ is e-total. Prove that $\mathcal{O}_\phi$ is creative in $\mathscr{C}_{\mathsf{dhe}}$.

5.5.7. Prove Corollary 5.5.6.2 using the result in Exercise 4.5.21.

5.5.8. Consider the function h in the proof of Theorem 5.5.4, and assume x and y are elements of $\mathcal{O}_g$. Prove the following:
(a) If $|x|_g < |y|_g$ then $|h(x)|_\phi < |h(y)|_\phi$.
(b) If $|x|_g = |y|_g$ then $|h(x)|_\phi \leq |h(y)|_\phi$.
(c) $|x|_g = |y|_g$ if and only if $|h(x)|_\phi = |h(y)|_\phi$.

5.5.9. Let $\mathscr{C}$ and $\mathscr{C}'$ be FRSs such that $\mathscr{C} \cap \mathscr{C}' \subseteq \mathscr{C}_{\mathsf{dh}\#}$. Assume there is a set A in the class $(\mathscr{C}_{\mathsf{dhe}} - \mathscr{C}_{\mathsf{ha}}) \cap \mathscr{C}'_{\mathsf{ha}}$. Prove that $|\mathscr{C}| < |\mathscr{C}'|$.

5.5.10. Let c be a total function, $\mathscr{C} = \mathsf{RC}(c)$, and $\mathscr{C}' = \mathsf{RC}(\{c, A\})$ $(= \mathsf{RC}(\{c, \chi_A\}))$, where the set A is $\mathscr{C}_{\mathsf{d}}$-hyperenumerable. Prove that the following conditions are equivalent:
(a) $|\mathscr{C}| < |\mathscr{C}'|$.
(b) A is not $\mathscr{C}$-hyperarithmetic.
(c) $\mathscr{C}_{\mathsf{dhe}} \subseteq \mathscr{C}'_{\mathsf{ha}}$.

5.5.11. Let $\mathscr{C}$ and $\mathscr{C}'$ be FRSs, where $\mathscr{C} \subseteq \mathscr{C}'$, $\mathscr{C}$ is e-total, and $\mathscr{C}'$ is closed under quantification. Prove that $\mathscr{C}_{\mathsf{ha}} \subseteq \mathscr{C}'_{\mathsf{d}}$.

5.5.12. Let c be a unary total function. Prove that $\mathsf{RC}_{\mathsf{ha}}(c) = \mathsf{HC}_{\mathsf{d}}(c) = \mathsf{HC}_{\mathsf{ha}}(c)$.

Notes

The basic result in this section is Theorem 5.5.6, due to Kleene [11], which characterizes the relation between $\mathscr{C}_{\mathsf{dh}\#}$ and $\mathcal{O}_\phi$, where $\mathscr{C}$ is a reflexive structure with internal interpreter (ϕ, S). Roughly, it means that $\mathcal{O}_\phi$ contains $\mathscr{C}_{\mathsf{dh}\#}$. On the other hand, if $\mathscr{C}$ is e-total then $\mathcal{O}_\phi$ is in $\mathscr{C}_{\mathsf{dh}\#}$, so we have a kind of self-referential relation. In fact, in this case $\mathscr{C}_{\mathsf{dh}\#}$ is a FRS with internal interpreter $(\mathcal{O}_\phi, \mathsf{S})$.

A given reflexive structure $\mathscr{C}$ determines a segment $|\mathscr{C}|$ of ordinals, which provides a constructive version of Cantor second class. While the segment is determined by $\mathscr{C}$, note that in order to have $|\mathscr{C}| < |\mathscr{C}'|$ it is not sufficient to have $\mathscr{C} \subset \mathscr{C}'$, but rather $\mathscr{C}_{\mathsf{dh}\#} \subseteq \mathscr{C}'$.

The fundamental selector property given by Corollary 5.5.6.1 is proved here in the frame of inductive definability. A different impredicative proof is given in Kreisel [13].

§6. Hyperhyperenumeration

In this section we extend barred quantification and allow two function variables to be simultaneously barred by one numerical variable. We call this form type 2 barred quantification, and the original form with only one function variable we call type 1 barred quantification. The usual unbounded numerical quantification can be thought of as type 0 barred quantification.

As usual there are two versions of type 2 barred quantification, one existential and the other universal. Formally, there are introduced as follows.

Existential Type 2 Barred Quantification. Let Q be a $(k + 2)$-ary predicate. We introduce a k-ary predicate P such that

$$P(\mathbf{x}) \equiv \exists\alpha\forall\beta\exists y Q(\mathbf{x}, \bar{\alpha}(y), \bar{\beta}(y)).$$

We say that P is obtained from Q by *existential type 2 barred quantification.*

Universal Type 2 Barred Quantification. Let Q be a $(k + 2)$-ary predicate. We introduce a k-ary predicate P such that

$$P(\mathbf{x}) \equiv \forall\alpha\exists\beta\forall y Q(\mathbf{x}, \bar{\alpha}(y), \bar{\beta}(y)).$$

We say that P is obtained from Q by *universal type 2 barred quantification.*

These forms follow the usual rules, which related existential and universal quantification. For example,

$$\sim \exists\alpha\forall\beta\exists y Q(\mathbf{x}, \bar{\alpha}(y), \bar{\beta}(y)) \equiv \forall\alpha\exists\beta\forall y \sim Q(\mathbf{x}, \bar{\alpha}(y), \bar{\beta}(y))$$

$$\sim \forall\alpha\exists\beta y Q(\mathbf{x}, \bar{\alpha}(y), \bar{\beta}(y)) \equiv \exists\alpha\forall\beta\exists y \sim Q(\mathbf{x}, \bar{\alpha}(y), \bar{\beta}(y)).$$

Theorem 5.6.1. *There are unary elementary functions* d_1, d_2, d_3, and d_4 *such that whenever* P *is a* $(k + 4)$-ary predicate then

$$\exists\gamma\exists\beta\forall\alpha\exists w\forall\beta'\exists y P(\mathbf{x}, \bar{\gamma}(w), \bar{\alpha}(w), \bar{\beta}(y), \bar{\beta}'(y))$$
$$\equiv \exists\beta\forall\alpha\exists y P(\mathbf{x}, d_1(\bar{\beta}(y)), d_2(\bar{\alpha}(y)), d_3(\bar{\beta}(y)), d_4(\bar{\alpha}(y))).$$

PROOF. We introduce the functions as follows:

$$d_1(u) = \left(\prod_{v<(\ell u \dot{-} 1)_0} \exp(\mathrm{p}_v, ((u)_v \dot{-} 1)_0 + 1)\right) \dot{-} 1$$

$$d_2(u) = \left(\prod_{v<(\ell u \dot{-} 1)_0} \exp(\mathrm{p}_v, ((u)_{\exp(2,v)} \dot{-} 1)_0 + 1)\right) \dot{-} 1$$

$$d_3(u) = \left(\prod_{v<(\ell u \dot{-} 1)_1} \exp(\mathrm{p}_v, ((u)_v \dot{-} 1)_1 + 1)\right) \dot{-} 1$$

$$d_4(u) = \left(\prod_{v<(\ell u \dot{-} 1)_1} \exp(\mathrm{p}_v, ((u)_{\exp(2,(\ell u \dot{-} 1)_0)\times\exp(3,v)} \dot{-} 1)_1 + 1)\right) \dot{-} 1.$$

To prove the equivalence we assume first the left side; hence there are total unary functions γ and β such that

$$\forall\alpha\exists w\forall\beta'\exists y P(\mathbf{x}, \bar{\gamma}(w), \bar{\alpha}(w), \bar{\beta}(y), \bar{\beta}'(y)).$$

We set $\beta'(v) = (\exp(2, \gamma(v)) \times \exp(3, \beta(v))) \dot{-} 1$, and fix α in order to prove

$$\exists y P(\mathbf{x}, d_1(\bar{\beta}'(y)), d_2(\bar{\alpha}(y)), d_3(\bar{\beta}'(y)), d_4(\bar{\alpha}(y))).$$

We set $\alpha'(v) = (\alpha(\exp(2, v)))_0$, and for every number w we introduce a function β_w such that $\beta_w(v) = (\alpha(\exp(2, w) \times \exp(3, v)))_1$. From the left side assumption it follows that there are numbers w' and y' such that

$$P(\mathbf{x}, \bar{\gamma}(w'), \bar{\alpha}'(w'), \bar{\beta}(y'), \bar{\beta}_{w'}(y')).$$

We put $y = \langle w', y'\rangle$, and note that $d_1(\bar{\beta}'(y)) = \bar{\gamma}(w')$, $d_2(\bar{\alpha}(y)) = \bar{\alpha}'(w')$, $d_3(\bar{\beta}'(y)) = \bar{\beta}(y')$, and $d_4(\bar{\alpha}(y)) = \bar{\beta}_{w'}(y')$. This proves that the right side of the equation holds.

Now we assume the left side is false, to prove that the right side is also false. This means that

$$\forall\gamma\forall\beta\exists\alpha\forall w\exists\beta'\forall y \sim P(\mathbf{x}, \bar{\gamma}(w), \bar{\alpha}(w), \bar{\beta}(y), \bar{\beta}'(y)).$$

To prove that the right side is false we fix β' and show that

$$\exists\alpha\forall y \sim P(\mathbf{x}, d_1(\bar{\beta}'(y)), d_2(\bar{\alpha}(y)), d_3(\bar{\beta}'(y)), d_4(\bar{\alpha}(y))).$$

We set $\gamma(v) = (\beta'(v))_0$ and $\beta(v) = (\beta'(v))_1$. Hence there is a function α' such that for every w we can associate a function β_w, and

$$\forall w\forall y \sim P(\mathbf{x}, \bar{\gamma}(w), \bar{\alpha}'(w), \bar{\beta}(y), \bar{\beta}_w(y)),$$
$$\forall y \sim P(\mathbf{x}, \bar{\gamma}((y \dot{-} 1)_0), \bar{\alpha}'((y \dot{-} 1)_0), \bar{\beta}((y \dot{-} 1)_1), \bar{\beta}_{(y \dot{-} 1)_0}((y \dot{-} 1)_1)).$$

Now we set $\alpha(v) = (\exp(2, \alpha'((v \dot{-} 1)_0)) \times \exp(3, \beta_{(v \dot{-} 1)_0}((v \dot{-} 1)_1))) \dot{-} 1$ and note that $d_1(\overline{\beta}'(y)) = \overline{\gamma}((y \dot{-} 1)_0)$, $d_2(\overline{\alpha}(y)) = \overline{\alpha}'((y \dot{-} 1)_0)$, $d_3(\overline{\beta}'(y)) = \overline{\beta}((y \dot{-} 1)_1)$, and $d_4(\overline{\alpha}(y)) = \overline{\beta}_{(y \dot{-} 1)_0}((y \dot{-} 1)_1)$. □

Theorem 5.6.2. *There are unary elementary functions d_5 and d_6, such that whenever P is a $(k + 4)$-ary predicate then*

$$\exists\gamma\forall\alpha\exists w\exists\beta\forall\beta'\exists y P(\mathbf{x}, \overline{\gamma}(w), \overline{\alpha}(w), \overline{\beta}(y), \overline{\beta}'(y))$$

$$\equiv \exists\gamma\exists\beta\forall\alpha\exists w\forall\beta'\exists y \colon \overline{\alpha}(w) = [\ell\overline{\beta}(y)]_1 \wedge P(\mathbf{x}, \overline{\gamma}(w), \overline{\alpha}(w), d_5(\overline{\beta}(y)), d_6(\overline{\beta}'(y))).$$

PROOF. We put

$$d_5(u) = \left(\prod_{v < [\ell u]_2} \exp(\mathrm{p}_v, (u)_{\langle [\ell u]_1, v\rangle})\right) \dot{-} 1$$

$$d_6(u) = \left(\prod_{v < [\ell u]_2} \exp(\mathrm{p}_v, (u)_v)\right) \dot{-} 1.$$

To prove the relation we assume first that the left side is true; hence there is a total function γ such that

$$\forall\alpha\exists w\exists\beta\forall\beta'\exists y P(\mathbf{x}, \overline{\gamma}(w), \overline{\alpha}(w), \overline{\beta}(y), \overline{\beta}'(y)).$$

For every number w we associate a function β_w as follows. If there is a function β such that

$$\forall\beta'\exists y P(\mathbf{x}, \overline{\gamma}(\ell w), w, \overline{\beta}(y), \overline{\beta}'(y)),$$

we take $\beta_w = \beta$. Otherwise we take $\beta_w = I_1^1$. Note that the following relation is satisfied:

$$\forall\alpha\exists w\forall\beta'\exists y P(\mathbf{x}, \overline{\gamma}(w), \overline{\alpha}(w), \overline{\beta}_{\overline{\alpha}(w)}(y), \overline{\beta}'(y)).$$

In order to prove that the right side is true we introduce a total function β such that $\beta(v) = \beta_{[v]_1}([v]_2)$. Since $d_5(\overline{\beta}(\langle\overline{\alpha}(w), y\rangle)) = \overline{\beta}_{\overline{\alpha}(w)}(y)$ and $d_6(\overline{\beta}'(\langle\overline{\alpha}(w), y\rangle)) = \overline{\beta}'(y)$, it follows that for every α there is w such that

$$\forall\beta'\exists y \colon \overline{\alpha}(w) = [\ell\overline{\beta}(y)]_1 \wedge P(\mathbf{x}, \overline{\gamma}(w), \overline{\alpha}(w), d_5(\overline{\beta}(y)), d_6(\overline{\beta}'(y))).$$

Now we assume the right side is true, in order to prove that the left side is true. It follows that there are functions γ and β such that for every function α there is w such that

$$\forall\beta'\exists y \colon \overline{\alpha}(w) = [\ell\overline{\beta}(y)]_1 \wedge P(\mathbf{x}, \overline{\gamma}(w), \overline{\alpha}(w), d_5(\overline{\beta}(y)), d_6(\overline{\beta}'(y))).$$

If we put $\beta_1(v) = \beta(\langle\overline{\alpha}(w), v\rangle)$ and assume that $[y]_1 = \overline{\alpha}(w)$, it follows that $d_5(\overline{\beta}(y)) = \overline{\beta}_1([y]_2)$, and furthermore $d_6(\overline{\beta}'(y)) = \overline{\beta}'([y]_2)$ holds for any function β'. It follows that

$$\forall\beta'\exists y P(\mathbf{x}, \overline{\gamma}(w), \overline{\alpha}(w), \overline{\beta}_1(y), \overline{\beta}'(y)).$$

This completes the proof. □

Corollary 5.6.2.1. *Let P be a $(k + 4)$-ary predicate, and let P' be the $(k + 2)$-ary predicate such that*

$$P'(\mathbf{x}, y_1, y_2) \equiv d_2(y_2) = [\ell d_3(y_1)]_1 \wedge P(\mathbf{x}, d_1(y_1), d_2(y_2), d_5(d_3(y_1)), d_6(d_4(y_2))),$$

where d_1, d_2, d_3, d_4, d_5, and d_6 are the elementary functions of Theorem 5.6.1 and Theorem 5.6.2. Then

$$\exists\gamma\forall\alpha\exists w\exists\beta\forall\beta'\exists y P(\mathbf{x}, \bar{\gamma}(w), \bar{\alpha}(w), \bar{\beta}(y), \bar{\beta}'(y)) \equiv \exists\beta\forall\alpha\exists y P'(\mathbf{x}, \bar{\beta}(y), \bar{\alpha}(y)).$$

Proof. Immediate from Theorem 5.6.1 and Theorem 5.6.2. □

Theorem 5.6.3. *Let P be a $(k + 3)$-ary predicate and d_5, d_6 the elementary functions of Theorem 5.6.2. Then*

$$\forall y < z\exists\beta\forall\alpha\exists w P(\mathbf{x}, y, \bar{\beta}(w), \bar{\alpha}(w)) \equiv \exists\beta\forall y < z\forall\alpha\exists w\colon y = [\ell\bar{\beta}(w)]_1 \wedge P(\mathbf{x}, y, d_5(\bar{\beta}(w)), d_6(\bar{\alpha}(w))).$$

Proof. We assume first that the left side is true; hence for every number y there is a function β_y such that

$$\forall y < z\forall\alpha\exists w P(\mathbf{x}, y, \bar{\beta}_y(w), \bar{\alpha}(w)).$$

We set $\beta(v) = \beta_{[v]_1}([v]_2)$, and note that $d_5(\bar{\beta}(\langle y, w\rangle)) = \bar{\beta}_y(w)$ and $d_6(\bar{\alpha}(\langle y, w\rangle)) = \bar{\alpha}(w)$; hence the right side is true.

Now we assume the right side is true; hence there is a function β such that

$$\forall y < z\forall\alpha\exists w\colon y = [\ell\bar{\beta}(w)]_1 \wedge P(\mathbf{x}, y, d_5(\bar{\beta}(w)), d_6(\bar{\alpha}(w))).$$

For any given y we set $\beta_y(v) = \beta(\langle y, v\rangle)$. If $[w]_1 = y$ it follows that $d_5(\bar{\beta}(w)) = \bar{\beta}_y([w]_2)$ and $d_6(\bar{\alpha}(w)) = \bar{\alpha}([w]_2)$; hence the left side is true. □

Corollary 5.6.3.1. *Let P be a $(k + 3)$-ary predicate, and let P' be a $(k + 3)$-ary predicate such that*

$$P'(\mathbf{x}, y, w_1, w_2) \equiv y = [\ell d(y, w_1)]_1 \wedge P(\mathbf{x}, y, d_5(d(y, w_1)), d_6(d(y, w_2))),$$

where d_5, d_6 are the elementary functions in Theorem 5.6.3, and d is the binary elementary function of Theorem 5.1.2. Then

$$\forall y < z\exists\beta\forall\alpha\exists w P(\mathbf{x}, y, \bar{\beta}(w), \bar{\alpha}(w)) \equiv \exists\beta\forall\alpha\exists w\forall y < z P'(\mathbf{x}, y, \bar{\beta}(w), \bar{\alpha}(w)).$$

Proof. Immediate from Theorem 5.6.3 and Theorem 5.1.2. □

Let $\mathscr{P}$ be a class of predicates. We say that a k-ary predicate P is *$\mathscr{P}$-hyperhyperenumerable* if there is a $(k + 2)$-ary predicate Q in $\mathscr{P}$ such that

$$P(\mathbf{x}) \equiv \exists\beta\forall\alpha\exists y Q(\mathbf{x}, \bar{\beta}(y), \bar{\alpha}(y)).$$

The class of all predicates that are $\mathscr{P}$-hyperhyperenumerable is denoted by $\mathscr{P}_{\mathsf{hhe}}$. We put $\mathscr{P}_{\mathsf{hhu}} = \mathscr{P}_{\mathsf{nhhen}}$, so $\mathscr{P}_{\mathsf{hhu}}$ denotes the class of all predicates which can be obtained by universal type 2 barred quantification from predicates in the class $\mathscr{P}$.

The predicates in the class $\mathsf{RC}_{\mathsf{dhhe}} = \mathsf{RC}_{\mathsf{pdhhe}}$ are said to be *recursively hyperhyperenumerable*. The predicates in $\mathsf{RC}_{\mathsf{dhhe}}(\mathscr{C})$ are *recursively hyperhyperenumerable* in $\mathscr{C}$, and the predicates in $\mathsf{RC}_{\mathsf{pdhhe}}(\mathscr{C})$ are *partially recursively hyperhyperenumerable* in $\mathscr{C}$.

Theorem 5.6.4. *Let $\mathscr{P}$ be a class of predicates closed under substitution with total $\mathscr{C}$-computable functions, where $\mathscr{C}$ is closed under elementary operations. Assume $\mathsf{EL}_{\mathsf{d}} \subseteq \mathscr{P}$. Then*

(i) *$\mathscr{P} \subseteq \mathscr{P}_{\mathsf{hhe}}$ and $\mathscr{P}_{\mathsf{hhe}}$ is closed under substitution with total $\mathscr{C}$-computable functions.*
(ii) *If $\mathscr{P}$ is closed under conjunction then $\mathscr{P}_{\mathsf{hhe}}$ is closed under existential type 2 barred quantification, under conjunction, under existential type 1 barred quantification, under existential and universal unbounded quantification, under existential bounded quantification, and under universal type 1 barred quantification.*
(iii) *If $\mathscr{P}$ is closed under disjunction then $\mathscr{P}_{\mathsf{hhe}}$ is closed under disjunction.*
(iv) *If $\mathscr{P}$ is closed under conjunction and universal bounded quantification then $\mathscr{P}_{\mathsf{hhe}}$ is closed under universal bounded quantification.*

PROOF. Part (i) is trivial. In part (ii) we note first that closure under existential type 2 barred quantification follows from Corollary 5.6.2.1. Closure under conjunction follows as usual by exportation of quantifiers. Closure under existential type 1 barred quantification is trivial because this is just a special case of type 2. This means $\mathscr{P}_{\mathsf{hhehe}} = \mathscr{P}_{\mathsf{hhe}}$; hence we can apply Theorem 5.1.3 (ii) and (iii), which gives closure under existential and universal unbounded quantification and under existential bounded quantification. Finally, we get closure under universal type 1 barred quantification by the following relation:

$$\exists\beta\forall y Q(\mathbf{x}, \bar{\beta}(y)) \equiv Q(\mathbf{x}, 0) \wedge \exists\beta\forall\alpha\exists y\colon \ell\bar{\alpha}(y) \neq 0 \wedge \ell\bar{\alpha}(y) = (\bar{\alpha}(y))_0 \wedge Q(\mathbf{x}, \bar{\beta}(y)).$$

which holds for any $(k + 1)$-ary predicate Q.

In part (iii) closure under disjunction follows in the usual way by exportation of quantifiers and closure under existential type 2 barred quantification. Finally, part (iv) follows from Corollary 5.6.3.1. □

Corollary 5.6.4.1. *Let $\mathscr{P}$ be a class of predicates closed under substitution with elementary functions, conjunction, disjunction, and existential type 2 barred quantification. Assume $\mathsf{EL}_{\mathsf{d}} \subseteq \mathscr{P}$. Then $\mathscr{P}$ is closed under existential and universal type 1 barred quantification, under existential and universal unbounded quantification, and under existential and universal bounded quantification.*

Proof. Immediate from Theorem 5.6.4 noting that $\mathscr{P}_{\mathsf{hhe}} = \mathscr{P}$. Closure under universal bounded quantification is derived using closure under disjunction. □

Example 5.6.1. If $\mathscr{C}$ is a class of functions closed under elementary operations then $\mathscr{C}_{\mathsf{d}}$ satisfies all the assumptions in Theorem 5.6.4; hence $\mathscr{C}_{\mathsf{dhhe}}$ is closed under existential type 2 barred quantification, type 1 barred quantification, unbounded numerical quantification, and bounded numerical quantification. Note that $\mathscr{C}_{\mathsf{d}} \subseteq \mathscr{C}_{\mathsf{de}} \subseteq \mathscr{C}_{\mathsf{dhe}} \subseteq \mathscr{C}_{\mathsf{dhhe}}$; hence $\mathscr{C}_{\mathsf{dhhe}} \subseteq \mathscr{C}_{\mathsf{dehhe}} \subseteq \mathscr{C}_{\mathsf{dhehhe}} \subseteq \mathscr{C}_{\mathsf{dhhehhe}} = \mathscr{C}_{\mathsf{dhhe}}$.

Example 5.6.2. Let $\mathscr{C}$ be a class closed under recursive operations. It follows that $\mathscr{C}_{\mathsf{pdhhe}} = \mathscr{C}_{\mathsf{pdehhe}}$, where $\mathscr{C}_{\mathsf{pde}}$ is closed under conjunction and disjunction. It follows that $\mathscr{C}_{\mathsf{pdhhe}}$ is closed under existential type 2 barred quantification, type 1 barred quantification, unbounded quantification, bounded quantification, conjunction, and disjunction.

The theory of the classes $\mathscr{P}_{\mathsf{hhe}}$ is quite similar to the theory of the classes $\mathscr{P}_{\mathsf{he}}$. We give the fundamental definitions and leave the results to be proved in the exercises.

Let $\mathscr{P}$ be a class of predicates. If h is a k-ary function and G_h is $\mathscr{P}$-hyperhyperenumerable, we say that h is *$\mathscr{P}$-hyperhyperenumerable*. The class of all functions that are $\mathscr{P}$-hyperhyperenumerable is denoted by $\mathscr{P}_{\mathsf{hh}\#} = \mathscr{P}_{\mathsf{hheg}}$.

The functions in $\mathsf{RC}_{\mathsf{dhh}\#} = \mathsf{RC}_{\mathsf{pdhh}\#}$ are said to be *recursively hyperhyperenumerable*. A function in $\mathsf{RC}_{\mathsf{dhh}\#}(\mathscr{C})$ is *recursively hyperhyperenumerable in $\mathscr{C}$*, and a function in $\mathsf{RC}_{\mathsf{pdhh}\#}(\mathscr{C})$ is *partially recursively hyperhyperenumerable in $\mathscr{C}$*.

Let $\mathscr{C}$ be a class of functions. If $\mathscr{C}$ is closed under recursive operations and $\mathscr{C} = \mathscr{C}_{\mathsf{pdhh}\#}$, we say that $\mathscr{C}$ is closed under *hyperhyperenumeration operations*. For example, it can be proved that if $\mathscr{C}$ is closed under recursive operations then $\mathscr{C}_{\mathsf{pdhh}\#}$ is closed under hyperhyperenumeration operations.

Let $\mathscr{C}$ be a class of functions. If $\mathscr{C}$ is closed under recursive operations and $\mathscr{C} = \mathscr{C}_{\mathsf{dhh}\#}$, we say that $\mathscr{C}$ is hhe-*total*. This notation refers to the fact that in such a case $\mathscr{C}$ is determined by its total functions and hyperhyperenumeration. For example, it can be proved that if $\mathscr{C}$ is closed under elementary operations then $\mathscr{C}_{\mathsf{dhh}\#}$ is hhe-total.

Let $\mathscr{C}$ be a class of functions. We say that a predicate P is *$\mathscr{C}$-hyperhyperarithmetical* if both P and $\bar{P}$ are $\mathscr{C}_{\mathsf{d}}$-hyperhyperenumerable. We denote by $\mathscr{C}_{\mathsf{hha}}$ the class of all $\mathscr{C}$-hyperhyperarithmetical predicates. Clearly, we have $\mathscr{C}_{\mathsf{hha}} = \mathscr{C}_{\mathsf{dhhe}} \cap \mathscr{C}_{\mathsf{dhhen}} = \mathscr{C}_{\mathsf{dhhe}} \cap \mathscr{C}_{\mathsf{dnhhu}}$. If $\mathscr{C}_{\mathsf{d}}$ is closed under negation we have $\mathscr{C}_{\mathsf{hha}} = \mathscr{C}_{\mathsf{hhe}} \cap \mathscr{C}_{\mathsf{hhu}}$. If h is a k-ary function and G_h is $\mathscr{C}$-hyperhyperarithmetical, we say that h is *$\mathscr{C}$-hyperhyperarithmetical*. The class of all $\mathscr{C}$-hyperhyperarithmetical functions is denoted by $\mathscr{C}_{\mathsf{hhag}}$. Note that in case $\mathscr{C}_{\mathsf{hha}} = \mathscr{C}_{\mathsf{hhae}}$ we have $\mathscr{C}_{\mathsf{hhag}} = \mathscr{C}_{\mathsf{hhaeg}} = \mathscr{C}_{\mathsf{hha}\#}$.

The predicates in the class $\mathsf{RC}_{\mathsf{hha}}$ are said to be *hyperhyperarithmetical*, and the predicates in $\mathsf{RC}_{\mathsf{hha}}(\mathscr{C})$ are said to be *hyperhyperarithmetical* in $\mathscr{C}$. A function in the class $\mathsf{RC}_{\mathsf{hha}\#}$ is *hyperhyperarithmetical*, and the functions in $\mathsf{RC}_{\mathsf{hha}\#}(\mathscr{C})$ are *hyperhyperarithmetical* in $\mathscr{C}$.

Theorem 5.6.5. *Let $\mathscr{P}$ be a* RRS. *Then $\mathscr{P}_{\mathsf{hhe}}$ and $\mathscr{P}_{\mathsf{ehhu}}$ are both* RRSs.

PROOF. Similar to that of Theorem 5.1.13. □

As an application of this theory we consider a GFRS $\mathscr{F}$ with internal interpreter (Φ_c, S). We recall that $\mathscr{F}_0$ is the class of all ordinary functions in the class $\mathscr{F}$; hence $\mathscr{F}_{0\mathsf{d}}$ is the class of all predicates that are $\mathscr{F}_0$-decidable, and similarly for $\mathscr{F}_{0\mathsf{pd}}$.

We know that $\mathscr{F}_0$ is closed under recursive operations. Furthermore, if c is a fixed unary function, then $\mathscr{F}_c$ is a FRS with internal interpreter (Φ_c, S), and we can use the notations $<_{\Phi_c}$, $\mathscr{O}_{\Phi_c}$, $+_{\Phi_c}$, and $|y|_{\Phi_c}$ with the meaning explained in Section 5. But we prefer to use the notations $<_c$, $\mathscr{O}_c$, $+_c$, and $|y|_c$. This is adequate as long as we consider the GFRS $\mathscr{F}$ and the interpreter (Φ_c, S) to be fixed.

On the other hand, in some situations we want to restrict c to a total unary function. In these cases we use a function variable, say β, and write $<_\beta$, $\mathscr{O}_\beta$, $+_\beta$, and $|y|_\beta$. Finally, we set $y \in \mathscr{O}_\Phi \equiv \exists \beta y \in \mathscr{O}_\beta$.

Theorem 5.6.6. *Let $\mathscr{F}$ be a* GFRS *with internal interpreter* (Φ_c, S). *There is a total binary $\mathscr{F}_0$-computable function $+_0$ such that $+_c = +_0$ whenever c is a fixed unary function.*

PROOF. We return to the specification of the function $+_c (= +_\phi)$ in Section 5. The function g given by the abstraction property is $\mathscr{F}_0$-computable, for S is $\mathscr{F}_0$-computable, and satisfies uniformly (i.e., for any fixed unary function c) the relation

$$\Phi_c^1(g(z, x, w), v) \simeq \Phi_c^2(z, x, \Phi_c^1(w, v)).$$

It follows that the function f is also $\mathscr{F}_0$-computable, and there is a number z (given by the RT) that satisfies uniformly the relation

$$f(\mathsf{z}, x, y) \simeq \Phi_c^2(\mathsf{z}, x, y).$$

Hence we take $x +_0 y = f(\mathsf{z}, x, y)$. □

In relation with the predicates $\mathfrak{D}_g^{g'}$ and $\mathfrak{E}_g^{g'}$ of Section 4, we shall use the notations $\mathfrak{D}_c^{c'}$ (or $\mathfrak{D}_\beta^\gamma$) and $\mathfrak{E}_c^{c'}$ (or $\mathfrak{E}_\beta^\gamma$), where $g = \Phi_c^1$ (or $g = \Phi_\beta^1$) and $g' = \Phi_{c'}^1$ (or $g' = \Phi_\gamma^1$).

Theorem 5.6.7. *Let $\mathscr{F}$ be a* GFRS *with internal interpreter* (Φ_c, S), *where the function* S *is strict. If Q is a $(k + 2)$-ary predicate that is $\mathscr{F}_0$-decidable, there is* *a total $\mathscr{F}_0$-computable function h such that*

$$\forall\alpha\exists yQ(\mathbf{x}, \bar{\beta}(y), \bar{\alpha}(y)) \equiv h(\mathbf{x}) \in \mathcal{O}_\beta$$

holds for arbitrary total function β.

PROOF. We give a uniform version of Theorem 5.5.6. We introduce a $(k + 4)$-ary general function q_c such that

$$q_c(z, \mathbf{x}, u, w, 0) \simeq \Phi_c^{k+2}(z, \mathbf{x}, u \,\square\, \langle c(\ell u)\rangle, w \,\square\, \langle 0\rangle)$$

$$q_c(z, \mathbf{x}, u, w, v + 1) \simeq q_c(z, \mathbf{x}, u, w, v) +_0 \Phi_c^{k+2}(z, \mathbf{x}, u \,\square\, \langle c(\ell u)\rangle, w \,\square\, \langle v + 1\rangle).$$

It follows that there is a total $(k + 3)$-ary $\mathscr{F}_0$-computable function g such that

$$q_c(z, \mathbf{x}, u, w, v) \simeq \Phi_c^1(g(z, \mathbf{x}, u, w), v)$$

holds for all unary functions c.

Using total definition by cases we introduce a total $(k + 3)$-ary $\mathscr{F}_0$-computable function f such that

$$\begin{aligned} f(z, \mathbf{x}, u, w) &= 1 && \text{if } Q(\mathbf{x}, u, w) \\ &= \exp(3, g(z, \mathbf{x}, u, w) + 1) && \text{otherwise.} \end{aligned}$$

We fix $\mathbf{z}$ uniformly by the RT such that the relation

$$f(\mathbf{z}, \mathbf{x}, u, w) = \Phi_c^{k+2}(\mathbf{z}, \mathbf{x}, u, w)$$

holds for all unary functions c. We set $h'(\mathbf{x}, u, w) = f(\mathbf{z}, \mathbf{x}, u, w)$, $q_c'(\mathbf{x}, u, w, v) \simeq q_c(\mathbf{z}, \mathbf{x}, u, w, v)$, and $d(\mathbf{x}, u, w) = g(\mathbf{z}, \mathbf{x}, u, w)$ and note that if c is restricted to a total function β then q_β' is always total. From the above specifications we derive the following relations, where β is an arbitrary total unary function:

$$\begin{aligned} q_\beta'(\mathbf{x}, \bar{\beta}(u), w, 0) &= h'(\mathbf{x}, \bar{\beta}(u + 1), w \,\square\, \langle 0\rangle) \\ q_\beta'(\mathbf{x}, \bar{\beta}(u), w, v + 1) &= q_\beta'(\mathbf{x}, \bar{\beta}(u), w, v) +_0 h'(\mathbf{x}, \bar{\beta}(u + 1), w \,\square\, \langle v + 1\rangle) \\ h'(\mathbf{x}, \bar{\beta}(u), w) &= 1 && \text{if } Q(\mathbf{x}, \bar{\beta}(u), w) \\ &= \exp(3, d(\mathbf{x}, \bar{\beta}(u), w) + 1) && \text{otherwise,} \end{aligned}$$

where the total $\mathscr{F}_0$-computable function d satisfies the condition

$$\Phi_\beta^1(d(\mathbf{x}, \bar{\beta}(u), w), v) = q'(\mathbf{x}, \bar{\beta}(u), w, v).$$

The rest of the proof is exactly as in Theorem 5.5.6. First, we show that if $Q(\mathbf{x}, \bar{\beta}(u), w) \equiv \mathsf{F}$ and $h'(\mathbf{x}, \bar{\beta}(u), w) \in \mathcal{O}_\beta$, then $h'(\mathbf{x}, \bar{\beta}(u + 1), w \,\square\, \langle v\rangle) \in \mathcal{O}_\beta$ and $|h'(\mathbf{x}, \bar{\beta}(u + 1), w \,\square\, \langle v\rangle)|_\beta < |h'(\mathbf{x}, \bar{\beta}(u), w)|_\beta$ for every v. From this it follows that in case $h'(\mathbf{x}, 0, 0) \in \mathcal{O}_\beta$ then $\forall\alpha\exists yQ(\mathbf{x}, \bar{\beta}(y), \bar{\alpha}(y))$.

Second, we show that if $Q(\mathbf{x}, \bar{\beta}(u), w) \equiv \mathsf{F}$ and $h'(\mathbf{x}, \bar{\beta}(u + 1), w \,\square\, \langle v\rangle) \in \mathcal{O}_\beta$ for every v, then $h'(\mathbf{x}, \bar{\beta}(u), w) \in \mathcal{O}_\beta$. From this it follows that whenever $h'(\mathbf{x}, 0, 0) \notin \mathcal{O}_\beta$ then $\exists\alpha\forall y \sim Q(\mathbf{x}, \bar{\beta}(y), \bar{\alpha}(y))$.

If we set $h(\mathbf{x}) = h'(\mathbf{x}, 0, 0)$ it follows that

$$\forall\alpha\exists y Q(\mathbf{x}, \bar{\beta}(y), \bar{\alpha}(y)) \equiv h(\mathbf{x}) \in \mathcal{O}_\beta$$

holds whenever β is a total unary function. □

Corollary 5.6.7.1. *Let $\mathcal{F}$ be a* GFRS *with internal interpreter* (Φ_c, S), *where* S *is strict. If P is a k-ary predicate, which is $\mathcal{F}_{0\mathsf{d}}$-hyperhyperenumerable, there is a total k-ary $\mathcal{F}_0$-computable function h such that*

$$P(\mathbf{x}) \equiv h(\mathbf{x}) \in \mathcal{O}_\Phi.$$

PROOF. Immediate from the definitions and Theorem 5.6.7. □

EXAMPLE 5.6.3. The preceding result applies to the class GRC of Chapter 4, Section 6, with internal interpreter (Φ_c, S) induced by the general function δ_c. Note that $\mathsf{GRC}_0 = \mathsf{RC}$ and $\mathsf{RC}_{\mathsf{dhhe}}$ is the class of all predicates that are recursively hyperhyperenumerable.

EXAMPLE 5.6.4. Another example is the class GRC(c_0), where c_0 is a fixed unary function, not necessarily total. An interpreter for GRC(c_0) of the form ($\Phi_{c_0,c}$, S_δ) is obtained by the procedure outlined in Example 4.6.3. In this case $\mathsf{GRC}_0(c_0) = \mathsf{RC}(c_0)$ and $\mathsf{RC}_{\mathsf{dhhe}}(c_0)$ is the class of all predicates that are recursively hyperhyperenumerable in c_0.

EXAMPLE 5.6.5. Another interesting example is the class GHC of Section 3, with interpreter (Φ_c, S_δ) induced by the function Δ_c.

Theorem 5.6.8. *Let $\mathscr{C}$ be a* FRS, *and assume $\mathscr{C}$ is* e-*total. There is a* GFRS *$\mathcal{F}$ with internal interpreter* (Φ_c, S_δ) *such that:*

(i) *$\mathcal{F}_0 = \mathscr{C}$ and there is a $\mathscr{C}$-decidable* 4-*ary predicate Q_1 such that*

$$\Phi^1_\beta(z, x) \simeq v \equiv \exists u Q_1(z, x, v, \bar{\beta}(u))$$

holds whenever β is a total function.

(ii) *There is a $\mathscr{C}_{\mathsf{d}}$-enumerable* 3-*ary predicate Q_2 such that*

$$x <_\beta y \equiv \exists w Q_2(x, y, \bar{\beta}(w))$$

holds whenever β is a total function.

(iii) *There is a $\mathscr{C}$-decidable* 3-*ary predicate Q_3 such that*

$$y \in \mathcal{O}_\beta \equiv \forall\alpha\exists w Q_3(y, \bar{\beta}(w), \bar{\alpha}(w))$$

holds whenever β is a total function.

(iv) *There is a $\mathscr{C}$-decidable* 5-*ary predicate Q_4 such that*

$$\mathfrak{D}^\gamma_\beta(x, y) \equiv \forall\alpha\exists w Q_4(x, y, \bar{\beta}(w), \bar{\gamma}(w), \bar{\alpha}(w))$$

holds whenever β and γ are total functions.

(v) *There is a $\mathscr{C}$-decidable* 5-*ary predicate Q_5 such that*

$$\mathfrak{E}^\gamma_\beta(x, y) \equiv \forall\alpha\exists w Q_5(x, y, \bar{\beta}(w), \bar{\gamma}(w), \bar{\alpha}(w))$$

holds whenever β and γ are total functions.

PROOF. If $\mathscr{C} = \mathsf{RC}(c_0)$ where c_0 is a total unary function we take $\mathscr{F} = \mathsf{GRC}(c_0)$ with the interpreter described in Example 4.6.3. The predicate Q_1 is given by

$$Q_1(z, x, v, u) \equiv Q(\mathsf{S}_\delta(z, x), v, \bar{c}_0(\ell u), u),$$

where Q is the elementary predicate of Theorem 4.6.4.

To prove (ii) we introduce by finitary induction a predicate Q_2 which satisfies the condition. In the specification we use the predicate Q_1 obtained in part (i). We assume Q_1 satisfies the following monotonic condition: if $Q_1(w, z, v, \bar{\beta}(u))$ and $u < u'$ then $Q(w, z, v, \bar{\beta}(u'))$. It is easy to modify Q_1 in such a way that this condition is satisfied. The specification of Q_2 is as follows:

1. If $\exists w$: $y = \mathsf{exp}(2, w) \wedge (x = w \vee Q_2(x, w, u))$ then $Q_2(x, y, u)$.
2. If $\exists w \exists z \exists v$: $y = \mathsf{exp}(3, w + 1) \wedge Q_1(w, z, v, u) \wedge Q_2(x, v, u)$ then $Q_2(x, y, u)$.

Note that the monotonicity of Q_1 extends to Q_2. Using this property it is easy to prove that

$$x <_\beta y \equiv \exists u Q_2(x, y, \bar{\beta}(u))$$

holds whenever β is a total function.

To prove part (iii) we rewrite the specification of $\mathcal{O}_\beta$ (i.e., $\mathcal{O}_g$ where $g = \Phi^1_\beta$) as in Section 4, but replace occurrences of $<_g$ with the expression in part (ii), and occurrences of G_g with the expression in part (i). Terms of the form $\bar{\beta}(u)$ can be eliminated noting that in general

$$\exists u Q(\mathbf{x}, \bar{\beta}(u)) \equiv \exists u\colon Q(\mathbf{x}, u) \wedge \forall v < \ell u\colon [u]_{v+1} = \beta(v) + 1.$$

This gives a specification where β appears as a free variable. Using the method of Theorem 5.2.4. we get an A-term U such that

$$y \in \mathcal{O}_\beta \equiv U,$$

and β occurs free in U. From Corollary 5.2.3.1 it follows that there is a 3-ary $\mathscr{C}$-decidable predicate Q_3 such that

$$U \equiv \forall \alpha \exists w Q_3(y, \bar{\beta}(w), \bar{\alpha}(w))$$

holds whenever β is a total function.

The predicates in part (iv) and part (v) are obtained by the same method used in part (iii). □

Theorem 5.6.9. *Let $\mathscr{C}$ be a* FRS *and assume $\mathscr{C}$ is* e*-total. Then,*

(i) *$\mathscr{C}_{\mathsf{dhhe}}$ is a* RRS.
(ii) *If $\mathscr{F}$ is the* GFRS *of Theorem 5.6.8 with internal interpreter $(\Phi_c, \mathsf{S}_\delta)$ then $\mathcal{O}_\Phi$ is $\mathscr{C}$-creative in $\mathscr{C}_{\mathsf{dhhe}}$.*
(iii) *$\mathscr{C}_{\mathsf{dhh\#}}$ has the selector property relative to $\mathscr{C}_{\mathsf{dhhe}}$.*
(iv) *$\mathscr{C}_{\mathsf{dhh\#}}$ is a* FRS.

PROOF. The proof of (i) is as in Theorem 5.1.13. Part (ii) follows from Theorem 5.6.7 and Theorem 5.6.8 (iii). To prove (iii) we use again the GFRS $\mathscr{F}$ of Theorem 5.6.8. Consider a $(k + 1)$-ary $\mathscr{C}_{\mathsf{d}}$-hyperhyperenumerable predicate P, so there is a $(k+1)$-ary $\mathscr{C}$-computable function h such that

$$P(\mathbf{x}, y) \equiv h(\mathbf{x}, y) \in \mathcal{O}_\Phi.$$

We denote by U the following expression:

$$P(\mathbf{x}, y) \wedge \forall\gamma\forall u\colon (u < y \wedge \mathfrak{E}^\gamma_\beta(h(\mathbf{x}, u), \mathsf{exp}(2, h(\mathbf{x}, y))))$$
$$\vee\ (u \geq y \wedge \mathfrak{E}^\gamma_\beta(h(\mathbf{x}, u), h(\mathbf{x}, y))).$$

We can eliminate the occurrences of $\mathfrak{E}^\gamma_\beta$ via Theorem 5.6.8 (v), and the barred terms can be eliminated in the usual way. In this way U becomes an A-term with free variables $\mathbf{x}$, y, and β. It follows that there is a $\mathscr{C}$-decidable predicate Q such that

$$U \equiv \forall\alpha\exists w Q(\mathbf{x}, y, \bar{\beta}(w), \bar{\alpha}(w))$$

holds whenever β is a total function. Now we introduce a $(k+1)$-ary predicate P' such that

$$P'(\mathbf{x}, y) \equiv \exists\beta\forall\alpha\exists w Q(\mathbf{x}, y, \bar{\beta}(w), \bar{\alpha}(w)),$$

and it follows that P' is $\mathscr{C}_\mathsf{d}$-hyperhyperenumerable.

To complete the proof we use the method as in Corollary 5.5.6.1. The predicate P' is single-valued, hence there is a k-ary function f such that $\mathsf{G}_f = P'$. The function f is $\mathscr{C}_\mathsf{d}$-hyperhyperenumerable and also a selector function for P.

To prove (iv) we obtain an internal interpreter for $\mathscr{C}_{\mathsf{dhh}\#}$ using the method of Corollary 5.5.6.2. □

Corollary 5.6.9.1. *Let $\mathscr{C}$ be a class of functions closed under recursive operations. Then $\mathscr{C}_{\mathsf{dhh}\#}$ has the selector property relative to $\mathscr{C}_{\mathsf{dhhe}}$.*

PROOF. Same proof as in Corollary 5.5.6.3. □

Let $\mathscr{F}$ be a GFRS with internal interpreter (Φ_c, S). We introduce a binary predicate $\mathfrak{E}_\Phi$ such that

$$\mathfrak{E}_\Phi(x, y) \equiv \exists\beta\colon y \in \mathcal{O}_\beta \wedge \forall\gamma\mathfrak{E}^\gamma_\beta(x, y).$$

Furthermore, for any number y we introduce a set $\mathcal{O}_{\Phi,y}$ such that

$$x \in \mathcal{O}_{\beta,y} \equiv {\sim}\mathfrak{E}_\Phi(x, y).$$

Theorem 5.6.10. *Let $\mathscr{F}$ be a* GFRS *with internal interpreter* (Φ_c, S). *If $y \in \mathcal{O}_\Phi$ then*

$$x \in \mathcal{O}_{\Phi,y} \equiv \mathfrak{E}_\Phi(y, \mathsf{exp}(2, x)).$$

PROOF. Assume the left side holds and $y \in \mathcal{O}_\beta$ where $|y|_\beta$ is minimal. It follows that there is γ such that $x \in \mathcal{O}_\gamma$ and $|x|_\gamma < |y|_\beta$. To prove $\mathfrak{E}_\Phi(y, \mathsf{exp}(2, x))$ assume there is β' such that ${\sim}\mathfrak{E}^{\beta'}_\gamma(y, \mathsf{exp}(2, x))$. This means $y \in \mathcal{O}_{\beta'}$ and $|y|_{\beta'} \leq |x|_\gamma < |y|_\beta$ which contradicts the minimality of $|y|_\beta$.

Now assume $\mathfrak{E}_\Phi(y, \mathsf{exp}(2, x))$ and let γ be such that $\mathsf{exp}(2, x) \in \mathcal{O}_\gamma$ and $\forall\beta\mathfrak{E}^\beta_\gamma(y, \mathsf{exp}(2, x))$. On the other hand, from $\mathfrak{E}_\Phi(x, y)$ it follows that there is β such that $y \in \mathcal{O}_\beta$ and $\mathfrak{E}^\gamma_\beta(x, y)$, that is, $|y|_\beta > |x|_\gamma \geq |y|_\beta$ which is a contradiction. □

Corollary 5.6.10.1. *Let $\mathscr{C}$ be a* FRS *and assume $\mathscr{C}$ is* e-*total. Let $\mathscr{F}$ be the* GFRS *of Theorem 5.6.8 with internal interpreter $(\Phi_c, \mathsf{S}_\delta)$. If $y \in \mathcal{O}_\Phi$ then $\mathcal{O}_{\Phi,y}$ is $\mathscr{C}$-hyperhyperarithmetical.*

Proof. Immediate from the specification of $\mathfrak{E}_\Phi$, Theorem 5.6.8 (v), and Theorem 5.6.10. □

Theorem 5.6.11. *Let $\mathscr{C}$ be a* FRS. *Assume $\mathscr{C}$ is* e-*total. Let $\mathscr{F}$ be the* GFRS *of Theorem 5.6.8 with internal interpreter $(\Phi_c, \mathsf{S}_\delta)$. Assume P is a k-ary predicate such that $\bar{P}$ is $\mathscr{C}_\mathsf{d}$-hyperhyperenumerable, and h is a k-ary total $\mathscr{C}$-computable function such that whenever $P(\mathbf{x})$ holds then $h(\mathbf{x}) \in \mathcal{O}_\Phi$. There is $y \in \mathcal{O}_\Phi$ such that whenever $P(\mathbf{x})$ holds then $h(\mathbf{x}) \in \mathcal{O}_{\Phi,y}$.*

Proof. Assume there is no such y. It follows that

$$y \notin \mathcal{O}_\Phi \equiv \forall\mathbf{x}\colon \sim P(\mathbf{x}) \vee \mathfrak{E}_\Phi(y, \mathsf{exp}(2, \mathbf{x})),$$

and this means that $\mathcal{O}_\Phi$ is $\mathscr{C}$-hyperhyperarithmetical, contradicting Theorem 5.6.9 (ii). □

Corollary 5.6.11.1. *Let $\mathscr{C}$ be a* FRS. *Assume $\mathscr{C}$ is* e-*total. Let $\mathscr{F}$ be the* GFRS *of Theorem 5.6.8 with internal interpreter $(\Phi_c, \mathsf{S}_\delta)$. A k-ary predicate P is $\mathscr{C}$-hyperhyperarithmetical if and only if there is a number $y \in \mathcal{O}_\Phi$ and a total k-ary $\mathscr{C}$-computable function h such that*

$$P(\mathbf{x}) \equiv h(\mathbf{x}) \in \mathcal{O}_{\Phi,y}.$$

Proof. The condition is sufficient by Corollary 5.6.10.1, and it is necessary by Corollary 5.6.7.1 and Theorem 5.6.11. □

Exercises

5.6.1. Let $\mathscr{P}$ be a class of predicates closed under substitution with total $\mathscr{C}$-computable functions, where $\mathscr{C}$ is closed under elementary operations. Assume $\mathsf{EL}_\mathsf{d} \subseteq \mathscr{P}$. Prove the following:
(a) $\mathscr{P} \subseteq \mathscr{P}_\mathsf{hhu}$ and $\mathscr{P}_\mathsf{hhu}$ is closed under substitution with total $\mathscr{C}$-computable functions.
(b) If $\mathscr{P}$ is closed under disjunction, then $\mathscr{P}_\mathsf{hhu}$ is closed under universal type 2 barred quantification, disjunction, universal type 1 barred quantification, universal and existential unbounded quantification, universal bounded quantification, and existential type 1 barred quantification.
(c) If $\mathscr{P}$ is closed under conjunction, then $\mathscr{P}_\mathsf{hhu}$ is closed under conjunction.
(d) If $\mathscr{P}$ is closed under disjunction and existential bounded quantification, then $\mathscr{P}_\mathsf{hhu}$ is closed under existential bounded quantification.

5.6.2. Let $\mathscr{P}$ be a class of predicates closed under substitution with elementary functions and conjunction. Assume $\mathsf{EL}_\mathsf{d} \subseteq \mathscr{P}$. Prove the following:

(a) $\mathscr{P}_{\mathsf{he}} \subseteq \mathscr{P}_{\mathsf{hhe}} \subseteq \mathscr{P}_{\mathsf{ehhe}} \subseteq \mathscr{P}_{\mathsf{hehhe}} = \mathscr{P}_{\mathsf{hhee}} = \mathscr{P}_{\mathsf{hhehe}} = \mathscr{P}_{\mathsf{hhehhe}}$.

(b) $\mathscr{P}_{\mathsf{h\#d}} \subseteq \mathscr{P}_{\mathsf{hh\#d}} \subseteq \mathscr{P}_{\mathsf{hh\#de}} \subseteq \mathscr{P}_{\mathsf{hh\#pd}} = \mathscr{P}_{\mathsf{hh\#pdhe}} = \mathscr{P}_{\mathsf{hh\#pdhhe}} = \mathscr{P}_{\mathsf{hhe}}$.

(c) $\mathscr{P}_{\mathsf{hh\#d\#}} \subseteq \mathscr{P}_{\mathsf{hh\#pdh\#}} = \mathscr{P}_{\mathsf{hh\#pdhh\#}} = \mathscr{P}_{\mathsf{hh\#}}$.

(d) If $\mathscr{P}_{\mathsf{hhe}}$ is closed under disjunction, then $\mathscr{P}_{\mathsf{hh\#}}$ is closed under recursive operations and $\mathsf{RC}_{\mathsf{pdhh\#}}(\mathscr{P}_{\mathsf{pd}}) = \mathscr{P}_{\mathsf{hh\#}}$.

5.6.3. Let $\mathscr{P}$ be a class of predicates closed under substitution with elementary functions and boolean operations. Assume $\mathsf{EL}_\mathsf{d} \subseteq \mathscr{P}$. Prove the following:

(a) $\mathscr{P} \subseteq \mathscr{P}_{\mathsf{hhe}} \cap \mathscr{P}_{\mathsf{hhu}} = \mathscr{P}_{\mathsf{hh\#d}}$.

(b) $\mathscr{P}_{\mathsf{hh\#dhhe}} = \mathscr{P}_{\mathsf{hh\#pd}} = \mathscr{P}_{\mathsf{hh\#pde}} = \mathscr{P}_{\mathsf{hh\#pdhe}} = \mathscr{P}_{\mathsf{hh\#pdhhe}} = \mathscr{P}_{\mathsf{hhe}}$.

(c) $\mathscr{P}_{\mathsf{hh\#dhh\#}} = \mathscr{P}_{\mathsf{hh\#pd\#}} = \mathscr{P}_{\mathsf{hh\#pdh\#}} = \mathscr{P}_{\mathsf{hh\#pdhh\#}} = \mathscr{P}_{\mathsf{hh\#}} = \mathsf{RC}_{\mathsf{dhh\#}}(\mathscr{P}_\mathsf{d})$.

5.6.4. Let $\mathscr{C}$ be a class of functions closed under elementary operations such that $\mathscr{C}_{\mathsf{dhh\#}} \subseteq \mathscr{C}$. Prove the following:

(a) $\mathscr{C}_{\mathsf{dhhe}} \subseteq \mathscr{C}_{\mathsf{pd}}$.

(b) $\mathscr{C}_{\mathsf{dhh\#}} \subseteq \mathscr{C}_{\mathsf{pd\#}}$.

(c) $\mathscr{C}_\mathsf{d} = \mathscr{C}_{\mathsf{dhhe}} \cap \mathscr{C}_{\mathsf{dhhu}} = \mathscr{C}_{\mathsf{dhh\#d}}$.

(d) $\mathscr{C}_\mathsf{d}$ is closed under existential and universal type 1 barred quantification.

5.6.5. Let $\mathscr{C}$ be a class of functions closed under hyperhyperenumeration operations. Prove the following:

(a) $\mathscr{C}_{\mathsf{dhh\#}} \subseteq \mathscr{C}$.

(b) $\mathscr{C}_{\mathsf{dhhe}} \subseteq \mathscr{C}_{\mathsf{pd}} = \mathscr{C}_{\mathsf{pde}} = \mathscr{C}_{\mathsf{pdhe}} = \mathscr{C}_{\mathsf{pdhhe}}$.

(c) $\mathscr{C}_\mathsf{d} = \mathscr{C}_{\mathsf{dhhe}} \cap \mathscr{C}_{\mathsf{dhhu}}$.

(d) $\mathscr{C}_\mathsf{d}$ is closed under type 1 barred quantification.

(e) A k-ary predicate P is $\mathscr{C}$-decidable if and only if both P and $\bar{P}$ are $\mathscr{C}_{\mathsf{pd}}$-hyperhyperenumerable.

5.6.6. Let $\mathscr{C}$ be a class of functions. Assume $\mathscr{C}$ is hhe-total. Prove the following:

(a) $\mathscr{C}_{\mathsf{dhhe}} = \mathscr{C}_{\mathsf{pdhhe}}$.

(b) $\mathscr{C}_{\mathsf{dhh\#}} = \mathscr{C}_{\mathsf{pdhh\#}}$.

(c) $\mathscr{C}$ is closed under hyperhyperenumeration operations.

5.6.7. Let $\mathscr{C}$ be a class containing only total functions. Prove the following:

(a) $\mathsf{EL}_{\mathsf{dhhe}}(\mathscr{C}) = \mathsf{RC}_{\mathsf{dhhe}}(\mathscr{C}) = \mathsf{RC}_{\mathsf{pdhhe}}(\mathscr{C})$.

(b) $\mathsf{EL}_{\mathsf{dhh\#}}(\mathscr{C}) = \mathsf{RC}_{\mathsf{dhh\#}}(\mathscr{C}) = \mathsf{RC}_{\mathsf{pdhh\#}}(\mathscr{C})$.

5.6.8. Let $\mathscr{C}$ be a class of functions closed under recursive operations. Prove that the following conditions are equivalent:

(a) $\mathscr{C}$ is hhe-total.

(b) $\mathscr{C} = \mathsf{RC}_{\mathsf{dhh\#}}(\mathscr{C}_{\mathsf{dd}})$.

(c) There is a class $\mathscr{C}'$ containing only total functions such that $\mathscr{C} = \mathsf{RC}_{\mathsf{dhh\#}}(\mathscr{C}')$.

(d) There is a class $\mathscr{C}'$ containing only total functions such that $\mathscr{C} = \mathsf{EL}_{\mathsf{dhh\#}}(\mathscr{C}')$.

(e) There is a class $\mathscr{C}'$ closed under elementary operations such that $\mathscr{C} = \mathscr{C}'_{\mathsf{dhh\#}}$.

5.6.9. Let $\mathscr{C}$ be a class of functions closed under elementary operations. Prove the following:

(a) $\mathscr{C}_{\mathsf{ha}} \subseteq \mathscr{C}_{\mathsf{hha}}$.

(b) $\mathscr{C}_{\mathsf{hha}} = \mathscr{C}_{\mathsf{dhh\#d}}$.

(c) $\mathscr{C}_{\mathsf{hha}}$ is closed under boolean operations, bounded quantification, unbounded quantification, and type 1 barred quantification.
(d) $\mathscr{C}_{\mathsf{hhahhe}} = \mathscr{C}_{\mathsf{dhhe}} = \mathscr{C}_{\mathsf{hhahhu}} = \mathscr{C}_{\mathsf{hhu}}$.
(e) An infinite set A is $\mathscr{C}$-hyperhyperarithmetical if and only if there is a total $\mathscr{C}_{\mathsf{d}}$-hyperhyperenumerable function f such that $A = \mathsf{R}_f$ and $f(y) < f(y+1)$ holds for all $y \in \mathbb{N}$.

5.6.10. Let $\mathscr{C}$ be a FRS. Assume $\mathscr{C}$ is e-total and $\mathscr{F}$ is the GFRS of Theorem 5.6.8 with internal interpreter $(\Phi_c, \mathsf{S}_\delta)$. Prove the following:
(a) $\mathcal{O}_\Phi$ is creative in $\mathscr{C}_{\mathsf{dhhe}}$.
(b) If β is a total unary function there is $y \in \mathcal{O}_\Phi$ such that $\mathcal{O}_\beta \subseteq \mathcal{O}_{\Phi,y}$.

5.6.11. Let $\mathscr{C}$ be a class of functions closed under recursive operations. Assume A and B are $\mathscr{C}_{\mathsf{d}}$-hyperhyperenumerable sets. Prove there are $\mathscr{C}_{\mathsf{d}}$-hyperhyperenumerable sets A' and B' such that $A' \subseteq A$, $B' \subseteq B$, $A' \cup B' = A \cup B$, and $A' \cap B' = \varnothing$.

5.6.12. Let $\mathscr{C}$ be a class of functions closed under elementary operations. Assume A is an infinite $\mathscr{C}_{\mathsf{d}}$-hyperhyperenumerable set. Prove there is an infinite $\mathscr{C}$-hyperhyperarithmetic set B such that $B \subseteq A$.

Notes

Hyperhyperenumeration generalizes hyperenumeration and preserves many of the properties of the latter. However, this theory requires uniform reflexive structures, which contain functionals rather than functions. This suggests that we have reached the limit of first-order computability, and further extensions must be carried out in the frame of a higher-order theory. Hinman [8] provides an extensive treatment of such a theory, but restricted to total functions. On the other hand, Tourlakis [29] considers functionals with partial functions as arguments. This appears to us to be the right approach if recursion is to be accounted for in the theory.

Most results in this section are generalizations of results on hyperenumeration and are due to Moschovakis (see Rogers [20] for reference).

References

[1] N. J. Cutland, *Computability. An Introduction to Recursive Function Theory*. Cambridge University Press, Cambridge, 1980.

[2] M. Davis, *Computability and Unsolvability*. McGraw-Hill, New York, 1958.

[3] J. Fenstad, *General Recursion Theory*. Springer-Verlag, New York, 1980.

[4] R. M. Friedberg, A criterion for completeness of degrees of unsolvability. *The Journal of Symbolic Logic*, **22**, 159–160 (1957).

[5] H. Friedman, Axiomatic recursive function theory. In *Logic Colloquium* '69 (R. O. Gandy and C. E. M. Yates, eds.). North-Holland, Amsterdam, 1971.

[6] P. Gabrovsky, Models for an axiomatic theory of computability. Ph.D. dissertation, Syracuse University, 1976.

[7] K. Gödel, Über formal unentscheidbare Sätze der Principia Mathematica und verwandten Systeme, I. *Monatschefte fur Mathematik und Physik*, **38**, 173–198 (1931).

[8] P. G. Hinman, *Recursion-Theoretic Hierarchies*. Springer-Verlag, New York, 1978.

[9] S. C. Kleene, *Introduction to Metamathematics*. North-Holland, Amsterdam, 1952.

[10] S. C. Kleene, Hierarchies of number-theoretic predicates. *Bulletin of the American Mathematical Society* **61**, 193–213 (1955).

[11] S. C. Kleene, On the forms of the predicates in the theory of constructive ordinals (second paper). *American Journal of Mathematics* **77**, 405–428 (1955).

[12] S. C. Kleene, Recursive functionals and quantifiers of finite type, I. *Transactions of the American Mathematical Society* **91**, 1–52 (1959).

[13] G. Kreisel, The axiom of choice and the class of hyperarithmetic functions. *Indagationes Mathematics* **24**, 307–319 (1962).

[14] J. D. Monk, *Mathematical Logic*. Springer-Verlag, New York, 1976.

[15] Y. N. Moschovakis, Abstract first order computability, I, II. *Transactions of the American Mathematical Society* **138**, 427–464, 465–504 (1969).

[16] J. Myhill Creative sets. *Zeitschrift fur mathematische Logik und Grundlagen der Mathematik* **1**, 97–108 (1955).

[17] E. L. Post, Recursively enumerable sets of positive integers and their decision problems. *Bulletin of the American Mathematical Society* **50**, 284–316 (1944).

[18] H. G. Rice, Classes of recursively enumerable sets and their decision problems. *Transactions of the American Mathematical Society* **74**, 358–366 (1953).
[19] H. E. Rose, *Subrecursion. Functions and Hierarchies.* Clarendon Press, Oxford, 1984.
[20] H. Rogers, *Theory of Recursive Functions and Effective Computability.* McGraw-Hill, New York, 1967.
[21] L. E. Sanchis, Hyperenumeration reducibility. *Notre Dame Journal of Formal Logic* **19**, 405–415 (1978).
[22] L. E. Sanchis, Reducibilities in two models for combinatory logic. *The Journal of Symbolic Logic* **44**, 221–234 (1979).
[23] L. E. Sanchis, Reflexive domains, in *To H. B. Curry: Essays on Combinatory Logic, Lambda-Calculus and Formalism* (J. R. Hindley and J. P. Seldin, eds.). Academic Press, Orlando, FL, 1980.
[24] L. P. Sasso, A survey of partial degrees. *The Journal of Symbol Logic* **40**, 130–140 (1975).
[25] J. R. Shoenfield, *Degrees of Unsolvability*, American Elsevier Publishing Co., 1971.
[26] R. M. Smullyan, *Theory of Formal Systems*, Ann. Math. Studies No 47. Princeton University Press, Princeton, 1961.
[27] C. Spector, Recursive well-orderings. *The Journal of Symbolic Logic* **20**, 151–163 (1955).
[28] C. Spector, Inductively defined sets of natural numbers. In *Infinitistic Methods*, Pergamon Press, New York, 1961.
[29] G. J. Tourlakis, *Computability*. Reston Publishing Reston, VA, 1984.
[30] A. M. Turing, On computable numbers with an application to the Entscheidung problem. *Proceedings of the London Mathematical Society* **43**, 544–546 (1937).
[31] E. G. Wagner, Uniform reflexive structures: On the nature of Gödelization and relative computability. *Transactions of the American Mathematical Society* **144**, 1–41 (1969).

Index